COMPETITION VERSUS PREDATION IN AVIATION MARKETS

Competition versus Predation in Aviation Markets

A Survey of Experience in North America, Europe and Australia

Edited by
PETER FORSYTH
DAVID W. GILLEN
OTTO G. MAYER
HANS-MARTIN NIEMEIER

Published in Association with the
German Aviation Research Society (GARS)

ASHGATE

Published by
Ashgate Publishing Limited
Gower House
Croft Road
Aldershot
Hampshire GU11 3HR
England

Ashgate Publishing Company
Suite 420
101 Cherry Street
Burlington, VT 05401-4405
USA

Ashgate website: http://www.ashgate.com

British Library Cataloguing in Publication Data
Competition versus predation in aviation markets : a survey
of experience in North America, Europe and Australia. -
(Ashgate studies in aviation economics and management)
1.Aeronautics, Commercial - Congresses 2.Predatory pricing
- Congresses 3.Airlines - Management - Congresses
4.Competition, Unfair - Congresses
I. Forsyth. P. (Peter)
387.7'42'068

Library of Congress Cataloging-in-Publication Data
Competition versus predation in aviation markets : a survey of experience in North America, Europe and Australia / edited by Peter Forsyth ... [et al.].
p. cm.
ISBN 0-7546-4195-3
1. Airlines. 2. Predatory pricing. 3. Airlines--North America. 4. Predatory pricing--North America. 5. Airlines--Europe. 6. Predatory pricing--Europe. 7. Airlines--Australia. 8. Predatory pricing--Australia. I. Forsyth, P. (Peter)

HE9776.C66 2005
387.7'1--dc22

2005007350

ISBN 0 7546 4195 3

Printed and bound in Great Britain by MPG Books Ltd, Bodmin, Cornwall

Contents

Acknowledgements *vii*
Editors and Contributors *ix*
Introduction *xiii*

Part A: Competition and Predation – Surveys from Different Perspectives

1. Strategic Behaviour of Incumbents
Kai Hüschelrath 3

2. Predatory Pricing: Still a Rare Occurrence?
Ashish Lall 37

3. Competition and Predation
Ruwantissa Abeyratne 57

Part B: Australia and New Zealand

4. Predatory Behaviour in Australian Aviation Markets
Peter Forsyth 81

5. When does Airline Competition become Predation?
David Greig 95

Part C: North America

6. Predation in Aviation: The North-American Divide
David W. Gillen and Ashish Lall 105

7. Airline Predation in Canada
Scott M. Butler 129

8. Most Low-Cost Airlines Fail(ed): Why Did Southwest Airlines Prosper?
Andreas Knorr and Andreas Arndt 145

Part D: Europe

9. The Lufthansa-Germania Case at a Glance
Kai Hüschelrath 173

10. Predatory Pricing in the Airline Industry as a Challenge to Competition Law Enforcement
Christian Ewald 183

11. The European 'No-Frills'-Aviation Market: Current and Future Developments
Bernard Michael Gilroy, Elmar Lukas and Tobias Volpert 203

Part E: Theoretical and Policy Aspects of Predatory Pricing and Competition

12. A Case of Beer and Pretzels
William G. Morrison 237

13. Investigating Airline Managers' Perception of Route Entry Barriers: A Questionnaire-Based Approach
Mirko C.A. Schnell 249

14. Why do Low Cost Carriers Arise and How can they Survive the Competitive Responses of Established Airlines?: A Theoretical Explanation
Markus Reisinger 269

15. A Capacity Lock-In Rule
David Starkie 289

16. Wide Angle Preying in Land Modes? The Problem of Missing Modes
Marc Gaudry 305

17. An Economic Framework for Assessing Predation in Air Services: Markets, Barriers to Entry, Market Power and Tests
Colin A. Gannon 311

Acknowledgements

This book is a compilation of selected papers presented at two workshops organized by the German Aviation Research Society. We would like to thank the Berlin School of Economics (FHW), HWWA – Institute of Economic Research, Hamburg, and the University of Applied Sciences, Bremen for acting as hosts. We are also grateful to Hamburg Airport, Lufthansa and the Wolfgang-Ritter-Stiftung for providing financial support.

John Hindley of Ashgate Publishers was encouraging with his active promotion of GARS during the critical years of its foundation.

Bill Morrison reviewed several papers and gave us valuable comments on many issues. We thank him very much for his advice and help.

Our thanks also go to Natalie Chantal McCaughey for carefully revising the English style as well as to Andreas Arndt, Jürgen Müller, Wolfgang Strehl, and Harald Wiese for their support.

Editors and Contributors

Ruwantissa Abeyratne is Co-Ordinator, Air Transport Programs at the International Civil Aviation Organization and Chairman, Advisory Joint Appeals Board of the Organization. He is also Professor of Aeropolitical Law and Policy at Concordia University, Montreal. Dr. Abeyratne holds the degrees DCL from McGill University, PhD from the University of Colombo, LLM from Monash University, Victoria, Australia, and LLB from the University of Colombo. He has published numerous books and over 250 leading journal articles on international law and air law. He is a Fellow of both the Royal Aeronautical Society and the Chartered Institute of Logistics and Transport. He is also a member of the International Law Association, in which he serves as member of both the Association's International Trade Law Committee and Space Law Committee. He can be contacted through his web page at www.abeyratne.com.

Andreas Arndt has been Research Fellow at the Institute for World Economics and International Management at the University of Bremen, Germany, since June 2000. He received his PhD in Economics from the University of Bremen in 2004. His main fields of research are air transport economics, competition policy and environmental economics.

Scott M. Butler has been at WestJet since 1999. His current position is Director, Corporate Planning and Real Estate. Previous roles included helping prepare WestJet's submissions for a Competition Bureau investigation of anti-competitive conduct. He holds a Bachelor of Commerce from the University of Victoria.

Christian Ewald, Economist, is a case-handler in the 9th Decision Division of the German national competition authority, the Bundeskartellamt (Federal Cartel Office (FCO), which is responsible for competition law enforcement in the airline industry. Furthermore, Mr. Ewald currently is one of the co-ordinators of the Task Force 'Economic Concepts' (ÖKON) set up in the FCO in 2001.

Peter Forsyth has been Professor of Economics at Monash University, Australia since 1997. Prior to this he held posts at Australian National University and the University of New England. He holds degrees from the University of Sydney and the University of Oxford. He has specialized in the economics of transport, especially aviation, privatization and regulation, and the economics of tourism. Most recently, he has been paying particular attention to the privatization and regulation of airports, and to the use of computable general equilibrium models in evaluating the economic impacts of tourism. He has recently published an edited volume of classic articles in the economics of air transport (Edward Elgar).

Colin A. Gannon is an Economic Adviser to the Australian Competition and Consumer Commission covering primarily transport, including airports, air navigation, mergers and anti-competitive conduct. Prior to this he was with the World Bank as Lead Economist and Principal Economist and worked on lending operations, strategic reviews, project evaluations, and policy with developing countries in East Asia, South America, Africa, the Pacific Islands, and the Caribbean, predominantly in the transport sector on all modes, including airports and air services. Prior to this he was Director of the Australian Bureau of Transport Economics, Economic Policy Adviser in International Aviation, Department of Transport, and Associate Professor of Economics, Monash University. Email: cgannon@bigpond.com.

Marc Gaudry is Professor of Economics at the University of Montreal (Montreal, Canada) and Associate Researcher at the *Institut National de Recherche sur les Transports et leur Sécurité* (Arcueil, France). Educated in North America (Canada and the USA) and in Europe (Britain and France), and teaching on both sides of the Atlantic for the last 20 years in Montreal and Karlsuhe or Strasbourg, he has published extensively in Economics and Transportation. His work, including public and fully documented statistical estimation programs available on the web site of the Agora Jules Dupuit (www.ajd.umontreal.ca), has been recognized in particular by an *Alexander von Humboldt* Research Award (*Forschungspreis*) and by more than 1300 references in the scientific literature.

David W. Gillen is YVR Professor of Transportation Policy in the Sauder School of Business at the University of British Columbia, Canada. He is also Director of the Centre for Transportation Research at UBC. He also holds the position of Research Economist, Institute of Transportation Studies, University of California-Berkeley. He obtained his PhD in Economics from the University of Toronto in 1975. He has published over 80 articles and 15 books in the areas of transportation economics, transportation management, aviation and airport economics, industrial organization and management strategy. His current research covers airline strategies, airport performance measurement and strategy, network economics and competition issues in airlines and airports.

Bernard Michael Gilroy has been a Professor of International Trade and Macroeconomics at the University of Paderborn since 1996. An American by birth, he has worked in Germany, Switzerland, and England since 1979. He holds degrees from Uppsala University (N.J., U.S.A.), the University of Constance (Germany), and the University of St. Gallen (Switzerland). He has published over 40 articles in scholarly journals and 5 books in the areas of international trade policy and theory, macroeconomics, multinational enterprises, integration processes and network economics.

David Greig is Executive Director with the Australian economic consulting firm ACIL Tasman. His interest in aviation began with work in the New Zealand Treasury in the 1980s when the then government-owned Air New Zealand was in financial distress. He was active in the subsequent deregulation of domestic and international aviation in New Zealand. In the 1990s in Melbourne he established the unit that privatized the railway and tram services. His consulting work has included pricing and investment policy in transport and other sectors, and newspaper op-ed articles on airline competition. He was educated at Victoria and Canterbury universities in NZ, and at MIT (airline management) and Harvard.

Kai Hüschelrath is a research and teaching assistant at the Institute for Industrial Organization at WHU Graduate School of Management in Vallendar, Germany. He holds degrees in both Economics and Management from the University of Würzburg, Germany. He has specialized in competition policy and transport economics. Most recently, he has been paying particular attention to predatory pricing strategies as well as concepts of market delineation.

Andreas Knorr has been Professor for Economic Policy at the Deutsche Hochschule für Verwaltungswissenschaften (DHV) Speyer, Germany, since October 2004. Prior to this he held a professorship in international economics at the University of Bremen, Germany as well as two temporary professorships at the Technical University Ilmenau and the Technical University Dresden. He received his PhD in Economics from the University of Bayreuth in 1993. His main fields of research are air transport economics, international trade in services, European economic integration, competition policy and environmental economics.

Ashish Lall teaches Economics and Business Strategy at Nanyang Business School in Singapore. Prior to this, Ashish worked in Canada with the Royal Commission on National Passenger Transportation and as a consultant with TNS Canadian Facts. Ashish has written extensively in the areas of aviation and transport, competition policy and the measurement of productivity.

Elmar Lukas is a Lecturer and Research Assistant at the Department of Business Administration, Economics and Business Computing, University of Paderborn, Germany. He graduated in physics and industrial management and holds a PhD in economics from the University of Paderborn. In 1996 he was a visiting scholar at the University of California-Berkeley. His main fields of research are international economics, foreign direct investment and financial economics.

Otto G. Mayer is Vice-President and Head of the Presidential Department of the Hamburg Institute of International Economics (HWWA). He is also managing editor of *Wirtschaftsdienst* and of *Intereconomics*. From 1995-2000 he was member of the board of the Gesellschaft für Wirtschafts- und Sozialwissenschaften – Verein für Socialpolitik (German Economic Association).

William G. Morrison is an Associate Professor of Economics at the School of Business and Economics at Wilfrid Laurier University, Waterloo, Ontario. His academic research combines both theoretical works in the fields of dynamic evolutionary game theory and experimental economics with more applied work, in the field of transportation economics. Dr. Morrison is the co-author of several research reports to government, including a study on Canada's airport system for the Canada Transportation Act Review Board and a study of demand-price sensitivity estimates for air travel, commissioned and published by the Federal Department of Finance. His recent published research includes articles on predatory pricing in air transportation, airline network competition and airport policy.

Hans-Martin Niemeier is Professor of Transportation Economics and Logistics at the University of Applied Sciences, Bremen. He received his PhD in economics at the University of Hamburg and worked in the aviation section of the State-Ministry of Economic Affairs of Hamburg. His research focuses on airport regulation and management.

Markus Reisinger has been Research and Teaching Assistant at the Department of Economics, Ludwig-Maximilians-University, Munich from 2000 till 2004. Currently he is visiting scholar at the Graduate School of Business, University of Chicago. His research interests are industrial organization and applied game theory, especially imperfect competition, price discrimination, and two-sided markets.

Mirko C.A. Schnell is currently working with BSL Management Consultants, a transportation management consultancy in Hamburg, Germany. His career began at an US-American top-management consultancy. Thereafter he joined the Deutsche VerkehrsBank AG as a transportation specialist. He has specialized in the economics and management of transport, especially aviation, public transport and railways. He received his PhD from Cambridge University (UK) and holds further degrees from the Technical University of Dresden, Germany.

David Starkie is a Director of Economics-Plus Ltd and Co-ordinator of Transport Programmes at the Regulatory Policy Institute, Oxford. He has held positions at the Universities of Leeds, Reading and Adelaide and at the Institute for Fiscal Studies. He has also worked in the Western Australian Public Service and been a director of a number of private sector companies. A former aviation adviser to the UK House of Commons Select Committees he currently advises the Irish Commission for Aviation Regulation, the UK CAA and Office of Fair Trading. His degrees are from the London School of Economics.

Tobias Volpert is the Managing Director of the Department of Business Administration, Economics and Business Computing, University of Paderborn, Germany. He holds a PhD in economics from the University of Paderborn. His main fields of research are industrial organization, the economics of patents and property rights and economic policy.

Introduction

Churchill's verdict about economics 'Ask five economists and you will receive six different answers' continues to hold even now. For roughly 50 years economists have debated predation, namely what it is (and if it exists at all), how to assess whether it is taking place, and how to stop it. No broad agreement has so far been reached. In practice the picture becomes even more complex as courts and legal scholars are involved when charges of predatory behaviour are finally settled in court. Predation is not confined to the aviation industry. Indeed the famous Matsushita and Brooke Group cases, which made the disagreement among economists and legal experts obvious to a wide public, stem from the TV manufacturing and cigarette industries. However, we believe that the aviation industry has currently come to a stage in which predation becomes a feature worthwhile of study.

There are several reasons for this. Firstly, the legal response to allegations of predation have been diverse – for example, there are two recent contradictory court rulings. Lufthansa and American Airlines reacted very similarly to the entry of low cost carriers (LCC), both matching prices and increasing frequency. While the German courts found Lufthansa guilty of predation, the US courts were not convinced, and ruled against the allegation. Secondly, in the current environment, LCCs are gaining more and more market share in the US, Europe, South America, Australia, and now, in Asia. For full service airlines (FSAs) it is vital to find a strategy against LCC in order to survive – what price response is safe from being deemed predatory?

In recent years it has been observed that whenever an LCC enters a market dominated by FSAs, prices fall. The price reaction is especially strong if the LCC goes into head to head competition with the FSA by flying into the hub of the FSA. The FSA very often matches the fare of the LCC, increases capacity and enhances service quality by offering bonuses on loyalty points programmes. In many, though not in all, cases the LCC cannot stand the fierce competition and leaves the market, complains about predation and often after some time has elapsed; the consumer has to pay ticket prices at or even above the initial level. While price reductions as a reaction to competition are no doubt a feature of a well functioning market, the interpretation of the effects of the market exit of the new entrant and the higher prices is controversial. Has the incumbent lowered the price below its cost to monopolize the market? Is such strategy successful and profitable? Should the competition authorities forbid such pricing? Why is fare matching or pricing below cost deemed to be a potentially successful strategy by incumbent carriers?

The story of predatory behaviour in aviation in recent years is very much the story of new LCCs entering markets and challenging well-established incumbents. All, or almost all, of the allegations of predation are made by LCCs against incumbents. This need not have been the case. There could have been cases of FSAs entering new route markets being subject to strong competitive responses from incumbents on the route, which could have been predatory. To this end, we give a lot attention to LCCs in the book, analysing their advantages, disadvantages and strategies.

It is not surprising that allegations of predation occur when new LCCs challenge established FSAs. When a FSA enters a new route and commences to compete with another FSA, it is quite probable that it will have similar costs and pricing policies to the incumbent. FSAs may have long pockets – while the financial troubles of major airlines recently may make one question this, a large FSA can normally withstand a price war on a route which accounts for only a small proportion of its traffic. It is also likely to compete with the incumbent on other routes, and be able to retaliate against the incumbent if it reacts too vigorously. New entrants, by contrast, are likely to be a lot smaller than the incumbent, and often they have only modest financial reserves. Thus they are not able to withstand a long period of loss making. They are also less likely than FSAs to be competing with the incumbent in other markets, and to be able to retaliate. To this extent, incumbents may be more willing to try predatory tactics against them.

This said, however, the strength of the position in which incumbent FSAs find themselves should not be overestimated. Many incumbents are afraid of competition from LCCs. The latter have lower costs, and the ability to pitch their product such that demand for it grows very rapidly. The incumbents face a dilemma – if they do not match fares they lose market shares rapidly. On the other hand, if they do match fares, they are likely to be setting fares below costs (since the LCCs' costs are lower than theirs) and they risk being found to have indulged in predatory conduct. Since it is usually very difficult for them to reduce their costs quickly, they find it difficult to choose a response to LCC entry which is both viable in the long run and not risking charges of predation. In this situation, it is not surprising that many have set up their own LCC subsidiaries, in spite of the difficulties in making this work effectively.

The LCCs, for their part, also fear the responses of the incumbents. Over the history of deregulated air transport markets, there are more examples of failures of LCCs than successes, and there is a multitude of examples of LCCs being forced from markets. LCCs have had to develop strategies to cope with vigorous competition from the FSAs, whether this is predatory or not. Airlines such as Southwest avoided head to head competition with the FSAs until they became large and successful enough to fight off the challenges. Successful LCCs such as Ryanair have sought to develop secondary markets in which they do not have to face direct competition from strong incumbents. LCC entrants will seek to appeal to competition authorities to put a break on the behaviour of the incumbents; however, as the experiences in this book show, they have rarely obtained much protection from this quarter. Significantly, the entrants surveyed by Schnell do not seem to regard predation as a significant entry barrier.

Competition and Predation: Overall Perspectives

In Part A we begin by looking at the incumbent. Hüschelrath starts with an overview of the strategic behaviour of incumbents. He argues that predation is just one option out of many an incumbent has to deter entry. According to Hüschelrath the incumbent has three types of potential advantages:

- *Structural advantages* ranging from economies of scale and scope, over product differentiation, to customer loyalty and the presence of switching costs. The incumbent FSAs have built their strategy around it by developing hub and spoke networks, frequent flyer programmes and yield management.
- *Strategic advantages of positioning before entry* such as deterrence through vertical integration or limit pricing. Although airlines typically do not own airports they sublease gates and possess grandfather rights to slots which can be quite effective as barriers when trading is restricted.
- *Strategic advantages of reaction after entry* like raising rival's cost and in particular, predation. For example, they can offer bonus frequent flyer points on routes where incumbent and LCCs compete.

Lall analyses the concept of predation further. Predation is defined as selling a product below costs to induce the rival to exit the market and to gain monopoly profits afterwards. Lall reviews the debate on predation between the 'Chicago school' (as articulated by McGee, 1958) and the 'Post-Chicago school'. The Chicago school views predation as irrational because the predator can never regain the losses from below-cost pricing and the prey can survive. Even when entrants are forced out of a market, more will appear. The Post-Chicago school uses modern industrial organization theory based on game theory to show that under certain circumstances predation can be rational. It argues that in modern complex market structures, predation can occur because the predator has a long purse so that it can drive out the new entrant and regain its monopoly position. The incumbent might try to establish a reputation as a ruthless competitor. By showing that it has the ability to force an entrant from market A, it might prevent the entrant and others from entering markets B, C and so on. Another rational strategy is to bluff, that is for a predator it might be profitable to signal predation so that the new entrant might exit, although in fact it might be profitable to stay. Under these circumstances, subsequent entry may be met with less vigorous response for fear of raising the attention of competition authorities. Although these models show that predation is a potentially rational strategy, they have not convinced courts in general. Lall points out two weaknesses of these models. Firstly, they are not robust as the results depend on some specific assumptions which might or might not be fulfilled in the actual market. Secondly, the empirical support of the models is 'at best, mixed' (Lall).

How should competition policy deal with predation? The fundamental problem for competition policy is, according to Lall, that predatory pricing can resemble vigorous price competition. Hüschelrath stresses the same point for strategic behaviour. There are three approaches for competition policy. The first one is the Chicago School approach, which Hüschelrath characterizes as the do nothing approach since it removes all constraints from the behaviour of the firm. This follows from the view that predation is irrational and is thus not likely to exist. The conclusion is less convincing if one is of the opinion that predation might occur, but only in rare, hard-to-prove cases. Nevertheless one can argue that active competition policy is doing more harm than good. The second option is the per se approach which prohibits specific behaviour generally. The approach has the advantage that it deters the incumbent from the specified uncompetitive behaviour. The rule presupposes that these types of behaviour are identifiable, which in the case of predation, as well as in the case of strategic behaviour, is generally not the case. The third policy option is the rule of reason approach. Each case should be judged on its own merits as to whether it distorts competition or not. While having the advantage of focusing on the identification of predation and uncompetitive strategic behaviour, it has the disadvantage that a complete analysis of each single case leads to relatively high costs. Thus several filters are used to find out if predation might have occurred in actual cases (for a discussion of these, see Ordover and Saloner, 1989).

One filter is the Areeda-Turner test, which defines predation as pricing below short-run marginal cost or in cases in which these costs are too difficult to measure, average variable costs. The rationale for this test is that a profit-maximizing firm under significant [perfect] competition would not price below short marginal costs. This is very appealing but it loses a lot of its theoretical precision in those settings in which predation might be profitable, namely markets with networks and asymmetric information. Several other tests have been developed, but as Lall analysis shows the Areeda-Turner test is 'generally accepted' (Lall). Another filter, applied in some jurisdictions, is a recoupment test which examines whether the predator profits from predation. In the case of American Airlines the court denied the existence of structural barriers to entry in the Dallas-Fort Worth market so that the incumbent could not reasonably expect to regain the losses from predation. In his review Lall concludes that in the United States 'the Chicago view will be liable to prevail' which, however, is not the case in other jurisdictions, especially in Europe. As Abeyratne notes in his paper, many believe air transport are far too complex to be treated on the basis of perfect competition. Gannon in section 5 of his chapter provides an excellent outline of the challenges of formulating an economic test for predation. He then argues that the 'most favoured' economic loss test using avoidable costs should be strengthened.

Predatory behaviour can take many forms. Most of the cases of alleged predation noted in this book take either the classic form of pricing below cost, or the related form of adding extra capacity to a market. Other forms could be relevant in the air transport context. One of these could include setting up a new LCC by a FSA – such an airline could be intended as a means of forcing independent LCCs out of the market, by making it impossible for the LCC to cover its costs. As Huschelrath and Ewald indicate, providing a higher quality/higher cost service at the same price as the entrant could also be construed as predatory.

Predation is most easily understood in the context of a profit maximizing firm that sets prices below cost to preserve market power and maximize profits in the long run. What if the firm is not a profit maximizer? In some sectors of transport, pricing below cost is the norm – rail and other surface modes are often heavily subsidized, as Gaudry documents. Is this predatory? It may have the effect of forestalling entry, though it is not a response to new entry. Systematic direct subsidies to airlines are not common (though unsystematic subsidies are: this is evidenced by the large subsidies to a select group of US carriers over the past three years and the use of Chapter 11 bankruptcy provisions as well as government assistance for fleet modernization for both Air France and Alitalia). Of relevance to the air transport industry is the possibility that large carriers may act as size or revenue maximizers (as Butler argues is the case for Air Canada). Such airlines may be more willing to risk the consequences of being convicted for predatory conduct when profit is reduced by such conduct because profit is not their objective. It may be that airlines operate as if they are revenue maximizers since they treat their network and capacity as fixed in the short to medium term and set fares to maximize revenue per flight.

Predation in Aviation Markets

In Parts B, C and D we survey how predation in aviation markets has been handled in three regions – Australia, North America and Europe. The identification of predatory behaviour poses major problems for competition policy as well as the appropriate measures against such uncompetitive behaviour. Furthermore, legislation differs across jurisdictions so what may be deemed predation in one case may not in another, and the requirements to obtain a conviction for predatory behaviour differs, as Abeyratne points out. Part B, C and D analyse the experience with predation in three major aviation markets. The papers seek to assess:

- *The incumbents' behaviour.* Did the incumbent airlines use predatory tactics to keep the market for their own? How effective was this strategy in keeping the market?
- *The new entrant behaviour.* What was the reaction of the new entrants to predation? What are successful strategies to enter the market given the structural and strategic advances of incumbents?
- *The policy reaction.* What was the reaction of the competition authorities? How did the courts rule?

These are summarized in Table 1.

Until recently, with the blooming of competition in Europe, competition in the US market has seemed to be more intense than in the other markets. Forsyth observes only two intense periods of competition in the Australian market and in each period predation was alleged. In 1990 to 1991 Compass Mark I entered and exited the market claiming that the incumbent airlines, Ansett and Australian Airlines (now absorbed into Qantas) had employed predatory tactics.

The established airlines offered a limited amount of cheap fares to match Compass's prices. Yield management is, as Greig points out, an 'aggressive competitive device' because it ensures that yields do not fall too much and because it is hard to prove to be predatory. In the case of Compass the Australian competition authority came to the conclusion that fares were above variable costs and that predation was not present. Compass's exit was due to management failure.

Table 1 Predation in major aviation markets

	Australia	US	Canada	Europe
Incumbent	Price matching with cheap seats against Compass Mark I; Qantas: Price matching plus capacity adding against Virgin Blue.	American Airlines matched or undercut prices and added capacity against LCC.	Air Canada matched fares against WestJet plus added capacity.	Lufthansa matched or undercut fare of Germania.
New entrant	Exit of Compass Mark I; Virgin Blue remained in the market.	LCC like Vanguard exited market. In other markets SouthWest overcame predation and other entry barriers.	Canjet exited the market but for reasons other than competitive action of Air Canada. WestJet maintained and improved its market and financial position, in spite of the action by Air Canada.	Germania continues operating the market. LCC airlines are successfully entering intra European markets.
Policy reaction	Inquiries with Areeda-Turner Test and below profit-maximizing price criteria. No case of predation demonstrated.	Court decision with Areeda-Turner Test plus recoupment found American Airlines not guilty of predation.	Fares below avoidable cost.	Federal Cartel Office found Lufthansa guilty of predation and enforced a minimum price distance.

During the second period of intense competition, from 2000 to 2001, Qantas matched the fares of the LCC Virgin Blue and increased frequency on a specific route, between Brisbane and Adelaide. Forsyth argues that while the first reaction is a competitive response the later reaction could be predatory, but it remains unclear if the additional flight generated enough demand to cover the costs, or led to a situation in which the revenues to Qantas from the route covered the costs. The analysis is made more complex by the fact that much Brisbane–Adelaide traffic travels indirectly via the Sydney and Melbourne hubs. Virgin Blue was not driven out of the market. However, it was helped to survive by the collapse of Australia's second major airline Ansett. The initial response of the authorities was to use the Areeda-Turner test of average variable cost, but in addition, to define predation as any price below the profit maximizing price given the entrant's price. Forsyth notes that predation is difficult to identify in such a context and notes that competition policy, as spelt out in the legislation, gives little guidance on predation. Greig shares much of the criticism and adds that new entrants should not count on too much from competition policy, but expect an aggressive and potential predatory reaction from the incumbent airlines.

Gannon in his paper provides a discussion of the Australian experience but uses it to provide context in which he discusses the underlying feasibility and prospects for predatory conduct. He asks the basic question, what type of market characteristics are more or less likely to see predation.

In North America there are differences between the Canadian and US legislation and requirements to obtain a conviction. The American Airlines (AA) and Air Canada cases provide an interesting contrast in the application of the respective law for each country, reflecting the approach to controlling predatory behaviour. In both cases, full service carriers faced entry by LCCs and responded by increasing capacity and cutting prices. In the US the court held that the government did not prove that AA had monopoly power on the routes in question, thus ruling out the feasibility of supra-competitive pricing in the post-predation period, an essential element in obtaining a conviction. The calculations also failed to show the recovery of losses incurred during the alleged period of predation, another essential element. In the US the average variable cost test is generally acceptable but the avoidable cost test or a test based on incremental cost may also be acceptable if it is implemented correctly. The cost measure should not include attributed or common costs or count forgone profits as costs – in other words, an opportunity costs test will not do. Prices must be below the appropriate measure of cost. Failure to maximize profits does not constitute predatory sacrifice and neither does failure to take advantage of more profitable opportunities. It is essential to show a reasonable likelihood of recoupment through use of monopoly power after the entrant has left the market.

In contrast the Competition Tribunal was faced with implementing a new section to the Competition Act which had been passed as a consequence of significant changes in the airline industry in Canada. The changes made were not just aimed at a specific industry but at a specific firm, something unheard of in other jurisdictions. The changes included definitions as to what were avoidable costs and what constituted predatory behaviour.

In the course of the case involving alleged predation by Air Canada against two LCCs, WestJet and CanJet the Tribunal accepted that 90% of Air Canada's fully allocated costs were avoidable and no recognition was given to beyond revenue. In essence it accepted the Commissioner's argument that the time period over which cost become avoidable was irrelevant – either costs were avoidable or they were not. The Commissioner and the Tribunal failed to acknowledge the network aspects of the airline business. The Commissioner's version of the avoidable cost test included costs that could be shed outright and increased profits from [full] recapture and profitable redeployment.[1] The Canadian decision may be viewed as controversial because some have argued government duplicity in not allowing, on the one hand, Air Canada to operate as a competitive airline and yet on the other, of claiming it was too aggressive. A significant element of Canadian competition law was the introduction of a penalty, a fine, should a firm be found guilty. Prior to this it was simply a cease and desist order, hardly an incentive to not be overly aggressive.

Knorr and Arndt focus on the strategy of the classic successful new entrant Southwest. They argue that the business model of Southwest gives such a great competitive advantage that in the long run Southwest can enter almost any US market. Southwest was able to circumvent structural and strategic barriers to entry, with the exception of infrastructure bottlenecks such as gates and slots. Initially, Southwest had to overcome legal hurdles, created by incumbents, in order to be allowed to start up. These actions could be regarded as predatory. For many years, it avoided head to head competition with strong incumbents. Predation could not stop Southwest. Knorr and Arndt argue, 'the effectiveness of strategic barriers to entry has long been overestimated'. But while this is true for Southwest, it does not have to be for other LCCs. South West is no more a small new entrant, but it is the fourth largest carrier in the US with deep pockets. It has the ability to, and indeed now does, compete head to head with major FSAs such as Northwest – a kind of competition it had wisely avoided in its start-up phase.

The perspective from a successful LCC, which has been subject to allegedly predatory competition from a FSA, is not surprisingly different to that of Knorr and Arndt. Butler, from the Canadian LCC WestJet, takes a much broader view than the traditional approach focusing on pricing in a particular market. He argues that new entrants can succeed in the short haul market with a LCC business model. However government policy, competition law, the current industry structure and the strategies of the incumbent might effectively close this window of opportunity. Government policy, by imposing a flat rate user charge on airline customers and a security charge while subsidizing the competing rail competition, is in danger of making short haul flights unprofitable (on the topic of intermodal predation, see Gaudry in section E). Competition law is not effective in cases of predation, as it takes too long to reach a decision. The takeover of Canadian Airlines by Air Canada has created a monopoly with a market share of 75%; this has limited the ability of LCC to succeed. The normal assumption is that privately owned firms maximize profits – under this assumption, they are less likely to indulge in predatory behaviour. However Butler argues that the dominant Canadian airline is not acting as a profit maximizer. Rather, it appears to be acting as a revenue maximizer, and to be willing to remain in route markets even when it is not earning profits in them. Such behaviour makes it difficult for LCC entrants to compete, and it can be regarded as predatory.

---

[1] This is an opportunity cost test and one that was applied incorrectly because it assumed full recapture. This test and its application will probably be the most controversial aspects of this decision.

European deregulation of aviation is in many aspects different from the North American and Australian liberalization. Europe liberalized aviation in three stages from 1988 to 1997. Although the major airlines were competing on some dense routes, and there was always some rivalry between FSA and charter airlines, intense competition began with the rise of the LCCs, as Gilroy et al. point out. In Europe the LCC revolution started in the UK with the success of Ryanair and easyJet in the middle of the 90s. In Germany Deutsche BA was never a serious competitor to Lufthansa as it did not fly into the Lufthansa's hub and had a similar high cost base and fares. When BA decided to sell its loss making subsidiary, Lufthansa had a de-facto monopoly in the German market at the dawn of the new millennium. However, competition from new entrants soon began as the LCCs began to expand their market to Continental Europe and start-ups adopted the LCC business model. On 12 November 2001, the LCC Germania started its first flights from Berlin into Lufthansa's main hub, Frankfurt. Germania announced a one-way fare of 99 €. Lufthansa reacted by reducing its price from 243 € to 100 €. After a short price war the Federal Cartel Office stepped in and required Lufthansa to charge a price for a one-way ticket which was at least 35 € above Germania's price for the next two years. As Hüschelrath reports, the decision was based on the reasoning that Lufthansa did not only adjust to Germania's price but effectively undercut it as it offered a service of higher quality, by providing services such as its frequent flyer programme, more daily flight connections and the availability of lounges.[2] The low fare did not cover the average cost per passenger and made it difficult for Germania to compete. The German courts approved the Cartel Office decision.

This decision is the only court decision of predation against a major airline in Europe to date. Gilroy et al. report that there were some complaints by LCCs against predatory behaviour by established airlines and that this will continue to be an issue in Europe in the future. LCCs will rapidly gain market share from the FSAs and the charter airlines. Furthermore, Gilroy et al. predicts a strong consolidation among LCC with only two dominant carriers. Although so far no LCC has been accused of predation, this might change in the future.

The decision of the Federal Cartel Office is in another aspect a hallmark. It makes German competition policy very different from the US approach. It should be noted that other jurisdictions in Europe might take a different stand. Despite several attempts, no alleged predator in the US airline industry has been convicted so far. What are the reasons for the diverging conclusions of the courts? Ewald compares both approaches. His analysis shows that case law in the U.S. is deeply influenced by the theoretical framework of the Chicago School, thereby downplaying the structural peculiarities of the airline industry and their potential for predatory pricing strategies. This explains part of the obvious disagreements, but Ewald argues that there is a deeper division. The divergences are results from what he calls different 'models' of competition law. While the German model gives priority to the protection of competitive market structures and the chances of market entrants, the US model prioritizes the freedom of price setting.

---

[2] Looking under the decision one finds it was based on quite limited information as to what was the value of certain flight amenities. See the paper by Morrison.

In summary, the papers which survey predation in aviation markets around the world suggest several conclusions. While there have been many allegations of predation, these are rarely proved. This is often due to the inherent difficulty of proving predation in many jurisdictions. Furthermore, even when it is accepted by the courts that predation has taken place, as it was in the Germania-Lufthansa case, no penalties were imposed. Rather, constraints on future behaviour were imposed. As a consequence, the alleged predator did not lose from its action. Overall, this would suggest that competition policy has been ineffective in protecting airlines from predation if it occurs.

Theoretical and Policy Aspects

The theoretical survey in Part A shows how diverse the approach towards predation is, and this diversity is reflected in the actual practice, as the surveys in Parts B to D of predation in the major aviation markets highlight. There is no consensus on predation among economists. This provides a rich field for further research and the papers in part E explore some of these aspects.

Morrison looks critically at the standard procedure to prove predation. He goes through the standard procedure and shows how to address the problem of complementarities, product and price differentiation. The Areeda-Turner test does not allow for pricing below marginal costs, but below marginal cost pricing may be rational if for example the below marginal cost fare of the feeder service creates additional demand for the hub flight. Morrison also analyses if and 'how' price-matching should be adjusted for service – a topic on which the German and the American case law differs. While the US court refused to 'engage in a series of subjective price comparisons on intangible values' (Lall, p. 28) the German court did exactly this, and quantified the quality difference. Morrison's intention is to make predation more readily handled by competition policy without going through the complexities of proving intent.

Schnell focuses on empirical research on the nature of barriers to entry and predation in aviation. The rationale of predation depends very much on the question of whether the incumbent can recoup the temporary losses of predation. This in turn depends on whether the aviation market has a monopolistic structure with barriers to entry. Schnell argues that very much of our understanding of barriers to entry depends on theoretical models. Barriers might be high or low, and are typically difficult to measure. Schnell argues that what matters to firms are *perceived* barriers of entry. His questionnaire based results show airline managers perceive different barriers to entry as more or less effective. According to Schnell predation does matter but its importance should not be overstated. In many respect his work is the other side of the coin of Hüschelrath's survey on strategies of incumbents.

LCCs are offering, for many short-haul markets, a superior business model – as Knorr and Arndt, and Gilroy et al. point out. They are overcoming the structural and strategic barriers to entry in many markets. Some incumbents have reacted by starting a LCC, but so far with mixed success. The widely shared view is that it is difficult for a FSA to operate a separate LCC. However, FSAs are adopting repeatedly this strategy, especially in the US, Australia and now, Asia. This raises the question why they adopt what seems to be a losing strategy. Reisinger tries to solve this puzzle from a theoretical point of view. He shows in a model that it can be rational for FSA to build a low cost subsidiary even if it has to be withdrawn later on. In the case the low cost subsidiary is able to predate the new entrant LCC the FSA can keep the market. If the predation was unsuccessful the own LCC is a competitor to the new entrant and to the traditional business of the FSA so that the FSA finds it more profitable to lessen competition by closing the low cost subsidiary.

Gannon's paper not only makes a contribution to understanding the issues surrounding the application of predation legislation to cases in Australia but provides a thoughtful discussion of what markets would give rise to such conduct and then investigates the issue of an economic test for predation. In the first part of his paper he systematically examines the linkages from market definition to barriers to entry to market power. After having described the definitions and economic features of aviation markets, he then goes on to analyse specific entry barriers, distinguishing structural and strategic barriers, as well as primary or secondary barriers. This latter point is to make a distinction between what he terms fundamental barriers and incidental or derivative barriers to entry.

The challenges for developing a test for predation occupies part II of Gannon's chapter. He argues there is agreement among economists that a sound economic test should include four components; market power, intent, pricing below costs and recoupment. Each jurisdiction places more or less emphasis on these components; in some cases they are necessary conditions to obtain a conviction while in other cases they are not (see Morrison for a discussion). The fundamental point as he sees it is, did the carrier suffer real economic loses as a result of its pricing and capacity responses to an entrant. As he points out, in most cases, this loss test is the avoidable cost test. Such a test faces fundamental questions of unit of analysis, time period for the test and market definition. His central tenant is that avoidable costs are opportunity costs.

The two final papers by Gaudry and Starkie address two central problems of economic policy. The entrants (see Butler), but also the analysts of the actual predation cases (see Forsyth, Greig), note the ineffectiveness of the competition authorities to assist the entrants. Predation is hard to prove and it takes too long to decide. The second complaint is that other competing transport modes are neglected – point which Butler emphasizes strongly. The latter is analysed by Gaudry who argues for a more complete view of predation taking into account, in particular, the regulatory elimination by restricting market access and the suppression of intermodal competition. In many countries, railways are heavily subsidized and Gaudry notes two victims of unfair competition, namely scheduled intercity buses and collective taxis.

The first complaint is taken up by Starkie. The paper originates from the opinion of the American Department of Transportation that the rise of LCCs was slowed down by predation. In his correspondence with Alfred Kahn, one of the architects of deregulation of the US aviation industry, Starkie develops and refines his argument for adopting a simple rule to prevent predation, instead of proving predation in long court cases. The incumbent carrier would be obliged to freeze the post-entry capacity for a period in case of an entrant subsequently exiting the market. By locking in the capacity for a certain period, it would become more costly for the incumbent to offer excessive capacity to drive out the new entrant. The rule might remind some readers of Baumol's rule to freeze post entry prices, but Starkie's rule has one crucial advantage over it, which becomes very clear in the discussion with Kahn. While Baumol's rule might lead to a 'pervasive re-regulation' (as prices and quality of service have to be monitored, the enforcement of capacity is rather more straightforward, where most traffic travels on the direct route in question), and capacity determines the average price and yield. This rule was not adopted by the US DOT, so that we have no experience of how such a lock-in rule might alter behaviour, though there are plenty of cases of alleged predation.

Conclusions: Predation in Aviation Markets

Most of the problems, which normally arise in handling allegations of predation, are present in aviation markets. Predation is a notoriously difficult issue to handle. It is difficult to define precisely what it is, and what affects it has, and it is difficult to draw a dividing line between robust competition and predation. It is particularly difficult to prove its existence – tests can be devised, though they are essentially arbitrary, and they are not straightforward to implement. One appealing way of thinking about predation is to define it in terms of impacts on welfare – predation could be defined as behaviour which reduces welfare. Several tests of predation take this approach, but putting it into operation is problematic. In the light of this, governments and regulators have not found it easy to develop policies to address problems of alleged predation. In the light of this, definitive answers to the core questions about predation in aviation markets are not to be expected. Notwithstanding this, some clear themes do emerge from the studies in this book.

Does Predation Occur in Aviation Markets?

This is not an easy question to answer. There have been many allegations of predation, but few have been proven in courts or tribunals. This could be because many allegations are unfounded – after all, disgruntled entrants have an interest in alleging predation even if it has not really taken place. On the other hand, it could simply be that predation has been occurring, but that it has been difficult to prove. Indeed, it may be more prevalent than is alleged, because victims may not always report instances of it because they consider they have little expectation of gain even if they can prove it. The examples in this book, especially those reported in the surveys of actual aviation markets, essentially confirm the point that predation is difficult to prove. This is so across a variety of different legal frameworks, even with those, such as the Canadian one, which seek to simplify matters. While there may be gains from clarifying the legal treatment of potentially predatory behaviour, analysing it will remain inherently complex.

Is Predation a Problem in Aviation Markets?

Predation may or may not be taking place, but if it is, does it constitute a problem? It would be a problem if it were preventing competition from emerging.

Looking at the major aviation markets of the US and Europe, as they are at present, it would be difficult to conclude that predation was holding up the development of competition. In both of these markets, competition has been becoming stronger with the proliferation of LCCs. These airlines have been adding to competition, and they have been entering route markets throughout these regions with great success. It is the FSC incumbents which fear the LCC entrants rather than the other way around. It is a similar picture in the other major aviation market, that in Asia. Here LCCs are booming, and successfully entering route markets previously dominated by the FSC, even though they enjoy little or no protection against predation from competition policy.

This suggests that the current conditions, in the major markets of the US, Europe and Asia are not conducive to predation. Arguably they approximate the environment as envisaged by McGee and the Chicago School when they argued that predation was unlikely. There are several airlines which have the potential to enter most specific route markets. Airlines compete against one another in several route markets, and have the ability to retaliate against predatory behaviour in one route market by responding in others. Competition on most markets is sufficiently strong to keep profits in check – incumbents do not have large profits to defend should entrants appear, and recoupment of losses incurred during a predatory episode is not likely. Incumbents may be able to push one airline off a route, but others will appear if profits are to be made. Finally, and perhaps fundamentally, it is the LCC entrants which have a significant, cost based, advantage over the incumbents – this makes predatory behaviour much less likely to succeed. Thus, overall, the current environment in the major aviation markets is one which makes predation difficult to succeed and also less likely.

However, this has not always been the case in the major markets, and it may not be the case currently in smaller markets. In the US, in the past, there was concern about predation in aviation markets in the 1990s (see Starkie). In the 1980s there were many LCC entrants, but they were almost all forced out by the incumbents which were able to reduce costs and use their network advantages as a competitive weapon. Indeed, by the late 1980s it was considered that incumbents had such strong advantages over entrants and that they could employ predatory tactics cheaply and effectively (Levine, 1987). From the late 1980s to the mid 1990s in the US there were few LCCs which could enter markets readily, and competition on some markets was weak – profitable markets were worth defending. If, in future, incumbent airlines in the major markets are able to match the costs of the LCCs, and weaken the strength of this competitive advantage, predation might become more rational for them again.

Predation may also be more of a concern in small to medium sized markets. Significantly, in medium sized markets such as those of Australia and Canada, predation has been an issue. Actual competition on several route markets is not strong, and some routes may be quite profitable (and worth defending). Furthermore, there are not many potential entrants, and if an entrant is pushed out, others may not appear. After the failure of Compass II in Australia in 1992, it was not until the year 2000 that another airline entered.

The upshot of this is that it is not valid to lump all aviation markets together when discussing the potential for predatory behaviour. In those large markets where LCCs possess strong advantages and are making effective inroads, predation may not now be much of an issue. In other markets, where there are fewer airlines and fewer potential entrants on to any route, predation may still be a live issue. Even in large markets, if the cost advantages of the entrants are eroded, and fewer LCCs are present, predation could become more of a problem.

Are there Special Features of Aviation Markets which Make Analysis of Predation More Difficult?

As with other markets, there are a number of features of aviation markets which make analysis difficult.

One of these concerns airline cost structures. Airlines have major investments in capacity in the short to medium term, but these investments can be moved readily from route to route. Capacity is used such that it is not filled most of the time – this makes sense to enable benefits from higher frequencies. The ratio of demand to capacity varies considerably over the year, month and even day. These aspects make measurement of marginal and variable cost difficult. Cost measurement has been a critical aspect in several predation cases.

Network aspects are another complication with airlines. Airlines typically offer a range of complementary and substitute services, embodied in a network. This makes measurement of cost problematic, since capacity on one route can be used on another, several costs are common to all routes, and traffic on one route can add to revenues on another. It is also difficult to measure the capacity on a city pair, since capacity is provided both directly and indirectly, and capacity on indirect routes also serves other city pairs.

Finally, a pertinent complication is that created by the complexity of the product. A single flight will provide many distinct products, with different prices. Airlines operate sophisticated demand responsive pricing. While this gives them the ability to lower prices in selected markets to meet competition, it also means that it is difficult to determine whether prices are below cost for a particular market segment. There is also the problem of comparing like with like – is the product of a FSC comparable to that of a LCC? This is something which competition authorities have had to resolve, as in the Lufthansa-Germania case.

The experience of predation cases in aviation is that much effort has to be devoted by courts and competition authorities to resolving issues such as these.

Can the Policy Framework to Address Predation in Aviation Markets be Improved?

The ways in which predation issues in aviation have been handled around the world do not seem to have been very satisfactory. Predation has been difficult to prove, though this may be because behaviour has not really been predatory. Predation cases have been long and expensive. Where remedies could have been offered, it would have been too late.

The approaches taken across jurisdictions have differed markedly. In some countries, predation is handled through broad misuse of market power provisions, while in others, there are specific prohibitions of predation. In Canada, there are airline specific provisions. Some countries, such as Germany, give competition regulators the discretion to intervene and impose remedies. Other countries such as the US have considered implementation of conduct rules which would make predation costly.

One issue concerns the ease of proof of predation. Predation can be made easier to prove. For example, it is easier to prove predation under German law than it is under Australian law. Australia could change its laws to make predation more straightforward to demonstrate – indeed the competition authority has argued that this should be done. However this will also increase the risk that predation will be found even when it has not taken place. Determining where the balance should lie is inherently difficult given the problems of analysing the welfare consequences of market behaviour. Indeed we are not aware of any ex-post assessments of the welfare consequences of decisions which have been made in the context of aviation markets. Ultimately it is a matter for governments to decide how tough they want to be on potentially anti competitive behaviour – economic analysis cannot inform them as to what the right balance should be.

Another issue concerns the nature of the anti predation provisions – how specific should they be? The Canadian experience suggests that highly specific provisions, directed at aviation, do not resolve many of the key problems. As long as the meaning of predation is clear, and tests for it are accepted, there does not seem to be much merit in being more specific.

One of the problems with handling predation has been that of delay – cases drag on for years, often after the alleged victim has failed. This is an issue which can be addressed simply, though at a cost. Competition authorities can be given discretion to determine whether predation has taken place, and to institute remedies. This will lessen delay, and it may increase the likelihood of anti predatory action. However, quick action by regulators also means that the facts of the matter are not analysed as thoroughly and there is a greater chance of a wrong decision being made. How much discretion is to be given to regulators is also a judgement call for governments – there is no correct answer that can be determined by economic analysis.

Finally, there is the issue of which mechanisms or remedies will work best if predation is considered a risk or is proven. There have been some suggestions, such as that of Starkie, and some rules have been implemented, as in the Germania case. This is an issue which is more amenable to economic analysis, though thus far there has not been much investigation of the properties of these rules and remedies. It is comparatively easy to devise rules, but they often have unintended consequences. If regulators are to have discretion, it is important that they implement rules which are effective and contribute positively to efficiency.

References

Levine, M. (1987), Airline Competition in Deregulated Markets: Theory, Firm Strategy and Public Policy, *Yale Journal on Regulation* 29, pp. 393-494.

McGee, J. (1958), Predatory Price Cutting: The Standard Oil Case, in: *Journal of Law and Economics* 1 (1958), pp. 137-169.

Ordover, J. and G. Saloner (1989), Predation, Monopolization and Antitrust, in R. Schmalensee and R. Willig (eds), *Handbook of Industrial Organization*, Amsterdam, North Holland, pp. 537-596.

Part A:
Competition and Predation – Surveys from Different Perspectives

Chapter 1

Strategic Behaviour of Incumbents
Rationality, Welfare and Antitrust Policy

Kai Hüschelrath

'Look ahead and reason back.'
Dixit and Nalebuff (1991)*

1. Introduction

The liberalization of formerly regulated markets often leads to 'asymmetric' market structures characterized by one or a few large incumbent firms retaining considerable market power and several smaller new entrants. These new competitors threaten the incumbents in at least two ways. On the one hand, they take market share away reducing an incumbent's share of the profit pie. On the other hand, new entrants often intensify competition reducing the size of the profit pie. Against this background, it is not surprising that incumbent firms would like to impede market entry or at least reduce the competitive threat of entry. One option for dominant incumbents to reach this aim is to use some form of *strategic behaviour* aiming at discouraging entry or encouraging exit of rivals.

However, what exactly is meant by 'strategic behaviour'? What are necessary preconditions to make strategic moves possible and which different strategic options are available to incumbents? What are the possible welfare consequences of strategic behaviour directly related to this: what role should antitrust policy play with respect to such strategic moves? The aim of the present paper is to give some basic answers to these essential questions. A general introduction of possible strategies of incumbents appears to be important, especially in the light of the overarching topic of the present book: *predation* as one option out of the 'toolbox of strategic behaviour for incumbents'.

To reach the central objective of this paper, it is structured as follows: In Section 2 the foundations of strategic behaviour are assessed, which is followed by a simple analysis of the interrelation between strategic behaviour and entry in Section 3. Subsequently, Section 4 focuses on some main features of the relationship between strategic behaviour, welfare and antitrust policy. Section 5 concludes the paper.[1]

---

* Dixit and Nalebuff ((1991), p. 34).

[1] Conceptual, it should be further mentioned that throughout the whole paper, the general economic reasoning is clarified by examples from the airline industry. These examples can be found in the 'Focus'-boxes integrated in the text.

2. The Foundations of Strategic Behaviour

A fairly large number of definitions of 'strategic behaviour' exist. One reason for this might be that different disciplines (like economics, management or the political science) have used and examined this term with respect to their backgrounds and motivations (see e.g. Ansoff (1987) and Grundy and Wensley (1999) for 'strategic management' interpretations). As far as the field of industrial organization is concerned, Carlton and Perloff ((2000), p. 332f) provide a straightforward definition:

> Strategic behavior is a set of actions a firm takes to influence the market environment so as to increase its profits. The market environment comprises all factors that influence the market outcome (prices, quantities, profits, welfare), including the belief of customers and of rivals, the number of actual and potential rivals, the production technology of each firm, and the costs or speed with which a rival can enter the market. By manipulating the market environment, a firm may be able to increase its profits.[2]

This attempt to define 'strategic behaviour' is very general and is being largely confined (and therefore substantiated) by research of several game-theorists (see especially Schelling (1960)) and their formalized concept of a 'strategic move'.

What Constitutes a Strategic Move?

The basic idea behind a strategic move is that when making an optimal choice now, an incumbent must try to anticipate how his rivals will respond in the future. The incumbent's expectation is largely based on his perception of the rivals' payoffs and their perception of his payoffs. In the words of Dixit and Nalebuff ((1991), p. 34) the incumbent must '*[l]ook ahead and reason back*' to make an optimal (i.e. profit-maximizing) decision.

Given this first delineation, the more precise question of 'what are necessary attributes of strategic moves' immediately suggests itself. An essential element of the answer is the acceptance of short-term 'sacrifices'[3] by the incumbent aiming at obtaining long-term (discounted) gains, which at least outweigh the sacrifices. This first delineation clarifies that strategic moves are always dynamic phenomena (i.e. 'dynamics' requirement).

'Strategic moves' in the sense of Schelling (1960) are actions which benefit an incumbent firm indirectly via their effects upon the behaviour of the rival firm (see Church and Ware (2000), p. 461ff for a roundup). These actions 'induce the rival to adopt a course of action more favourable to the incumbent ...' (Vickers (1985), p. 33). Such 'strategic moves' are often based on some first mover advantage, thus a temporal edge of the incumbent in comparison with the entrant (i.e. 'first mover' requirement).

---

[2] Carlton and Perloff ((2000), p. 332ff) differentiate between non-cooperative and cooperative strategic behaviour. In the following the latter case is omitted.

[3] The term 'sacrifices' should be interpreted in a very broad sense, e.g. as investments which bond capital or as a sacrifice of current profits.

Another important feature of a strategic move is the non-revocation of decisions. In situations with perfect information this condition means that a successful strategic move is only possible if the incumbent changes the expectations of the rival in a credible way, e.g. by undertaking a binding and irrevocable decision (i.e. 'commitment' requirement), which the rival can observe (i.e. 'communication' requirement). In cases of imperfect information, the incumbent must be in a position to signal a credible threat of such a commitment.

Where can Strategic Moves Take Place?

Strategic moves depend on situations where strategic interactions take place. 'Strategic interaction' entails that pricing and production decisions of any one firm will affect overall industry price and production levels, and hence the performance of other firms. These characteristics however are only present in oligopolistic markets (i.e. 'market power' requirement). In a monopoly situation, there is (by definition), only a single supplier and entry is impossible. A monopolist has no actual or potential rival and therefore has no need to contemplate strategic behaviour. A firm in a perfectly competitive market is (by assumption) a price-taker, hence there is no payoff to strategic behaviour[4] (see Martin (2001) for a detailed analysis). The same conclusion is true for 'perfectly contestable markets' in the sense of the path breaking book by Baumol, Panzar and Willig (1982).

3. Strategic Behaviour and Entry

Free entry and exit is an essential assumption in the model of perfect competition. If the price of a good lies above average cost, the firm will realize abnormal profits. These abnormal profits are the central incentive for other firms wanting 'to join the party' (Saloner et al. (2001), p. 215). Over time, market entry will increase supply and will depress prices sufficiently for firms to return to normal economic profits. As abnormal profits are competed away, entry will cease and the market will reach its long-run equilibrium. This point coincides with the minima of the firm's average cost curve, the point where the firms use their resources in the most efficient manner.

This essential coherency between entry, competition and market performance is challenged by multiplicities of empirical studies (see Geroski (1995) for a survey), which show that incumbents in many industries are able to earn high abnormal profits without stimulating entry for a long time (see Focus 1 for evidence from the airline industry). In other words, these empirical results indicate that the model of perfect competition alone might be an insufficient description of many real markets.

[4] It seems obvious that the polar cases of 'monopoly' and 'perfect competition' must be interpreted as valuable starting points for further theoretical reasoning rather than descriptions of real markets.

Focus 1:
Entry, Exit, and Performance in Airline Markets

Joskow et al. (1994) examine quarterly data on major, non-stop city pairs in the U.S. between 1985 and 1987. They find that:

- entry generally is not induced by price levels substantially above the norm
- entry reduces fares and increases output, and exit increases fares and reduces output
- incumbents cut prices and maintain output in response to entry, and survivors increase both prices and output in response to exit.

The general significance of free entry and entry incentives for competition intensity and market performance is not restricted to the model of perfect competition. Similarly in standard Cournot oligopoly models, current abnormal profits are a central incentive for market entry. With a growing number of firms in the market, industry prices decline and per-firm output and profit ought to decrease (see Amir and Lambson (2000) for an exact formal treatment). With an infinite number of firms in the market, Cournot competition leads to perfectly competitive outcomes.[5]

In oligopoly models, compared to the model of perfect competition, especially the conditions surrounding entry (e.g. level of fixed cost, market size, cost differences between firms, heterogeneity of products etc.) become more important in determining the nature of market competition and market outcomes. Generally speaking, strategic interaction in oligopolistic markets complicates competition and makes market outcomes dependent on factors like players, actions, timing, information and repetition (see Saloner et al. (2001), p. 187ff).

What Drives a Firm's Entry Decision?

Imagine you are the CEO of a small low-cost airline and you have to decide if you should enter a certain route or not. The actual profits in the market might be an important signal guiding your decision, but the individual entry decision will depend particularly on two aspects: incentives and barriers. In other words, a CEO might first ask 'Is entry profitable?' and afterwards 'Is entry possible?'

[5] Besides the depicted effects of increased competition, entry also plays a creative role in markets, serving as a vehicle for the introduction and diffusion of innovations. Geroski ((1991), p. 219) therefore separates 'entry' into two components: imitative entry and innovative entry. Imitative entry occurs when the entrant can reap profits by copying the established firm's product or method of production. This type of entry is regarded as an equilibrium force in that it competes away excess profits to an equilibrium level. Innovative entry occurs when an entrant finds new ways of e.g. producing more cheaply. This type of entry is seen as a disequilibrium force, which propels the industry from one equilibrium state to another (for a detailed overview, see Lipczynski and Wilson (2001), pp. 153-154).

Is entry profitable? A profit-maximizing, risk-neutral firm should enter a market if the net present value of expected post-entry profits is greater than the sunk costs of entry.[6] As post-entry profits depend on post-entry competition, the entry decision therefore is connected to the expectations of the entrant about the conduct and performance of the firms after entry. Furthermore, the level of sunk costs incurred is a critical determinant of the entry decision (see Besanko et al. (1996), p. 396ff). The higher the necessary sunk costs to enter an industry are the higher is the risk of entry and the lower are the expected profits. Additionally, the entry condition above clarifies that profits immediately after entry are not necessary for a rational entry decision. It is sufficient that e.g. market growth expectations should promise sufficient profits in the future.

Is entry possible? A positive net present value (which at least outweighs sunk costs) is a necessary but not sufficient condition for entry. Imagine e.g. the U.S. Airline Industry 30 years ago. In spite of many entry applications by new carriers, the Civil Aeronautics Board did not approve one single market entry request of a new firm. This example clarifies, that 'barriers to entry' are sometimes high enough to completely deter entry.

'Barriers to entry' are defined by Stigler ((1968), p. 67) commonly as '... a cost of producing (at some or every rate of output) which must be borne by a firm which seeks to enter an industry but is not borne by firms already in the industry'.[7] The sources of such 'barriers to entry' are diverse. Besides legal entry barriers (like e.g. entry regulation by the state), the literature focuses on the so-called private entry barriers (see Focus 2 for some examples from the airline industry).

Focus 2:
Operating and Marketing Barriers in the Airline Industry

In a recent study, the U.S. GAO (2001) found the following operating and marketing barriers, which constrain new entry into dominated airline markets:

- Access to airport facilities such as gates, ticket counters, baggage handling and storage, take-off and landing slots,
- Frequent flyer programmes,
- Corporate incentive agreements,
- Travel agent commission overrides,
- Flight frequency,
- Network size and breadth.

---

[6] Furthermore, as investment capital is scarce, the entry decision for a special market depends on the existence and the profit expectations of other investment alternatives.

[7] See McAfee et al. (2004) for a discussion of this and other definitions of 'barriers to entry'.

These 'private entry barriers' are subdivided into *structural* and *strategic* barriers to entry. *Structural barriers to entry* are related to structural or technical characteristics of an industry (e.g. economies of scale, absolute cost advantages (e.g. favourable access to raw materials or a favourable geographic location), capital cost requirements, product differentiation advantages, etc.). *Strategic barriers to entry* are largely based on the notion of strategic entry deterrence, realizing that existing firms might deliberately behave in ways that decrease the probability of entry by other firms (see Waldman and Jensen (2000), p. 129).

Bearing this knowledge in mind, the entry decision of an entrant should be guided by an evaluation of the significance of entry barriers in a certain market. As many of these barriers simply reduce the expected post-entry profits in the market, the suggested separation between entry profitability and entry possibility does not seem to be very strict. Nevertheless, it might be helpful to structure the arguments in the course of the present essay.

Additionally, it should be mentioned that entry decisions are much more complex – in practice as well as in its examination by the economic profession – than described above by looking at 'incentives' and 'barriers'. Schulz (1995) provides an overview of the theoretical industrial organization literature. Geroski (1995) shows that the empirical evidence sometimes contradict the theoretical findings. Recent research in strategic management and marketing has exposed different options of entering a market. Montaguti et al. ((2002), p. 23) for example distinguish between strategies of penetration, compatibility, pre-announcing and external routes to market. The choice of a certain kind of entry strategy depends largely on factors such as technology characteristics (e.g. network externalities, appropriability, etc.), the competitive environment (e.g. industry concentration, level of incumbency) and firm-specific factors (e.g. reputation, multi-market contact, and order of entry). Focus 3 provides some insights from the airline industry.

Further important influences on the choice of an entry strategy – discussed in the marketing literature – are the expected responses by the competitors *and* assessments of how consumers adoption decisions can be influenced.

Focus 3:
Competitive Dynamics of Interfirm Rivalry in the California Commuter Airlines Market

Baum and Korn (1996) examined how firm-specific competitive conditions influence firms' patterns of market entry and exit. Their research focused on two features of firms' competitive conditions: market domain overlap, which measures the potential for competition, and multi-market contact, which measures the potential for mutual forbearance. Their results for the California commuter airlines' market between 1979 and 1984 show that increases in market domain overlap raised airlines' rates of market entry and exit, whereas increases in multi-market contact lowered them, especially in markets clearly dominated by a single airline. In other words, close competitors are not the most intense rivals: airlines that meet in multiple markets are less aggressive towards each other than those that meet in one or a few markets.

What Drives an Incumbent's Entry (Re)Action?

Now imagine you are the CEO of a big incumbent airline and you are confronted with an entry threat (on a certain route) by a potential entrant. What are possible options for (re)action and how should you choose between the different options?

Regarding the first question, the incumbent can try to reduce the expected level of profits that the entrant can hope to earn.[8] If it is assumed that the profit of the entrant is given by $\pi^E = pq - cq - F$, then the incumbent has four alternatives to negatively influence this expected profit and hence to induce exit or prevent entry: Firstly, to drop the price p, secondly to raise variable costs c, thirdly to raise fixed costs F or, finally as $p>c$, to drop quantity q. In short, the incumbent can act in a way that '*raises rivals' cost*' and/or '*reduces rivals' revenues*'.

The precise ways to reach this aim are diverse. Simply speaking, one option for the incumbent is to *raise the structural entry barriers* with the aim of making entry impossible or at least unprofitable. As this strategy might not be sufficient or too expensive to deter entry completely, the incumbent might consider *strategic moves* to complicate or even deter entry. The choice of a particular strategy again depends on its profitability and its probability to succeed.

Is strategic behaviour profitable? Strategic behaviour normally incurs costs. Therefore, the decision to behave strategically or not should depend on a comparison between the expected costs and revenues of such behaviour. Bain (1956) uses this essential interrelation for the design of a classification of reactions against potential market entry. His approach differentiates between blockaded, deterred and accommodated entry:

- Entry is *blockaded* according to Bain's taxonomy, if – although the incumbent behaves like a monopolist – entry is not profitable. In such a situation strategic behaviour is useless and therefore irrational.
- Entry is *deterred*, if the incumbent could not behave like a monopolist without causing entry and it therefore changes its behaviour in a way that discourages entry. In such settings strategic behaviour is rational and would deter entry completely as long as the cost of the deterrence is smaller than the additional profits in the less competitive market.
- Entry is *accommodated*, if structural barriers are low and the costs to the incumbent of deterring entry are greater than the benefits it could gain from repelling the entrant. In other words, firms do not want to deter market entry completely. Even so, strategic behaviour could be rational, as far as it commits to a market conduct more favourable to the incumbent after entry has occurred.

---

[8] For example, as particularly signalling models in predation theory show, an incumbent can act strategically to make the entrant expect a lower profitability of market entry (see Ordover and Saloner (1989)). Furthermore, an alternative way for the incumbent to deter entry would be to negatively influence the potential entrants' market growth expectations.

To illustrate the *cases of blockaded and deterred entry*, consider the following decision tree (Figure 1.1). The incumbent (I) can first choose between non-strategic and strategic behaviour. The decision tree clarifies that the best outcome for the incumbent is reached, when it behaves non-strategic and the entrant (E) stays out of the market: The incumbent reaps monopoly profits while the entrants profit is zero. Could the incumbent ever achieve this outcome? The answer is yes, if the structural entry barriers are high enough to make entry unprofitable or impossible for the entrant. This is a position of *blockaded entry* in the taxonomy of Bain (1956).

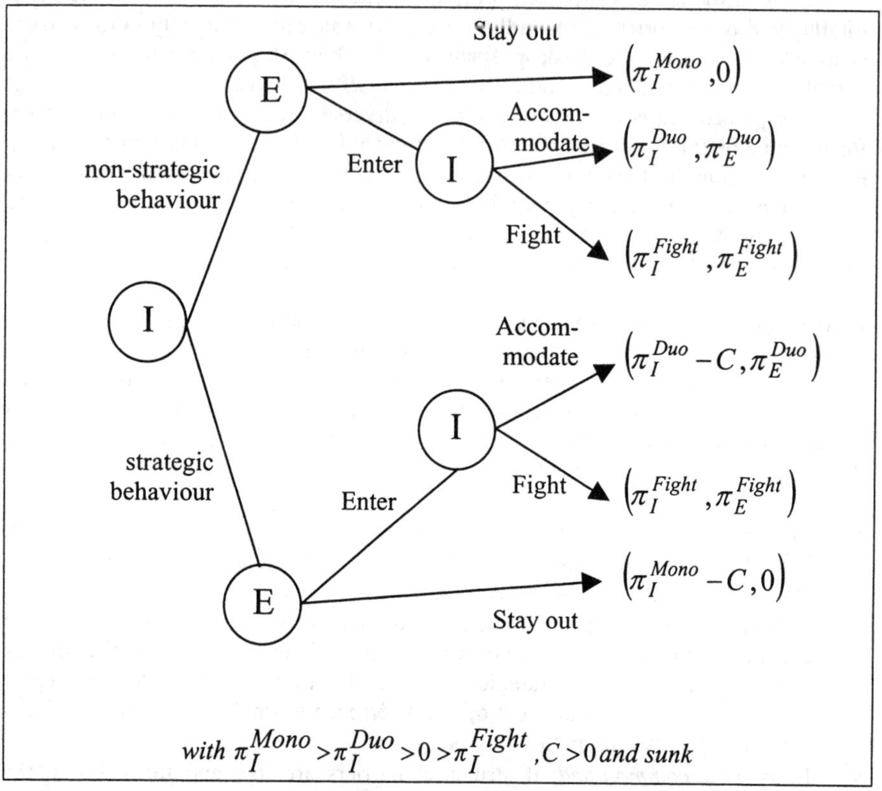

Figure 1.1 Strategic (entry deterring) behaviour of an incumbent

Now assume that the structural entry barriers are too low to deter entry and an entrant is in the position to enter the market. Can the incumbent deter entry (non-strategic) by simply threatening the entrant with a 'fight to death' in case of entry? The answer is no, as, once the entrant has entered the market, the incumbent has no longer an incentive to keep its price low. As the duopoly profits are higher than the profits during a fight, the incumbent would find it more profitable to raise prices and accommodate entry. However, such an attempt of '(non-) strategic behaviour' is not credible.

The central question therefore should be how the incumbent could credibly commit to a low post-entry price? Hence, imagine that the incumbent invests into some sunk costs C, e.g. excess production capacity (see Spence (1979) and Dixit (1980) for formal treatments). This 'commitment' requires that the incumbent consciously chooses a strategy generating additional costs in the current period. The incumbent undertakes this move before it is known if the rival enters the market or not. By undertaking this action, the incumbent wants to signal, that it has now an incentive to react more aggressively towards market entry.[9]

Under what condition is such a behaviour rational (i.e. profit-maximizing) for the incumbent and therefore a credible threat to the potential entrant? In case of a successful strategic move (which deters entry) the incumbent realizes a profit of $\pi_I^{Mono} - C$. The incumbent now has to compare this profit with the profits that could be earned if entry was accommodated. These are the duopoly profits π_I^{Duo} (without the sunk costs C). Thus, one condition for rational strategic behaviour is that $\pi_I^{Mono} - C > \pi_I^{Duo}$, meaning that the monopoly profits less C are bigger than the duopoly profits. The second condition requires that, in case the entrant does enter the market, aggression from the incumbent will be 'rewarded'. This is the case if $\pi_I^{Duo} - C < \pi_I^{Fight}$. Rearranging the inequalities delivers the following condition for rational entry-deterring strategic behaviour:

$$\pi_I^{Mono} - \pi_I^{Duo} > \mathrm{C} > \pi_I^{Duo} - \pi_I^{Fight}$$

If this condition is fulfilled, the incumbent will undertake the strategic investment in C. As the aggressive reaction of the incumbent is now credible, the rival will not enter the market for fear of incurring losses. If the above condition is not fulfilled, strategic entry deterrence is not profitable, because it is too expensive. In such a situation, the incumbent might have a continuing interest in strategic behaviour, which – although not deterring entry – could nevertheless improve its position after entry (see below and e.g. Tirole (1995), p. 326ff).[10]

Again, for completeness sake, it should be noted that the economic knowledge about entry and determinants of (re)actions by incumbents to entry exceeds the here presented characterization considerably. Gruca and Sudharshan (1995) review the strategic management and industrial organization literature and construct a framework based on their findings. They emphasize that entry decisions depend

---

[9] In technical terms, a commitment (e.g. an irreversible investment) makes an entry-deterring strategy credible because it alters the incumbent's best response function in the face of entry.

[10] Based on a simple two-stage game between a K_I-investing incumbent (with a first-mover advantage) and a potential entrant (with a profit of π^2 in the second period), Fudenberg and Tirole (1984) construct a whole taxonomy of business strategies dependent on the kind of investment (either tough ($d\pi^2/dK_I < 0$) or soft ($d\pi^2/dK_I > 0$) and the slope of the reaction function R' (strategic complements (R'>0) and strategic substitutes (R'<0)). Products are strategic complements, if an aggressive action in one product induces an aggressive reaction. Products are strategic substitutes when the reaction is dissimilar (see Bulow et al. (1985)).

particularly on three aspects: the realized and anticipated consequences for the incumbent, alternative entry deterrence strategies and the choice of the entry deterring strategy. The existing competitive environment (e.g. cost conditions, demand conditions, industry history, and legal climate) interacts with each of these factors. By way of illustration, Focus 4 highlights some empirical findings regarding incumbent reactions to entry in the U.S. airline industry.

Focus 4:
Incumbent Reactions to Entry in the U.S. Airline Industry

Lin et al. (2002) conducted an investigation of factors contributing to competitive reactions to entry by incumbent airlines in the short and longer runs. Using data on 889 incumbent reactions to entry between 1991 and 1997, the authors found several factors that have a significant impact on the level of incumbent price cuts in response to entry. They include

- the size of the entrant's price cut,
- the number of passengers carried by the new entrant on the route,
- the costs, size and number of complaints of the entrant.

Interestingly, Lin et al. (2002) found no evidence that incumbents respond more aggressively to small, low-cost carriers than to other carriers. Incumbents reserve their largest price cuts for larger new entrants with higher costs. The longer-run results of Lin et al. (2002) indicate that even if the entrant is forced to withdraw from a route, prices do not rise back up to pre-entry levels.

In a recent study, Ito and Lee (2003) examined incumbent responses to lower cost entry. Based on their analysis of 370 market events, the authors found that highly aggressive incumbent reactions (with respect to price and capacity) are more the exception than the rule. Furthermore, their study showed that the entrant's success or failure on a certain route cannot be explained by the incumbent's capacity or pricing decisions but rather depend on factors such as the entrant's capacity choice, pre-existing market density and the entrant's pre-entry presence at the endpoints of a market.

Is strategic behaviour possible? Successful strategic moves are dependent on specific conditions. As explained in Section 2, some type of first-mover advantage is a precondition for successful strategic behaviour. Furthermore, the rivals must be aware of the strategic move (and clearly understand) before they move. In other words, a successful strategic move requires some form of communication. Additionally the strategic move must credibly change the optimal behaviour of the incumbent in the future.

In this respect, recent marketing research on the question how managers (acting under uncertainty) interpret competitors' signals in various market contexts is of particular relevance. One basic finding of this strand of literature is that the effect of signals critically depends on the context in which it is used. In their empirical study, Prabhu and Steward ((2001), pp. 63-64) find that the entrant's perceptions of the incumbent's aggressiveness depends on the focus and strength of the incumbent's signals. The strength of the entrant's responses to the incumbent's signals depends on the incumbent's use of 'bluffs' and the informational cost regarding the factors driving the incumbent's signals.

Furthermore, strategic behaviour is only possible in oligopolistic markets. In other words, a certain degree of market power is a necessary precondition for strategic behaviour. Moreover, as shown by von Weizsäcker ((1980), p. 13), entry-deterring strategies are not available if there are no structural barriers to entry. Other factors which might influence the possibility of strategic behaviour include: informational asymmetries, managerial incentives, the expected market duration, the necessary amount of sunk cost investment, the presence of network effects or the role of antitrust enforcement.

It is important to also mention that strategic behaviour could have – besides the cost burden – other potential disadvantages. In this respect, literature particularly analyses the trade-off between 'Commitment and Flexibility' (see Boyer and Moreaux (1997) or Ghemawat and del Sol (1998)) or more generally 'First-Mover-*Dis*advantages' (see Lieberman and Montgomery (1998)). For example, in very dynamic markets, strategic commitments might incur a prohibitive risk, because the incumbent loses some of its flexibility to react to unexpected changes in the market environment and/or unexpected moves of (new) rivals.

What are Potential Advantages of Incumbents?

Advantages of incumbents (over entrants) generally stem from their persistent market power in conjunction with first-mover advantages and a profound general knowledge of the industry in question.[11] As incumbents often enjoy a 'quiet life' as a monopolist – protected from entry threats by *structural entry barriers* – they will therefore use their advantages foremost to protect such monopoly positions. In other words, the incumbent's primary aim is to raise structural entry barriers up to an entry-deterring level, e.g. by influencing politicians to introduce entry regulation (see e.g. Viscusi et al. ((1997), p. 331) for an example from the U.S. peanut industry), achieving learning economies, creating customer loyalty and switching costs or attempting to generally raise the amount of sunk costs necessary to start business in the industry (see Focus 5 for an example from the airline industry).

[11] Salop ((1981), pp. 18-19) refers to *information advantages* of the incumbent simply from its longer experience in the industry or the product segment and *strategic advantages*, stemming from the possibility of undertaking binding commitments. Following Spence ((1981), p. 60), it is also important to mention that the evolutionary question of how firms acquire *and* maintain positions of market power (which is essential for the understanding of strategic behaviour in a certain industry) is reduced in recently liberalized industries to the question of maintaining market power.

Focus 5:
A Structural Entry Barrier in the Airline Industry:
Frequent Flyer Programmes

As argued in detail in Hüschelrath ((1998b), pp. 277ff), frequent flyer programmes (FFP) can be interpreted as an attempt to create structural entry barriers in airline markets. By rewarding frequent flights with the same airline (alliance), the customers can secure private non-pecuniary benefits like free flights, luggage allowances etc. Such an incentive system creates customer loyalty and raises the switching costs for the customers. From the view of a potential entrant, FFP reduce the expected demand on a certain route and therefore may impede profitable service on the route in question. Furthermore, FFP also have a strategic potential, e.g. by raising the rewards on a certain route after entry has occurred ('marketing mix reaction after entry').

If the structural entry barriers are not high enough to prevent entry (e.g. the incumbent is too weak to achieve entry regulations via the 'regulatory policy process' or the formation of other sufficient barriers is too expensive), the incumbent often has the *opportunity and incentives* to introduce strategic behaviour. The opportunity, because the incumbent's position in the market often generally allows strategic behaviour, and the incentives, because (dependent on the costs and the expected reaction of the entrant) strategic behaviour pays for the incumbent in form of a higher net present value of the profit stream.[12]

With the aim of reducing the threat of entry and/or promoting exit by rivals, incumbent firms have at least two major possibilities to (re)act. On the one hand, if the entrant is not already in the market, the incumbent could try to deter *or* at least impede entry by negatively influencing the entry incentives of the potential entrant (*'positioning' or 'strategic action before entry'*). Examples for those strategies include 'limit pricing' and 'capacity investments' (see Besanko et al. (1996), p. 407ff).

On the other hand, if entry could not be prevented by those strategies, the incumbent could choose between different options to react to market entry (see Bunch and Smiley (1992) for empirical evidence) (*'reaction' or 'strategic action after entry'*). Predatory Pricing or Predation (by definition) is one of the 'after entry' reply options. Figure 1.2 summarizes some central advantages of

[12] It is often reasonable to assume that an incumbent is active in many product markets. As strategic behaviour might not be financially possible in every market, the incumbent has to decide which markets are most important to protect against entry. Following the strategic management literature, it seems obvious that the incumbent's aim is foremost to protect 'cash cow' and 'star' markets from market entry. 'Cash cow' markets are characterized by a high relative market share and a low industry growth ('high actual profits'), while 'star' markets combine a high relative market share and a high industry growth ('high potential of high future profits', see Oster (1994), p. 130). It is further worth mentioning, that the financing of market protection by the incumbent is sometimes possible with the revenues of 'rudiments of regulation', meaning that the incumbent can subsidize strategic moves with revenues from state-granted monopoly markets.

incumbents.[13] The arrows in Figure 1.2 should point up that interaction between the categories is expected. For example, if an incumbent reacts to market entry with predatory prices, this action might bring about a tough reputation. New potential entrants will take this reputation into account before deciding to enter the same (or a related) market.

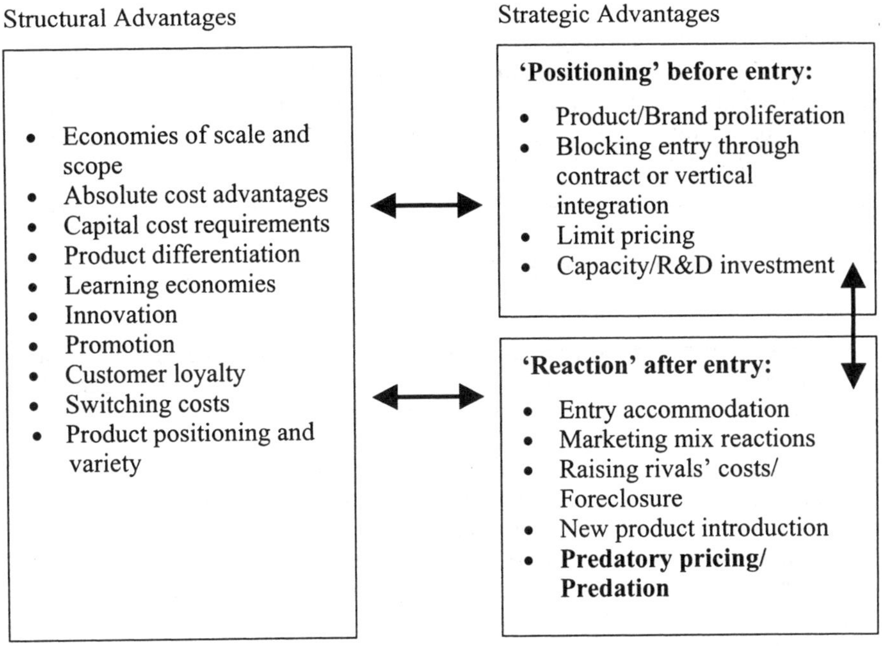

Figure 1.2 Potential structural and strategic advantages of incumbents

In general, 'after entry' strategies could have 'before entry' effects, because they influence the profit expectations of potential entrants. Furthermore, it is reasonable to assume that structural advantages influence the availability, the choice and the success of the incumbent's strategic instruments.[14]

---

[13] The characterization between 'positioning' for strategic moves before entry and 'reaction' to strategic moves after entry stems from Spence ((1981), p. 53). Please note further that Figure 1.2 only uses one possible way to structure advantages of incumbents. Another possibility might be a characterization by 'strategic behaviour through actions' and 'strategic behaviour through signals'. In an overview, Porter ((1981), p. 499ff) shows that signals might be a preferred strategic instrument, as its use is costless. On the other hand, some actions are still needed in order to make the signals credible and strategic behaviour possible.

[14] It is important to mention that – in the sense of Sutton (1991) – structural advantages (or entry barriers respectively) are often the result of strategic escalations on particular endogenous sunk cost dimension(s) like advertising or R&D. In other words – following Sutton's thinking – exogenous sunk costs interact with endogenous sunk costs in determining industrial structure (see especially Sutton (1991), pp. 45-81).

It is however beyond the scope of this paper to discuss in detail all the 'options out of the toolbox of strategic behaviour' available to incumbents presented in Figure 1.2 (see Saloner et al. (2001), Besanko et al. (1998), Gabszewicz (1999) or Grimm and Smith (1997) for further explanations). At first glance, it may be questionable, why accommodation of entry as such could be a strategic advantage. Ashiya (2000) gives an interesting answer to this question by showing the theoretical conditions under which 'weak entrants are welcome'. This basic idea is that the incumbent may allow entry of a weak firm in order to block other more threatening potential entrants. Salop and Scheffman (1987) examine a variety of strategies by which firms could disadvantage rivals by raising their (absolute or relative) costs (e.g. by raising switching costs or wages). Focus 6 presents an example from the airline industry.[15]

Focus 6:
A Strategic Entry Barrier in the Airline Industry:
Gate (Sub)Leases

As argued in detail in Hüschelrath ((1998b), pp. 263ff) the necessary access to airport gates for potential entrants could be used as a strategic (and structural) entry barrier by an incumbent airline. Before entry, airline ownership of gates (or long-term contracts between the airport and a hub airline) may impede or deter entry by other airlines (and might therefore lead to an 'overinvestment' in gate capacities). If entry occurred anyway, the entrant is often dependent on gate sublease contracts with the incumbent airline. This situation makes it possible for the incumbent airline to raise the input prices of the rival, e.g. by demanding high sublease fees. Dependent on the fee level, the market can even be foreclosed to the entrant. Chen and Ross (2000) provide a theoretical examination of these relationships.

It is important to further note that the different (strategic) options of incumbents in Figure 1.2 can reach the same aim of deterring or impeding market entry. In other words, the incumbent often faces a trade-off between the undertaking of (one or more) positioning investments (which are costly and which have the aim to create an incentive for the established firm to react towards the entrant in a way that is destructive to the latter) and the undertaking of (one or more) reaction(s) after the

[15] The results of the models of Salop and Scheffman (1987) suggest that cost-raising strategies can be an important anticompetitive instrument even if the 'predator' is a price-taker in the output market. Furthermore, they show that strategies designed to raise rivals' costs have a number of advantages over predatory pricing strategies: First, cost-raising strategies do not have a problem of credibility. Second, single firm market power is not essential for the success of cost-raising strategies. Therefore, raising rivals' costs is not a strategic move in the sense of the definition in Section 2. Nevertheless, such strategies can be an advantage of incumbents and are therefore part of Figure 1.2.

entry event (see Spence (1981), p. 56).[16] This trade-off is especially important for the following analysis of the relationship between strategic behaviour, welfare and antitrust policy. It indicates that antitrust action against one particular instrument (e.g. predation) might not lead to the intended aim of increasing competition in the market, but simply to a switch by the incumbent from one (prohibited) strategy to another (permitted) strategy (e.g. capacity investment (assuming suitable industry characteristics)).

4. Strategic Behaviour, Welfare and Antitrust Policy

The main purpose of the last section was to illustrate under which conditions strategic behaviour is rational – in the sense of profit maximizing – for an individual incumbent firm facing an entry threat. However, this knowledge is not sufficient in order to make a sound economic judgement on the necessity of public policy interventions in general and the application and/or creation of antitrust rules in particular. What is needed for economic judgements are assessments, first, of the essential features of antitrust policy, second, of the welfare effects of strategic behaviour and third, of approaches to intervene using antitrust tools which will enhance economic efficiency and welfare. In the following, all three areas will be mainly sketched in a *normative* way; that is intending to answer the general question of 'what policies are optimal?' from an economist's point of view.

What are Essential Features of Antitrust Policy Generally?

Despite the historical fact that antitrust laws have been motivated and legitimized by various arguments (brought forward by diverse interest groups), the economic profession currently sees the *central aim of antitrust policy* as being in maintaining and improving economic efficiency.

Given this primary objective, antitrust is largely concerned with two *types of offences*, which comprise the potential to harm competition and efficiency: exploitative and exclusionary behaviour (see Fox (2002) for an overview). 'Exploitative' implies that a number of firms have entered into some form of business together (e.g. form a cartel or a strategic alliance). 'Exclusionary' stands for the attempts of (usually) one firm aiming to achieve a higher degree of sole market power by harming rivals through certain forms of behaviour. For both types of behaviour, theoretical as well as empirical research has shown that they could have either pro- or anti-competitive motives and effects. While several forms of behaviour – like for example horizontal cartels or certain refusals to deal – are presumed to be detrimental to efficiency generally, others – like strategic alliances or certain tying agreements – could be either supportive or detrimental to competition and efficiency.

---

[16] LeBlanc (1992) examines this trade-off theoretically for the case of limit pricing (before entry) and predatory pricing (after entry). In his model setup, he finds that when the incumbent is likely to be strong relative to the entrant, predatory pricing is chosen. When the incumbent is likely to be weak relative to the entrant, limit pricing is chosen. For intermediate cases a strong incumbent may choose a combination of these two signalling strategies.

This observation led to the development of two types of *control strategies* (applied by the antitrust authority and probably the courts): per-se rules and the rule-of-reason. The per-se rule prohibits certain behaviour generally, because the detrimental effects are not only assumed to be dominant, but also almost certain and easily identifiable by the firms (see Carlton et al. (1997), p. 427). The rule-of-reason accommodates the more frequent case, where the anti-competitive effects of certain behaviours must be compared with the efficiency-enhancing effects on a case-by-case basis (see Black (1997) for an overview).

The existence of different control strategies immediately leads to the question of the *timing of control* (ex-ante or ex-post). This criterion is often used to discriminate between regulatory interventions (which normally take place before specific business actions) and antitrust interventions (which normally take place after the business actions). It is obvious that both approaches differ in many aspects, e.g. the information situation, the risk of regulatory capture or the enforcement cost burden (see Rey ((2001), pp. 42-48) for an overview). The fact that antitrust interventions sometimes must take place ex-ante (e.g. in the field of merger control) shows that the 'time of control' delineation between regulation and antitrust is not perfectly selective.[17]

All combinations of 'timing' and 'types' of control strategies contain the *danger of antitrust errors* in the form of wrong (and therefore welfare reducing) case decisions. Basically, two types of false inference (errors) are differentiated. On the one hand, a failure to condemn an instance of anti-competitive behaviour (Type 2 error) and on the other hand an erroneous condemnation of pro-competitive behaviour (Type 1 error). Interestingly, Gal ((2000), p. 93) identifies a third possible antitrust error (Type 3 error) which arises '[w]here a court correctly identifies anti-competitive conduct, but its mode of interference, i.e. its prescribed remedy, harms competition'. Consequently, with the aim of optimizing antitrust policy, the antitrust authority must take the trade-off into account that '[t]oo little enforcement leads to bad behaviour and monopoly power, whereas too vigorous enforcement may deter firms from pursuing desirable forms of competition for fear that this competition will be misinterpreted' (Carlton and Perloff (2000), p. 361).

What are the Welfare Consequences of Strategic Behaviour?

At first glance, it seems straightforward, that strategic behaviour has a strong tendency towards a welfare reduction.[18] Especially if the rival is completely

---

[17] Furthermore it is important to mention that from an institutional point of view, the aims and duties of regulation and antitrust authorities might conflict, at least as far as cases of 'liberalized industries with some form of regulatory oversight' (like e.g. in most telecommunication and post markets) are concerned.

[18] See Canoy et al. (2003) for a discussion of different welfare measurement concepts. In this paper the consumer welfare concept is mainly used. One central justification is that maximizing consumer surplus implies minimizing the deadweight loss (in a simple monopoly setting). An important disadvantage is that this relationship might not hold true in more complex settings. Furthermore such a simple measurement of welfare ignores important dynamic considerations (e.g. innovation and product choice) which normally have a value for consumers.

deterred from entry into a monopoly market, the expected negative welfare effects should be very high as the gains from competition are unachievable. Allocative inefficiency, inefficient rent seeking expenditures and x-inefficiencies remain persistent, leading to inefficiently high monopoly prices for a product with an inefficient (too high or too low) quality level. Due to these problems, the switch from monopoly to duopoly through market entry seems to be most desirable from a welfare point of view, even if the entrant is less efficient than the incumbent (see Edlin (2002)).

But even in such allegedly clear cases of a net welfare reduction due to strategic behaviour, there might be situations in which strategic behaviour enhances welfare. Generally speaking, these are situations in which strategic behaviour is used as an instrument to prevent socially inefficient entry. What would such situations look like? For example, if the cost structure and market demand interact in a specific way – technically, if the demand curve intersects the average cost curve of the monopolist in its sub-additive part – a natural monopoly is created (see Focus 7 for some evidence from the airline industry). In such situations market entry and competition reduces welfare, as the costs of competition for society (e.g. the fixed cost of the second firm) are bigger than the gains of competition (e.g. reduction in price).

Another prominent example is patent protection. Granting a monopoly position (with a patent) to an innovative firm might be necessary to preserve the innovation incentives in the future. As patents normally cannot provide perfect protection (e.g. consider the problem of 'innovating around a patent' by a rival), strategic moves by incumbents might be a substitute to reach the (socially efficient) degree of protection. A third example of 'too much entry' might be cases of high switching costs. Klemperer (1988) shows that in such cases, entry even into monopoly markets might be socially undesirable.

In the case of *entry deterrence into an already oligopolistic industry* comparable arguments concerning welfare are valid. As the performance in oligopoly markets (without collusion) is better than in the monopoly situation, it can be assumed that the social gains of market entry are (relatively) smaller (but still positive) than in the monopoly case above. Negative impacts on social gains of market entry can again be realized in cases where the efficient market structure is already reached ('natural oligopolies') and further competition would reduce welfare, e.g. in cases of excessive entry, created by the so-called business stealing effect (see Mankiw and Whinston (1986) for a basic model). This effect occurs, when the private entry incentives are larger than the social entry incentives.[19]

[19] As far as the arguments of 'inefficient entry' are concerned, it is important to mention two limitations of these theoretical findings. First, almost no empirical evidence concerning this question is existent. The presumably first empirical study is delivered by Berry and Waldfogel (1999). They find a welfare loss due to excessive entry in the radio industry. The generality of their findings is reduced by the very special cost structure of radio broadcasting (large fixed costs and zero marginal costs). In such situations, free entry is most likely to be inefficient. Second, the focus of the theoretical models is mainly static, with the consequence that positive dynamic welfare aspects – like the welfare-enhancing expansion of product choice – are not part of the welfare assessment.

Focus 7:
Natural Monopolies in the Airline Industry

Natural monopolies in the airline industry can be examined at different levels. For example, economies of scale on a *firm level* were analysed mainly prior to deregulation in the US. The central finding was that '... economies of scale are negligible or non-existent at the overall firm level' (White (1979), p. 564). If scale effects are analysed on a route level, scale economies are existent with respect to the aircraft size (because the average operating costs decrease) and with respect to the amount of flights and passengers (as the average operating costs on the ground decrease). This leads to the conclusion that dependent on aircraft technology and the traffic volume many route markets are natural monopolies in the sense that only one firm can operate profitably. Especially short-haul markets are therefore often natural monopoly markets. But it should be kept in mind that the extent of natural monopolies is limited '... as of 1988, only 11% of passenger-miles traveled were in markets with only one carrier, compared with 22% in 1979 ... This suggests that, while natural monopoly exists in US airline markets, its importance should not be overstated' (Keeler (1991), p. 127).

In *situations of accommodated entry*, the welfare implications of strategic behaviour become more ambiguous than in the two former cases. In order to show why, consider a two-period (strategic) game (t = 1, 2) with two firms α and β. In the first period (t = 1), there is no market demand ($P_1 = 0$), but firm α has the possibility to invest into some production capacity K (first-mover-advantage), which it can use in t = 2. This capacity must be interpreted as a commitment; hence the capacity cannot be sold in period 2. The inverse demand function in t = 2 is given by $P_2 = a - bQ$. The cost function is given by:

$$C(q,K) = \begin{cases} cq + lK & \text{if} \quad q \le K \\ \infty & \text{otherwise} \end{cases}$$

The profit π for firm α is then given by:

$$\pi^{\alpha} = -lK_1 + (a - b(q_{\alpha} + q_{\beta}))q_{\alpha} - cq_{\alpha} - l(K_2 - K_1) \quad \text{with} \quad q_{\alpha} \le K_2, K_1 \le K_2$$

The profit π for firm β is:

$$\pi^{\beta} = (a - b(q_{\alpha} + q_{\beta}))q_{\beta} - cq_{\beta} - lK_{\beta} \quad \text{with} \quad q_{\beta} \le K_{\beta}$$

Firm α uses its strategic advantage by taking the reaction function of the follower into account. After maximization and a few transformations the optimal capacity investment is:

$$K_1^* = K_2^* = q_\alpha^* = \frac{a-c-l}{2b}$$

By establishing the whole (irreversible) capacity in period 1, firm α can establish the credible threat that it will use this capacity in period 2. The rival realizes this threat and provides a 'new' optimal quantity:

$$K_\beta^* = q_\beta^* = \frac{a-c-l}{4b}$$

This optimal quantity in response to firm α's strategic move leads to decreasing profits of firm β:

$$\pi_\beta^* = \frac{(a-c-l)^2}{16b}$$

while firm α's profits are rising,

$$\pi_\alpha^* = \frac{(a-c-l)^2}{8b}$$

in each case compared to the symmetric profits of the Cournot outcome (in t = 2)

$$\pi_\alpha^c = \pi_\beta^c = \frac{(a-c-l)^2}{9b}$$

Therefore, as $\pi_\alpha^* > \pi_\alpha^C$, strategic behaviour is rational for firm α.

In order to derive the welfare consequences of the strategic behaviour of firm α, the total market outcome (Q_S) of the strategic game

$$Q_S = q_\alpha^* + q_\beta^* = \frac{3}{4} \cdot \frac{a-c-l}{b}$$

must be compared with the reference outcome (Q_C) of Cournot competition (in t = 2)

$$Q_C = q_\alpha^c + q_\beta^c = \frac{2}{3} \cdot \frac{a-c-l}{b}$$

As $Q_S > Q_C$, the consumers are better off in the situation with strategic behaviour than in the symmetric Cournot world. This example shows in a simple way, that strategic behaviour in situations of accommodated entry is not per se (consumer) welfare reducing.[20] It shows further that increased market power could lead to higher consumer welfare.[21] It therefore provides an intuition about the robust theoretical finding, that not the maximization but the *optimization of competition (and entry)* is congruent with the central aim of antitrust policy.

But this positive judgement of strategic behaviour is only one side of the (welfare) coin as Ordover and Saloner ((1989), p. 538) clearly state: 'Theoretical models ... provide a guarded support for the proposition that strategic choices made by dominant firms are not invariably consistent with the objective of welfare-maximization and that some constraints on firm behaviour may, in fact, increase welfare.' For example, if fixed costs are added to the cost function in the model above, we get a simple limit-pricing model with adverse effects on consumer welfare over a certain range.

Summing this section up, it was shown that strategic behaviour can reduce welfare by deterring or at least impeding market entry. Such behaviour forecloses the general gains from market competition to the consumers and the whole economy. Furthermore it was argued that strategic behaviour also has the potential to either *passively* increase welfare by maintaining a socially efficient degree of market power, or *actively* by leading to forms of 'asymmetric' oligopolistic competition (after accommodated entry with strategic behaviour of the incumbent), which reach better market performances than 'symmetric' oligopoly competition.

How should Antitrust Policy React to Strategic Behaviour?

Antitrust policy's interest lies largely in maintaining and improving economic efficiency. The conversion of this aim into policy actions targeting strategic behaviour is challenged by (at least) two forms of identification problems: First, it might generally be problematic to distinguish competitive from strategic behaviour, as '[r]eal world competition involves direct rivalry between competitors, and rivalry cannot exist without recognition of interdependence and without winners and losers (among the competitors)' (Holt and Scheffman (1989), p. 66). The empirical observation that firms which operate in largely competitive industries often use variants of most types of strategies reinforces the existence of

---

[20] In a general economic sense, this finding reminds of the '*second best problem*', meaning that an increase in market power for a firm that already has some market power does not necessarily reduce welfare. 'Therefore, if the strategist has market power absent of strategic activity, the theory of the second best shows that as a theoretical matter, it is very difficult to establish that aggregate (or even consumer) welfare is necessarily reduced by the strategic activity' (Holt and Scheffman (1989), p. 68).

[21] The market share of the leader in the strategic game is 2/3, while the follower gets 1/3 of the market. In the Cournot case, both firms would gain a market share of 1/2. Therefore, at least in this case, asymmetry in market shares corresponds with higher outputs and correspondingly higher consumer welfare.

this 'signal jamming' problem for antitrust policy.[22] Second, even if certain behaviour is proven to be strategic, antitrust regulation encounters another form of identification problem, as '[i]t is difficult to determine whether strategic behaviour raises or lowers welfare' (Carlton and Perloff (2000), p. 360). Both identification problems together increase the likelihood of wrong case decisions ('antitrust errors') hence raising the question what an optimal antitrust action against strategic behaviour would entail.

Given these challenges, antitrust policy has at least three possible ways to react against strategic behaviour: 'Doing nothing', implementing a 'per-se rule' or introducing some form of 'rule-of-reason' procedure. In the following, all three reaction possibilities will be outlined and discussed, taking into account Spence's general warning ((1981), p. 82) that '[s]ometimes the cure is worse than the disease. Indeed, the policy problem is to know or to learn what if any market processes are regulatable at reasonable cost and low risk, and by what means'.

The *'doing nothing' approach* can be justified by various arguments. For example, 'hard-core' supporters of the Chicago-School would argue that any antitrust action against strategic behaviour is unnecessary, because the market forces (in the long run) will 'automatically' select the most efficient firms and lead to efficient market outcomes. Others may support the opinion that strategic behaviour is generally irrational and that suspicious cases only represent the richness of normal competitive behaviour. In both cases, any need for antitrust action is rejected in principle due to a strong belief in the *power of market processes*.

A related view acknowledges the welfare-reducing potential of strategic behaviour in real markets, but maintains that it is impossible for an antitrust authority to separate between normal competitive behaviour and harmful strategic behaviour. Generally speaking, dynamic market behaviour is assumed to be too complex to be interpreted correctly by the antitrust authority (see Joskow and Klevorick (1979), p. 217). Interventions therefore contain the danger of errors and consequential socially harmful antitrust decisions.[23] Any antitrust action is rejected due to a strong belief in the *failure of the antitrust authority*.

Even in the hypothetical case that the antitrust authority is able to perfectly identify socially harmful strategic behaviour, this does not automatically mean that antitrust action is always the welfare-optimizing answer. Such a judgement should largely depend on some form of cost-benefit analysis, meaning that the optimal level of antitrust enforcement has to be chosen subject to cost considerations. If e.g. the detection and enforcement costs are extremely high and the benefits extremely low, a 'doing nothing' approach might be the optimal antitrust response.

---

[22] A further identification problem might arise between the intent of a certain (strategic?) move. Excess capacity e.g. can be an instrument to deter or impede entry but it could also be an instrument to stabilize a collusive outcome (see Phlips ((1995), pp. 151-172) for an analysis).

[23] In this respect it must be mentioned that antitrust rules (e.g. against predation) might demotivate incumbents to undertake welfare-enhancing activities (e.g. to drop prices) as they are concerned with being wrongly charged by the antitrust authority and the court(s).

Without denying the general importance of such cost-benefit assessments, extreme policy conclusions need to be handled with care. Suppose for example that a certain form of strategic behaviour has a rather rare occurrence in a certain industry, the negative welfare effects are assumed to be small and the antitrust enforcement costs are relatively high (see Focus 8 for an example from the airline industry).

Focus 8:
An Empirical Assessment of Predation in the Airline Industry

In a study *on behalf of United Airlines*, Bamberger and Carlton (1999) assessed in more detail whether predatory pricing is a widespread occurrence in the airline industry. The authors examined entry by low-fare carriers since 1990, the survival rates of low-fare carriers, the fare and capacity responses of incumbents to entry by low-fare carriers and the fare responses of incumbents to exit by low-fare carriers. Bamberger and Carlton conclude that the empirical evidence fails to support the growing concern that incumbent major carriers systematically engage in exclusionary or predatory conduct to force low-fare carriers out of the market. 'The evidence shows that the type of fare and capacity responses that concern the DOT and others are rare and also occur in response to entry by major carriers' (Bamberger and Carlton (1999), p. 14).

At first glance, such a situation appears to call for the application of a 'doing nothing' approach as the welfare-optimizing solution. In order to show that such a conclusion might be drawn too hastily, consider the fact above that 'a certain form of strategic behaviour has a rather rare occurrence'. Such a finding might state little about the importance and influence of a certain form of strategic behaviour on industry competition, because it neglects e.g. the important reputation effects, which are essential for several forms of strategic behaviour. In other words, some reputation-based forms of strategic behaviour (e.g. predatory pricing) would be really expensive, ineffective and therefore irrational, if they had to be frequently employed before or after market entry. The value of these strategies to the incumbent stems from its pre-entry deterrence effect, created by a combination of uncertainty and reputation. The incumbent uses such variables to reduce the profit expectations of the potential entrant, leading – as the case may be – to its decision not to enter at all. Consequently, observable cases of strategic behaviour should be relatively rare and could be interpreted as investments of the incumbent in the credibility of its strategic signals. The 'doing nothing' approach can hence not be simply justified by demonstrating that a certain strategic option is rarely observable in a certain industry.

In a nutshell, the 'doing nothing' approach might be regarded as a problematic general reaction to strategic behaviour, as it appears clear that some forms of strategic behaviour lead to evident negative welfare implications and are only used to protect monopoly power. Consequently, antitrust policy has at the very least the potential to improve welfare. Although the argument that antitrust policy contains the risk of errors is valid, it is not outstanding, as every policy action potentially faces similar problems. In general, a complete abandonment of all policy actions – which is equivalent to the maximization of Type 1 errors while minimizing Type 2 errors – is not a welfare-optimizing solution. Therefore, the most important insight of the 'doing nothing' approach is that costs and benefits should be taken into account when choosing and designing antitrust rules. Furthermore, the approach unintentionally points out that the economic profession should guide the analysis of antitrust authorities (e.g. by the development of analytical frameworks) to minimize the probability of erroneous case decisions.

The *'per-se rule' approach* prohibits certain behaviour generally. Accordingly, the antitrust authority only has to answer the question 'Did the incumbent engage in the proscribed practice?' If the answer is yes, the antitrust authority and afterwards the courts are committed to suppress the behaviour in question and eventually fine the relevant firm. The authority/court actions are furthermore independent of the question whether there has been an injury to competition or not (see above and Calvani (2001) for a more exact treatment). This 'per se' commitment to suppression and punishment creates an important deterrence effect for incumbent firms who consider the adoption of such prohibited strategies.

The central characteristics of the 'per-se rule' predestine the approach for types of behaviour that are clearly identifiable[24] (for the firms and the antitrust authority) and have clear (and almost certain) negative welfare consequences. However, assessing strategic behaviour generally faces the problems in achieving one or even both of these necessary conditions, as strategic behaviour is often difficult to distinguish from other competitive actions and it is often unclear whether strategic behaviour reduces or enhances welfare. Therefore, it appears impossible to design one single per-se rule equally applicable to the large number of different forms of strategic behaviour (at least without accepting the frequent emergence of all three possible errors of antitrust action). Consequently, strategic behaviour overall should not be treated by a 'per-se' rule approach.

Acknowledging this general judgement, it might be worth considering some form of *conditional 'per-se rule' approach*. Such a modified per-se rule would still directly prohibit certain behaviour by the incumbent, but *subject to* certain environments and attributes of the incumbent, the entrant and the entry attempt. Focus 9 outlines one example of such a conditional per-se rule approach.

---

[24] Note that the interpretation of a strategy as 'easily identifiable' might contain errors. For example, for a long time courts treated predatory pricing using a 'rule-of-thumb', stating that only if prices are below average variable costs, welfare-reducing predatory pricing is assumed as a matter of principle. However, economic research has shown that such a selection by rules of thumb is highly error-prone and hence often the source of wrong case decisions. For example, Edwards ((2002), p. 183ff) shows conditions in which pricing below average variable costs is welfare enhancing. Edlin (2002) discusses the antithetic problem of successful predatory pricing above average variable costs.

Focus 9:
An Ex-Ante 'Per-Se' Approach to Stop Above-Cost Predatory Pricing in the Airline Industry

In a recent paper, Edlin (2002) proposes an ex-ante per-se rule to reduce predatory pricing incentives to incumbents. The rule states

... that in markets where an incumbent monopoly enjoys significant advantages over potential entrants, and the firm decides to enter anyway charging substantially lower prices (at least a 20% discount), the monopoly cannot respond with lower prices or significant product enhancements for 12-18 months or until the entrant's share grows enough that the monopoly loses its dominance. If the entrant does not observe these structures, ... then the entrant under this proposal can sue successfully for predatory pricing without needing to demonstrate below-cost pricing or the opportunity for recoupment (Edlin (2002), p. 949).

In order to better understand the general idea of conditional 'per-se' approaches, consider for example a liberalized industry characterized by a large number of monopoly markets. As discussed above, entry into such monopoly markets is on one hand (normally) very desirable from a welfare perspective (it gives access to the general merits of competition), but on the other hand it is often relatively difficult (because entry deterrence and encouragement of exit might be relatively easy to achieve). In such an environment, it might be welfare enhancing to introduce a conditional per-se approach like the following: 'If the entry attempt fulfils certain conditions and takes place in a special environment (in other words if entry is most likely to be welfare enhancing), then the incumbent is prohibited from reacting in ways, which are highly likely to strongly inhibit entry incentives and/or entry attempts. In case the incumbent violates this rule, its behaviour will be automatically suppressed and fined'.

The general motivation for the design of such a conditional per-se rule is twofold: First, *pre-entry*, the incumbent is encouraged to price low, 'because otherwise entry will be attractive and the incumbent won't be able to respond' (Edlin (2002), p. 955). Second, *post-entry*, the rule will protect entrants from the particular form of strategic behaviour (e.g. predatory pricing above), which could otherwise lead to the failure of the entrant.

At first sight, such ex-ante approaches appear to be the silver bullet of antitrust policy, because they seem to reduce or even block the gains of strategic behaviour almost without cost. However, closer inspection unfortunately reveals that the practicability of such approaches is often quite low (see Hüschelrath (2003) for a discussion). In addition to the general disadvantages of per-se rules and ex-ante antitrust action (consider for example the high information needs, the low flexibility and the introduction of a Type 3 antitrust error), such approaches are particularly challenged by a *consistency problem*. This problem emerges, when one strategic option (out of the entire toolbox of strategic behaviour) is scrutinized by antitrust policy, while other options (which might be substitutes) are assumed to be independent.

This 'independency' assumption bears problems, as it denies the fact that selective changes in antitrust policy obviously affect the relative advantages of the mechanisms of strategic behaviour. Under the assumption that the strategic choices of a dominant firm grow primarily out of the specific industry structure, such selective antitrust rules would most likely lead to switches between anticompetitive strategies rather than the intended complete disappearance of strategic behaviour.

To illustrate the argument, imagine a scenario with a perfectly enforceable (but costly) per-se rule targeting one certain form of strategic behaviour (e.g. predation). Such a perfect rule might not lead to the intended aim of more competition in the market, but simply to a switch by the incumbent from the now prohibited and deterred strategy to a substitutive strategy (e.g. capacity investment), which is permitted or more difficult to identify by the antitrust authority. In the worst case scenario, antitrust action would lower overall welfare twice because the general enforcement costs are used to achieve the switch to a strategy which is more harmful to society (and the incumbent) than the previous (see Smith (1984), p. 521).

In a nutshell, the analysis illustrates that although the 'per-se' rule approach has some desirable general advantages, the essential conditions for its welfare-enhancing use are not present in each possible case of strategic behaviour. An exception to this could be those environments in which entry is most likely to be welfare-enhancing and the entrant promises 'some degree of efficiency'. In such cases a conditional per-se rule might be (at least in theory) a desirable commitment of the antitrust authority. If and how such rules can overcome the identified problems of practicability (foremost the expected counterproductive reactions of incumbents to such rules) should be the focus of further research.

The *'rule-of-reason' approach* stands for a case-by-case analysis of suspicious behaviour by the antitrust authority (and probably) the courts. Broadly, it is used when a per-se rule is not applicable; that is particularly when certain business behaviour cannot be clearly identified and/or its influence on welfare cannot be generally determined. Consequently, from a procedural point of view, the rule-of-reason allows the antitrust authority some considerable administrative discretion.

At first, such a comprehensive case-by-case analysis might appear like the optimal general antitrust reaction to strategic behaviour, as it has advantages such as fairness or the minimization of antitrust errors (see Black (1997), p. 152 for a further discussion). However, when enforcement costs and predictability matters are taken into account, it becomes obvious that an inclusive analysis of every single case at hand would be a very expensive and inefficient undertaking (see Joskow and Klevorick (1979) for a complete treatment). Consequently, with the aim of concentrating resources and efforts on the *likely* cases of anticompetitive strategic behaviour, the antitrust authority must filter out the unlikely cases at the earliest point possible during the investigation.

How would such a 'filtering procedure' take place, keeping the general finding in mind that the economic profession cannot 'provide a universally valid set of conditions that can be used by competition authorities as a safe checklist in their rule-of-reason analyses' (Ahlborn et al. (2003), p. 56)? The central idea is to use essential (robust) preconditions for strategic behaviour as well as suitable economic lines of argument (at best models). The questions investigated are as follows: is anticompetitive strategic behaviour in a certain case possible (first filter), plausible (second filter) and possibly offset by efficiency benefits (third filter). This basic structure of a policy framework for strategic behaviour will be outlined in the following (see Ahlborn et al. (2003) generally and for an application to the tying strategy).

First Filter: Is strategic behaviour possible? The first filter has the function to rule out cases, in which strategic behaviour is impossible or extremely unlikely. Here the robust theoretical knowledge concerning essential preconditions of successful strategic behaviour is largely applied. If such preconditions cannot be observed in the actual case, the business behaviour must be explained differently and the case can be closed (at least as far as strategic behaviour is concerned).

The most robust precondition for successful strategic behaviour is some degree of market power. Therefore, cases in which a firm is alleged to behave strategically but does not have a considerable degree of market power can be closed immediately without causing further costs. Other necessary preconditions for strategic moves are the existence of a first-mover advantage, information advantages, a credible commitment and the possibility to send/receive signals (see Section 2).

Besides these basic preconditions of strategic moves, other structural characteristics of the market might influence the general probability of strategic behaviour (see Scherer and Ross (1990), p. 5 for the following). Important basic conditions on the supply side could be technology, product durability, business attitudes or the legal framework. On the demand side, especially price elasticity, substitutes, growth rates or the cyclical or seasonal character of the industry should be considered. The market structure (in the narrower sense) can be examined via the assessment of entry barriers, cost structures, the general status of competition, product differentiation, vertical integration, attributes of (potential) entrants (e.g. the likelihood of exit and the abilities to react) and the maturity and dynamics of the market (particularly relevant for examining the role of reputation effects).[25]

In short, a case will have to face the second filter, if the antitrust authorities or courts can reasonably expect that anticompetitive strategic behaviour could occur and impose significant economic losses on society.

Second Filter: Is strategic behaviour plausible? The second filter has the function to find logical (theory-based) and likely (case-based) arguments making the observed (strategic?) behaviour '*plausible ex ante and probable ex post*' (Bolton et al. (2000), p. 2266). To reach this aim it is first necessary to generally assess market conduct (e.g. pricing behaviour, product strategy, research and innovation,

[25] It is beyond the scope of this paper to show more precisely how these market structure characteristics interact with the rationality and the success of strategic behaviour. Such assessments obviously depend on the kind of strategic behaviour used (see Bolton (2000) for an application to the predatory pricing strategy and Ahlborn et al. (2003) to the tying strategy).

plant investments, managerial incentives or legal tactics). Second, suitable lines of argument (at best suitable economic models) as well as supporting empirical evidence are needed to prove anticompetitive strategic conduct. In other words, the antitrust authority must present an 'explanation', which clarifies the general rationality, the actual occurrence and the anticompetitive impact of the presumed strategic mechanism. In this respect, the second filter follows the instinct of Spence ((1981), p. 58): '*My instinct as an economist is to study industries on a case-by-case basis, applying and adapting models as appropriate*' (Spence (1981), p. 58).

Third Filter: Are there offsetting efficiency benefits of strategic behaviour? Ideally only cases of anticompetitive strategic behaviour will have to face the third filter. Before such conduct is prohibited (and eventually fined), the antitrust authority has to examine if the behaviour generates efficiencies which offset the anticompetitive effects *and* which can only be achieved through anticompetitive strategic behaviour (the 'no less restrictive alternative' requirement (see Bolton et al. (2000), p. 2277)). It is obvious that the third filter uses the information collected in the previous two stages, supplemented by further internal information from the parties (e.g. cost data). Additionally, some projections (or scenarios) of the likely future development of the industry (with and without terminating strategic behaviour) and the (dynamic) effects on competition and consumer welfare would become necessary.

With respect to the discussed major problems faced by antitrust analysis of strategic behaviour, the 'rule-of-reason' approach has one important advantage: It directly addresses the central identification problems. This does not mean that the probability of antitrust errors is zero, but a diligent use of the approach minimizes the occurrence of such errors, assumed that the antitrust authority has sufficient economic knowledge and enforcement budgets. Furthermore, the 'rule-of-reason' has clear flexibility and fairness advantages compared to 'per-se' rules, as the peculiarities of every case are examined and estimated during the investigation. These flexible investigations lead to another advantage of the 'rule-of-reason': new economic knowledge is easier and quicker to adopt than in a 'per-se' approach. This characteristic reduces antitrust errors further and enhances antitrust decisions.

Some disadvantages of the approach – which should be taken into account while designing policy frameworks – are for example that the framework approach provides only a very broad structure for analysis, which must be substantiated for special cases in special industries (complexity problem). It is obvious that such substantiation exacerbates the main problems of the 'rule-of-reason' approach: it is often rather costly and time-consuming (enforcement cost problem). In particular the long processing time within the antitrust authority creates difficulties in situations where strategic moves are successful (e.g. induced market exit) before the antitrust authority is in the position to intervene (time lag problem). A further general problem lies in the large information needs of the concept. In order to make an informed framework-based decision, extensive industry as well as case data is necessary (information problem). Furthermore, as indicated above, the 'rule-of-reason' approach sets high standards as far as economic competence in the antitrust authority as well as general enforcement cost budgets are concerned (competence and budget problem). Besides, as the 'rule-of-reason' concept imparts the antitrust authority with vast administrative powers, a further channel of influences by political and private interest groups is opened (regulatory capture problem). This might reduce the 'optimality' of antitrust decisions.

In brief, despite the disadvantages of the 'rule-of-reason' concept as discussed, the major advantages (as well as the disadvantages of the other approaches discussed above) support the application of this concept to problems of strategic behaviour. Exceptions might be cases in which the antitrust authority lacks sufficient resources (especially knowledge and manpower) to perform 'rule-of-reason' analyses. In such cases, the application of a (conditional) 'per-se' rule approach might be preferable to simply 'doing nothing' against anticompetitive strategic behaviour.

Complementary to all three antitrust approaches, a further way to fight undesirable strategic behaviour and to improve market performance could be an approach which '*reduces entry barriers*'. Focus 10 discusses several examples of such entry stimulation from the airline industry.

Focus 10:
Examples of Entry Stimulation in the Airline Industry

One possibility to stimulate entry into airline markets is to reduce the loyalty effects of FFP, e.g. by taxation of these programmes. Such a step would reduce the value of FFP for the customers and might therefore increase the demand potential for entrants. Further possibilities of entry stimulation include steps to ease access to essential facilities like landing rights (e.g. by the introduction of slot trading in slot clusters, see Hüschelrath ((1998a), pp. 253ff)) or airport gates (e.g. by the introduction of entrepreneurial gates, see Hüschelrath ((1998b), pp. 259ff)).

Some researchers generally favour this approach, because it is easy to implement and it minimizes Type 1 errors as well as enforcement costs (see e.g. Porter (1981), p. 493ff). Besides the fact that reducing entry barriers has in most cases indeed the potential to promote competition by enhancing the possibility and probability of entry, it is questionable if such an approach can act as a perfect substitute to some form of antitrust intervention. One problem, which supports this sceptical judgement, is that the reduction of mainly structural entry barriers does not necessarily lead to a reduction of strategic behaviour, but rather to an increase. The reason is the following: The reduction of mainly structural entry barriers does not erase the incentives of incumbents to impede entry. Such moves might still be profit maximizing and incumbents will use the cheapest instrument to reach their aim. Therefore, if the structural entry barriers are reduced, a corresponding increase in strategic behaviour can be expected. Consequently, a 'reducing entry barriers' approach might have a positive effect on the growth of strategic behaviour in the market. This in turn would reinforce the call for some form of antitrust policy.

A further problem, which limits the influence of the 'reducing entry barriers' approach, is that antitrust authorities often do not have the executive powers to change policy parameters in favour of entry conditions. For example, the antitrust authority normally has only limited influence on tax levels, independent of the question if they act as an entry barrier in an industry or not. In conclusion, although it is undisputed that 'reducing entry barriers' is desirable in most industries, such an approach is not a substitution but rather complements the development and enforcement of antitrust rules.

5. Summary and Conclusion

What is meant by 'strategic behaviour'? – What are necessary preconditions for strategic moves and which different strategic options are available for incumbents? – What are the welfare consequences of strategic behaviour and – in direct connection – what role should antitrust policy play? The aim of this paper was to answer these fundamental questions. Such a general introduction of *strategic behaviour of incumbents* is important for both the better understanding of competitive interaction generally and for the understanding of the particular topic of the present book: *predation* as one option out of the 'toolbox of strategic behaviour for incumbents'.

After a short introductory section, *Section 2* concentrated on the initial characterization of strategic behaviour from an industrial organization point of view. It was shown that a successful strategic move is based on at least five necessary conditions: a dynamic context, a first mover advantage, some form of (credible) commitment, some form of communication and last but not least some degree of market power. If these conditions are present, an incumbent is able to undertake a successful strategic move, meaning that – by altering its own actions – it can induce the rival to adopt a course of action more favourable to itself.

The interrelation between strategic behaviour and entry was assessed in greater detail in *Section 3*. After reviewing the general importance of entry for competition and market performance, the first focus lay on an analysis of the drivers of a firm's entry decision as well as an incumbent's (re)action to such entry. In both cases, the decision is dependent on its 'profitability' and its 'possibility'. The entrant will enter the market if the net present value of the expected post-entry profits is greater than the sunk costs of entry and if entry is not blocked or deterred by structural entry barriers. The incumbent will generally react to entry threats by acting in a way that 'raises rivals' costs' or 'reduces rivals' revenues'. The precise ways to reach this aim are diverse. Simplifying, one option for the incumbent is to increase the structural entry barriers with the aim of making entry impossible or at least unprofitable. As this strategy might not be sufficient or too expensive to deter entry, the incumbent might think about further strategic moves to secure its monopoly profits. The rationality and success of such moves depend on the difference between the monopoly and the duopoly profits, the level of the sunk investment, and the difference between the duopoly profits and the profits during a fight (after entry of the rival). Even if the result of such a calculation is that entry deterrence is not profit maximizing, strategic behaviour might still act as an instrument to create an advantage in post-entry market competition.

Based on this notion of strategic behaviour, the second focus of Section 3 was an assessment of the potential advantages of incumbents. The results show that incumbents often have market power as well as information advantages, which allows them to raise structural barriers as well as to use some form of strategic behaviour. The strategic advantages of incumbents can be classified into pre-entry options (e.g. product proliferation, blocking entry through contract, limit pricing or capacity investment) and post-entry options (e.g. entry accommodation, marketing mix reactions, raising rivals' costs or predation). The incumbent can choose one or combinations of different options to reach its aim of maintaining monopoly profits.

After analysing the individual rationality of strategic behaviour for an incumbent, *Section 4* focused on the question of how antitrust policy should (re)act to such strategic moves. A sound economic answer to this question must contain assessments, first, of the essential features of antitrust policy, second, of the welfare effects of strategic behaviour and third, of approaches to intervene using antitrust tools which will enhance economic efficiency and welfare. With respect to the first emphasis, it was found that antitrust policy's central aim is maintaining and improving economic efficiency. Competition (as the central instrument to achieve economic efficiency) is endangered by exploitative as well as exclusionary behaviour of firms with considerable degrees of market power. Antitrust policy can react to these challenges with the implementation of different types of control strategies, which can be applied ex-ante or ex-post. However all potential options contain the possibility of antitrust errors and therefore welfare-reducing case decisions.

In order to decide which antitrust reaction to strategic behaviour is appropriate, an examination of the likely welfare effects of such behaviour is essential. In cases of entry deterrence in monopoly markets, negative welfare consequences are most likely, because such moves foreclose the social gains of competition. But even in such cases, phenomena of socially inefficient entry (e.g. natural monopolies, patent protection and high switching costs) prevent an unambiguous negative welfare judgement. It is straightforward that such a clear judgement cannot be reached either in cases of entry deterrence in oligopolistic industries or in cases of accommodated entry. Therefore two important conclusions of the welfare assessments are first, that the welfare implications of strategic behaviour are diverse and second, that entry into an industry is not always socially desirable.

These results are an important input requirement for the assessment of an appropriate antitrust reaction to strategic behaviour. Generally, three policy options are available: 'doing nothing', a 'per-se rule' or a 'rule-of-reason' approach. The decision between these options particularly depends on the characteristics of the strategy in question (especially its clear identification and welfare consequences), but also on an assessment of enforcement cost and antitrust errors. A discussion of the policy options showed that the 'doing nothing' approach is the least attractive option. Due to the diverse welfare effects and identification problems of strategic behaviour, a general 'per-se' rule can also be ruled out as an appropriate antitrust reaction. The theoretically attractive idea of some kind of 'conditional per-se' rule (subject to certain environments and certain attributes of the incumbent and the entrant) is also flawed, because such rules do not take the strategic responses of incumbents into account.

Partly due to the disadvantages of these approaches, the application of a 'rule-of-reason' (case-by-case) approach was identified as the appropriate antitrust reaction to strategic behaviour. In order to make this approach operable (especially to impede the expensive examination of every case at hand), the antitrust authority must promptly identify those cases in which anticompetitive strategic behaviour is unlikely. Such a 'filtering procedure' or 'framework approach' should contain the following three filters: 1. Is strategic behaviour possible?, 2. Is strategic behaviour plausible?, 3. Are there offsetting efficiency benefits of strategic behaviour? Additional to the compulsory assessments of market structure and market conduct in such a framework, the antitrust authority must develop an 'explanation', which clarifies the general rationality, the actual occurrence and the anticompetitive impact of the presumed strategic mechanism. If these anticompetitive effects are not offset by efficiency benefits, the antitrust authority must prohibit this form of anticompetitive strategic behaviour.

In summary, the paper showed that strategic behaviour can be a rational strategy for an incumbent firm facing an entry threat. As long as such strategic moves deter entry into monopoly markets, negative welfare effects are likely. In other markets, the probable welfare effects become unclear. When the identification problems of strategic behaviour are taken into account as well, the need for some antitrust action can best be met by the application of a 'rule-of-reason' (case-by-case) approach. Although the difficulties with this approach should not be neglected when designing policies, it is unlikely that they would lead to the introduction of error-prone 'per-se rules' or even the complete abandonment of any kind of antitrust action. On the contrary, the problems with the 'rule-of-reason' approach should provide a basic motivation for further economic research in antitrust economics. The general positive influence of such research efforts on the quality of antitrust decisions during the last decades is beyond controversy.

Acknowledgements

I'm indebted to Peter Forsyth and Natalie McCaughey for many valuable comments. Furthermore I would like to thank Fridrik Baldursson, Hans-Martin Niemeier and Martti Virtanen for helpful comments on earlier drafts of the paper.

References

Ahlborn, C., D. Evans and J. Padilla (2003), The Antitrust Economics of Tying, Unpublished Working Paper, AEI-Brookings Joint Center for Regulatory Studies, Washington D.C.

Amir, R., V. Lambson (2000), On the Effects of Entry in Cournot Markets, *Review of Economic Studies*, Vol. 67, 235-254.

Ansoff, I. (1987), The Emerging Paradigm of Strategic Behavior, *Strategic Management Journal*, Vol. 8, 501-515.

Ashiya, M. (2000), Weak Entrants are Welcome, *International Journal of Industrial Organization*, Vol. 18, 975-984.

Bain, J. (1956), *Barriers to New Competition*, Harvard University Press, Cambridge.

Bamberger, G. and D. Carlton (1999), *An Empirical Assessment of Predation in the Airline Industry*, Draft, Lexecon Inc. and University of Chicago.

Baum, J. and H. Korn (1996), Competitive Dynamics of Interfirm Rivalry, *Academy of Management Journal*, Vol. 39, 255-291.

Baumol, W., J. Panzar and R. Willig (1982), *Contestable Markets and the Theory of Industry Structure*, Harcourt Brace Jovanovich Publishers, San Diego et al.

Berry, S. and J. Waldfogel (1999), Free Entry and Social Inefficiency in Radio Broadcasting, *Rand Journal of Economics*, Vol. 30, 397-420.

Besanko, D., D. Dranove and M. Shanley (1996), *The Economics of Strategy*, John Wiley & Sons, New York.

Black, O. (1997), Per Se Rules and Rules of Reason: What are They?, *European Competition Law Review*, Vol. 18, 145-161.

Bolton, P., J. Brodley, M. Riordan et al. (2000), *Predatory Pricing: Strategic Theory and Legal Policy*, Georgetown Law Review, Vol. 88, 2239-2330.

Boyer, M. and M. Moreaux (1997), Capacity Commitment versus Flexibility, *Journal of Economics and Management Strategy*, Vol. 6, 347-376.

Bulow, J., J. Geanakoplos and P. Klemperer (1985), Multimarket Oligopoly: Strategic Substitutes and Complements, *Journal of Political Economy*, Vol. 93, 488-511.

Bunch, D. and R. Smiley (1992),Who Deters Entry? Evidence on the Use of Strategic Entry Deterrents, *Review of Economics and Statistics*, Vol. 74, 509-521.

Cabral, L. (2000), Introduction to Industrial Organization, MIT Press, Cambridge.

Calvani, T. (2001), Some Thoughts on the Rule of Reason, *European Competition Law Review*, Vol. 22, 201-207.

Canoy, M., P. Rey and E. van Damme (2003), Dominance and Monopolization, forthcoming in: Neumann, M. and J. Weigand, *International Handbook of Competition*, Edward Elgar, Cheltenham.

Carlton, D., R. Gertner and A. Rosenfield (1997), Communication among Competitors: Game Theory and Antitrust, *George Mason Law Review*, Vol. 5, 423-440.

Carlton, D. and J. Perloff (2000), *Modern Industrial Organization*, Addison-Wesley, Reading.

Chen, Z. and T. Ross (2000), Strategic Alliances, Shared Facilities, and Entry Deterrence, *Rand Journal of Economics*, Vol. 31, 326-344.

Church, J. and R. Ware (2000), *Industrial Organization: A Strategic Approach*, McGraw-Hill, Boston.

Dixit, A. (1980), The Role of Investment in Entry Deterrence, *Economic Journal*, Vol. 90, 95-106.

Dixit, A. and B. Nalebuff (1991), *Thinking Strategically*, Norton, New York.

Edlin, A. (2002), *Stopping Above-Cost Predatory Pricing*, Yale Law Journal, Vol. 111, 941-991.

Edwards, G. (2002), The Perennial Problem of Predatory Pricing, *Australian Business Law Review*, Vol. 30, 170-201.

Fox, E. (2002), What is Harm to Competition? Exclusionary Practices and Anticompetitive Effect, *Antitrust Law Journal*, Vol. 70, 372-411.

Fudenberg, D. and J. Tirole (1984), The Fat-Cat Effect, the Puppy-Dog Ploy, and the Lean and Hungry Look, *American Economic Review*, Vol. 74, 361-366.

Gabszewicz, J. (1999), Strategic Interaction and Markets, Oxford University Press, Oxford.

Gal, M. (2000), Harmful Remedies: Optimal Reformation of Anticompetitive Contracts, *Cardozo Law Review*, Vol. 22, 91-132.

Geroski, P. (1991), *Market Dynamics and Entry*, Oxford University Press, Oxford.

Geroski, P. (1995), What Do we Know about Entry?, *International Journal of Industrial Organization*, Vol. 13, 421-440.

Ghemawat, P. and P. del Sol (1998), Commitment versus Flexibility?, *California Management Review*, Vol. 40, 26-42.

Grimm, C. and K. Smith (1997), *Strategy as Action: Industry Rivalry and Coordination*, West Publishing, Cincinnati.

Gruca, T. and D. Sudharshan (1995), A Framework for Entry Deterrence Strategy: The Competitive Environment, Choices, and Consequences, *Journal of Marketing*, Vol. 59, 44-55.

Grundy, T. and R. Wensley (1999), Strategic Behavior: The Driving Force of Strategic Management, *European Management Journal*, Vol. 17, 326-334.

Holt, C. and D. Scheffman (1989), Strategic Business Behavior and Antitrust, in: Larner, R. and Meehan, J. (eds), *Economics and Antitrust Policy*, New York, 39-82.

Hüschelrath, K. (1998a), *Infrastrukturengpässe im Luftverkehr*, Deutscher Universitäts-Verlag (Gabler), Wiesbaden.

Hüschelrath, K. (1998b), *Liberalisierung im Luftverkehr*, Tectum Verlag, Marburg.

Hüschelrath, K. (2003), *How to Prey on Predators?*, unpublished Working Paper, Vallendar.

Ito, H. and D. Lee (2003), *Incumbent Responses to Lower Cost Entry: Evidence from the U.S. Airline Industry*, unpublished Working Paper, Providence.

Joskow, P. and A. Klevorick (1979), A Framework for Analyzing Predatory Pricing Policy, *Yale Law Journal*, Vol. 89, 213-270.

Joskow, A., G. Werden and R. Johnson (1994), Entry, Exit, and Performance in Airline Markets, *International Journal of Industrial Organization*, Vol. 12, 457-471.

Keeler, T. (1991), Airline Deregulation and Market Performance: The Economic Basis for Regulatory Reform and Lessons from the US Experience, in: Banister, D. and K. Button, *Transport in a Free Market Economy*, London, 121-170.

Klemperer, P. (1988), Welfare Effects of Entry into Markets with Switching Costs, *Journal of Industrial Economics*, Vol. 37, 159-165.

LeBlanc, G. (1992), Signalling Strength: Limit Pricing and Predatory Pricing, *Rand Journal of Economics*, Vol. 23, 493-506.

Lieberman, M. and D. Montgomery (1998), First-Mover Advantages, *Strategic Management Journal*, Vol. 9, 41-58.

Lin, J.-S., M. Dresner and R. Windle (2002), Determinants of Price Reactions to Entry in the U.S. Airline Industry, *Transportation Journal*, Vol. 41, 5-22.

Lipczynski, J. and J. Wilson (2001), *Industrial Organization*, Prentice Hall, Harlow.

Mankiw, N. and M. Whinston (1986), Free Entry and Social Inefficiency, *Rand Journal of Economics*, Vol. 17, 48-58.

Martin, S. (2001), *Industrial Organization: A European Perspective*, Oxford University Press, Oxford.

McAfee, P., H. Mialon and M. Williams (2004), *What is a Barrier to Entry?*, unpublished Working Paper, Pasadena.

Montaguti, E., S. Kuester and T. Robertson (2002), Entry Strategy for Radical Product Innovations: A Conceptual Model and Propositional Inventory, *International Journal of Research in Marketing*, Vol. 19, 21-42.

Ordover, J. and G. Saloner (1989), Predation, Monopolization, and Antitrust, in: Schmalensee, R. and R. Willig (eds), *Handbook of Industrial Organization*, Volume I, Amsterdam.

Oster, S. (1994), *Modern Competitive Analysis*, Oxford University Press, New York.

Phlips, L. (1995), *Competition Policy: A Game-Theoretic Perspective*, Cambridge University Press, Cambridge.

Porter, M. (1981), Strategic Interaction: Some Lessons from Industry Histories for Theory and Antitrust Policy, in: Salop, S. (ed.), *Strategy, Predation, and Antitrust Policy*, Washington D.C., 449-506.

Prabhu, J. and D. Steward (2001), Signaling Strategies in Competitive Interaction: Building Reputations and Hiding the Truth, *Journal of Marketing Research*, Vol. 38, 62-72.

Rey, P. (2001), *Towards a Theory of Competition Policy*, unpublished Working Paper, Toulouse.

Saloner, G., A. Shepard and J. Podolny (2001), *Strategic Management*, Wiley, New York.

Salop, S. (1981), Strategy, Predation, and Antitrust Analysis: An Introduction, in: Salop, S. (ed.), *Strategy, Predation, and Antitrust Policy*, Washington D.C., 1-42.

Salop, S. and D. Scheffman (1987), Cost-Raising Strategies, *Journal of Industrial Economics*, Vol. 36, 19-34.

Schelling, T. (1960), *The Strategy of Conflict*, Harvard University Press, Cambridge.

Scherer, F. and D. Ross (1990), *Industrial Market Structure and Economic Performance*, Boston.

Schulz, N. (1995), *Unternehmensgründungen und Markteintritt*, Heidelberg.

Smith, R. (1984), On Strategic Behavior as a Basis for Antitrust Action, *Antitrust Bulletin*, Vol. 29, 501-522.

Spence, M. (1979), Investment Strategy and Growth in a New Market, *Bell Journal of Economics and Management*, Vol. 10, 1-19.

Spence, M. (1981), Competition, Entry, and Antitrust Policy, in: Salop, S. (ed.), *Strategy, Predation, and Antitrust Policy*, Washington D.C., 45-88.

Stigler, G. (1968), *The Organization of Industry*, Homewood, Irwin.

Sutton, J. (1991), *Sunk Costs and Market Structure*, MIT Press, Cambridge.

Tirole, J. (1995), *The Theory of Industrial Organization*, MIT Press, Cambridge.

United States of America. General Accounting Office (2001), *Aviation Competition. Challenges in Enhancing Competition in Dominated Markets*, GAO/01-518T, Washington D.C.

Vickers, J. (1985), The Economics of Predatory Practices, *Fiscal Studies*, Vol. 6, 24-36.

Viscusi, W., J. Vernon and J. Harrington (1997), *Economics of Regulation and Antitrust*, Cambridge.

Von Weizsäcker, C. (1980), *Barriers to Entry. A Theoretical Treatment*, Springer, Berlin.

Waldman, D. and E. Jensen (2000), *Industrial Organization*, Addison Wesley, Boston.

White, L. (1979), Economies of Scale and the Question of 'Natural Monopoly' in the Airline Industry, *Journal of Air Law and Commerce*, Vol. 44, 545-573.

Chapter 2

Predatory Pricing: Still a Rare Occurrence?

Ashish Lall

> Ronald [Coase] said he had gotten tired of antitrust because when the prices went up the judges said it was monopoly, when the prices went down they said it was predatory pricing, and when they stayed the same they said it was tacit collusion. – William Landes

1. Introduction

There seems little agreement among economists about the structure of the aviation industry in the United States. Some claim that the big five carriers have increased their shares of passenger traffic over the last decade and this has come at the cost of independent low cost carriers. Concentration is even higher at the local and regional level and the large airports of the Nation are dominated by monopolies and tight oligopolies. Network carriers have used 'sharp' and 'tightly-focused' price cuts and predatory pricing to evict low cost entrants and then raised prices. This, and the practice of price discrimination is evidence of the market power of dominant incumbents. As a result, airfares have increased by more than the increase in the general price level.[1] Government prosecutors in the recent American Airlines and Air Canada predatory pricing cases painted a similar picture; one of an industry dominated by strong monopolists who use their mobile capacity and reputation for toughness to fight entrants.

Others argue that network carriers in the United States and in Canada have a serious fiscal problem. They support their claim with evidence which shows that over the past decade the market share of all network carriers in the United States has fallen from about 70% to 65% and that of the big five, from 60% to 55%. The share of independents has declined from about 22% to 19% but this is because Southwest has increased its share from 7% in 1990 to over 15% in 2001. Market shares at major U.S. airports show similar trends. The average price per mile for domestic air service has remained constant in nominal terms: 18.8 cents in 1991 and 18.6 cents in 2001. In real terms however it declined because consumer prices increased at about 2.8% per annum during the 1990s.[2] Further, they find no

---

[1] See Brock (2000).
[2] See Lee (2003) and Ito and Lee (2004).

evidence of predation or reputations effects. The response of incumbents to entry has been muted in general and exit by low cost carriers has more to do with thin markets and poor business models.

The stark difference in these views carries over to the debate on predatory pricing. The recent cases in the airline industry have resurrected old arguments and started new ones. The purpose of this paper is to provide an introduction to the predatory pricing debate. The overview is not comprehensive. Yet, it should provide the reader a reasonable appreciation of current controversies. For sake of completeness, Section 2 begins with reviewing the various tests for predation; these are all well known. Standards differ across jurisdictions and while Courts in the United States favour an Arreda-Turner type rule, competition regulators in Canada use Baumol's avoidable cost test in aviation cases. Section 3 introduces the multi-faceted debate on predation. The traditional divide in the predation debate is often characterized as 'Chicago' vs. 'post-Chicago'. The former suggest caution because they are sceptical about both the rationality of predation and frequency with which it may occur. The latter believe that predation is pervasive and takes many forms. Others view the debate as a divide between modern economics and the law. They claim that courts have not kept up with the developments in game theory and the strategic entry deterrence literature. Their critics in turn argue that this literature has little empirical support and provides no objective standards that courts can apply. The empirical evidence discussed in section 4 supports the latter view. There appears to be little empirical support for more sophisticated models and indeed recent work on the aviation industry in the United States finds no support for the argument that exit by low cost carriers is due to the aggressive capacity and price responses of incumbent firms. It shows instead that an overwhelming proportion of entrants in the last decade have been successful in entering hub markets – the probability of exit is inversely related to measures of concentration at hubs. Entry in thinner markets is more likely to be unsuccessful and the response of incumbents has no statistically discernable effect on the probability of exit by low cost carriers. At least in the aviation sector in the United States, it would appear that predation is a rare occurrence.

2. Tests for Predatory Pricing

One American legal scholar refers to predation as 'the most maligned area of monopolization law' in the United States.[3] To the extent that antitrust law is based on the principles of industrial organization economics, this observation may well be indicative of the state of the underlying economics. Though there are diverse views on what predation is and which tests courts should apply, the practice has the following five elements.

---

[3] Elhauge (2003b) at 11.

The first is monopoly or market power because predation is usually listed as an exclusionary practice under the monopolization or abuse of dominant position provisions of most statutes.[4] The second is an immediate sacrifice of profit by the incumbent, which comes from reducing prices below some reasonable measure of costs.[5] The third, at least in the extreme case, is exit by the entrant, or a dead body. The fourth is continuing dominance by the incumbent for a period of time long enough to allow recoupment of the predatory sacrifice.[6] Lastly, exit must be costly because without sunk costs the predator cannot harm the entrant.

The Areeda-Turner test is the most well known test for predation.[7] The test is based on elementary economics, which suggests that under perfect competition prices must equal marginal costs and under imperfect competition they must exceed marginal cost. So any price below short-run marginal costs must be deemed predatory. Since marginal costs are difficult to calculate in general and particularly from accounting data, average variable costs could be used as a proxy measure. This test has been criticized primarily on two counts. First, that average variable costs can be less than, equal to or greater than marginal costs depending on capacity utilization. For example, if a firm faces a capacity barrier, marginal costs will include the capital costs of adding capacity and therefore exceed average variable costs. The second criticism is that it is a static test, one that does not recognize the dynamic and strategic elements of predation. Models of reputation and signalling show that undermining the viability of entrants does not require prices to be below marginal or average variable costs. Regardless, this test continues to find favour with courts in the United States.

Posner suggests that using a cost floor is unavoidable because unless the incumbent charges a price below cost, they 'cannot exclude an equally or more efficient competitor'.[8] Posner would condemn pricing below short-run marginal costs and pricing below long-run marginal cost with intent to exclude the competitor. A firm must cover long-run marginal costs in order to stay viable, however in some circumstances a price below long-run marginal cost may be

---

[4] Posner (2001) at 209-210 suggests that predation is method of maintaining but not obtaining monopoly power. He argues that a monopolized market is more attractive to an entrant than a competitive one and a monopolist is likely to gain more from predation than a competitive firm. Using predation to obtain monopoly power would require building enough capacity in advance to supply the entire market.

[5] Temporary price reductions are not deemed predatory.

[6] A comparison across jurisdictions is beyond the scope of this paper. It is clear however, that the standards and indeed the goals of antitrust differ across jurisdictions. Recoupment is an essential component of the predation doctrine in the United States and in Canada for all sectors other than aviation. See Niels and TEN Kate (2000) for an international comparison covering North and South America, the UK and the EU. See Niblett et al. (2004) for a discussion of Australian cases including the recent *Boral* case.

[7] With the exception of Posner's work and Baumol's more recent contribution, the discussion here draws on the more extensive survey by Ordover and Saloner (1989).

[8] Posner (2001) at 205.

justified so Posner adds intent to his rule.[9] He acknowledges however that a doctrine based on intent would be difficult to apply in a uniform manner.[10]

The tests proposed by Williamson and Baumol are based on limit pricing models or models of strategic entry deterrence. In the static limit pricing model a firm chooses a price-quantity combination such that there is insufficient residual demand for profitable entry. In dynamic limit pricing models the incumbent sets prices or quantities over time so as to limit entry. Williamson would prevent the incumbent from expanding output in response to entry, for a period of twelve to eighteen months. Compared to the Areeda-Turner rule, the Williamson rule provides a weaker deterrent because residual demand for the entrant is larger. However the incumbent could expand capacity and use limit pricing to deter an equally efficient entrant and this would increase consumer welfare prior to entry. If entry takes place nonetheless, it will occur at a larger scale under an Areeda-Turner type rule than under the Williamson rule. Thus post-entry prices will be lower and welfare will be higher under an Areeda-Turner type rule.[11] Baumol's rule would allow the incumbent to reduce prices (and increase output) in response to entry, but not allow raising prices after exit. The price freeze after entry would prevent a reversal of the price cut and therefore prevent recoupment.

Joskow and Klevorick suggest a two-stage approach to testing for predation. The first stage of the inquiry would focus on market power and market structure to ensure that the market satisfies the structural preconditions for possible successful predation. The inquiry would move to the second stage only if the first-stage analysis shows a serious monopoly problem. In the second stage they would condemn a price below average variable costs since these represent avoidable costs. They would also condemn a price between average variable and average total cost unless the firm could show that it was minimizing losses. Temporary price reductions, or those reversed within two years would also be considered predatory even if the reduced price exceeded average total costs.

Ordover and Willig define predation as:[12]

> ... a response to a rival that sacrifices part of the profit that could be earned, under competitive circumstances, were the rival to remain viable, in order to induce exit and gain consequent additional monopoly profits.

This standard introduces the notion of examining alternatives available to the incumbent. Was there any other alternative response available to the incumbent, which would have been less harmful to the entrant? Is the aggressive response that results in the rival's exit more profitable than an alternative accommodating response? If the rival remains viable, does the aggressive response lead to lower

---

9 Posner (2001) provides an example at 215-216 where he suggests that depreciation expense would usually be included in long-run marginal cost, but if the seller does not intend to replace its plant it is of no value to force the seller to include a depreciation change in their price. In this case a price below long-run marginal cost (or one that does not include a depreciation charge) would not be considered predatory.

10 Posner (2001) at 214.

11 Scherer and Ross (1990), Chapter 12.

12 As quoted in Ordover and Saloner (1989) at 586.

profits for the incumbent than an alternative less harmful or more accommodating stance? Tests such as these introduce the problem of the counterfactual and therefore considerable speculation in antitrust cases. In addition they introduce notions of opportunity costs and forgone revenues which convert test for predation into tests of economic rationality or profit maximization – they ask: could the firm have done better by choosing an alternative action? Courts in the United States have not been receptive to tests of economic rationality – business cannot be penalized for failing to maximize profits.

Baumol (1996) proposes a multi-product generalization of the Areeda-Turner test. He defines a price as predatory if it satisfies the following three necessary and sufficient conditions. Firstly, the price must have no legitimate business purpose. Secondly, it should threaten the existence or entry of an equally efficient rival. Lastly, there should be a reasonable prospect of recoupment of at least the initial predatory sacrifice via subsequent monopoly profits.[13] The avoidable costs test uses cost concepts for multi-product developed by Baumol and others in the context of their work on contestability.

Consider a multi-product firm producing two products x and y. Then, the total cost of production can be written as $C(x, y)$. Stand-alone cost of x is the cost that would be incurred if x were produced by a single-product specialized firm. Stand-alone cost of x can be written as $C(x, 0)$; similarly stand-alone cost of y can be written as $C(0, y)$. Incremental cost of x can be derived as $IC(x) = C(x, y) - C(0, y)$. Incremental cost can be interpreted as the cost of adding a second output x to a single-product firm that is currently producing y. Incremental cost includes both product specific variable and fixed costs of production. Average incremental cost can be expressed as $AIC(x) = IC(x) / x$. Baumol refers to avoidable cost as *decremental* cost. In other words, if the firm were producing two goods to begin with and then stopped producing one of them. Avoidable or decremental costs would be less than incremental costs if some product specific costs were sunk. Thus, AAC AIC were AAC represents average avoidable costs.

Avoidable costs do not include common costs and product specific sunk costs, but as the period of time under consideration increases, more costs may become avoidable or escapable. Baumol's rule would consider a price below average avoidable cost as predatory. In other words, if incremental revenue exceeds avoidable costs either for one product or a combination of products there is no predation.[14] The test should be conducted for the period of the alleged predatory practice. Further, Baumol clarifies that avoidable costs should include the costs of inputs that the firm supplies to itself but not revenues forgone by charging the (lower) predatory price. This test has some of the same drawbacks as the Ordover and Willig proposal in that it requires examination of alternatives.

---

[13] Baumol (1996) at 52.

[14] Baumol (1996) at 58, states that the incumbent's price should be compared with the avoidable costs of the victim. This is because he defines predation as the ability to exclude an equally efficient competitor. However, if a price covers the incumbent's (own) avoidable costs then it would also cover the avoidable costs of an equally or more efficient rival. At 64, Baumol (1996) states: 'The point to be noted here is that the avoidable cost to be considered in the test is that of the defendant and that is the cost that has usually been examined in practice in carrying out the Areeda-Turner test.'

Though Baumol does not suggest the inclusion of revenues forgone due to charging a predatory price, presumably the opportunity cost of any resources diverted by the firm from other markets to the market where the alleged predation took place would be included in avoidable cost. This problem would affect what Ghemawat (1991) calls 'use flexible resources'. These resources can be used to respond to changes in the marketplace and they help the firm to revise its position in the product market. Since aircraft are an asset that the firm supplies to itself, Baumol's proposed test would presumably include the opportunity costs of any capacity response by an incumbent.[15] The major shortcoming of this approach is that courts have to deal with the problem of the counterfactual; which is speculative. Further, examining other opportunities usually results in a test for profit maximization or a test that asks if the incumbent could have done better by not adding capacity to the route in question. The speculative nature of examining alternatives makes the avoidable cost measure imprecise and imposes a higher burden on defendants. Thus there will be more convictions in jurisdictions that have blanket rules for applying the avoidable cost test and do not include a reasonable prospect of recoupment as an element of the offence. A test based on variable costs even though not ideal, is clearly more sensible and easier to implement and gives the benefit of doubt to business. The views on predation continue to evolve and as the next section shows, there is a movement away from the simple interpretation of below cost pricing toward strategic entry deterrence and toward multiple benefit arguments, where actions in one market may serve to deter entry in others.

3. The Continuing Debate

The predation debate is usually characterized as having two camps and it appears to be customary to refer to them as 'Chicago' and 'post-Chicago'. The Chicago approach is more cautionary and views predation as being either irrational or very unlikely. Post-Chicago scholars argue for aggressive enforcement because they view predation as being both rational and commonplace. Another way to view the scholastic divide is to equate post-Chicago with the strategic entry deterrence literature in industrial organization economics, which essentially relies on game theory to show that a wide range of exclusionary conduct including predation could occur in a variety of settings.

Legal scholars such as Bork, Posner and Easterbrook are cited most frequently as the proponents of the Chicago view.[16]

The hallmark of the Chicago approach to antitrust is skepticism. Doubt that we know the optimal organization of industries and markets. Doubt that government could use that knowledge, if it existed, to improve things ...

The Chicago view on predation is based on simple logic and continues to be influential. The argument begins with the assumption that the incumbent is a

[15] Aircraft are both use flexible and firm flexible.

[16] Easterbrook (1992) at 119.

dominant firm and the entrant is small.[17] Size is important because the entrant has to be large enough to be able to influence the market price and to supply the increased demand due to lower prices. The dominant incumbent evicts the entrant from the market by charging a price below some reasonable measure of cost. The incumbent hurts itself to hurt the entrant and loses more money than the entrant because it has a larger market share. The sacrifice from predation is both immediate and certain, but the benefit, or recoupment, is uncertain and comes later through future monopoly profits the incumbent expects to earn after evicting the entrant. Future benefits are uncertain because there may be other entrants and the incumbent may never have the opportunity to recover its losses.[18] Recoupment is the more important consideration because the sacrifice, or the price war between the entrant and the incumbent, is beneficial to consumers. Sacrifice without recoupment is not a concern for antitrust and successful recoupment requires the ability to exercise monopoly power for a reasonable period of time after the entrant leaves the market.[19]

We need a way to distinguish competition from exclusion without penalizing competition. *If* the practices are exclusionary, they will be profitable only if the aggressor can recoup. If the aggressor can not, there is no reason for antitrust concern.

Within the Chicago camp, Bork takes a particularly extreme stand and discards all tests of predation:[20]

> It seems unwise, therefore, to construct rules about a phenomenon that probably does not exist or which, should it exist in very rare cases, the courts would have very grave difficulty in distinguishing from competitive price behaviour.

Posner does not discard predation as being irrational, but subscribes to the view that it is often difficult to distinguish between predatory and efficient pricing.[21] Posner doubts the effectiveness of predation as a means of obtaining monopoly power. In his view predation may delay entry or increase the scale at which entry occurs because fighting large-scale entry is unlikely to yield a net benefit to the incumbent. Segmented markets may limit predatory sacrifice, but in Posner's view this still does not establish that predation is a profitable tactic.

Recall that Posner explicitly considers the role of sunk costs or 'resources that are specialized to the market'.[22] Harming the victim requires sunk costs, but sunk

---

[17] In aviation markets entrants do not, at least in the first instance, attempt to replicate the size or frequency of the incumbent.

[18] Since there is no difference between entry barriers and exit barriers, if entry, and subsequent exit takes place, it may indicate that barriers to entry are low in the industry, making recoupment less likely. In order for predation to be rational, the present discounted value of the future (uncertain) stream of monopoly profits must exceed the current (certain) loss. Another argument due to McGee is often attributed to the Chicago school. It says that predation is a very expensive way of disciplining or excluding rivals and it may be cheaper instead to acquire them. Modern merger law however closes this route.

[19] Easterbrook (1984) at 26.

[20] Bork (1978) at 154.

[21] Posner (2001) at 209.

[22] Posner (2001) at 213. The argument of capacity mobility was used in Canada to impose special rules on airlines. It was argued that dominant carriers can quickly move capacity to discipline entrants. However the capacity mobility argument can also be applied to entrants. Entrants can escape harm by moving capacity equally quickly.

costs are a barrier to exit. Barriers to exit are also barriers to entry. Moreover dominance or monopolization is a precondition for predation, so there must be some barrier to entry in the first place. Then how does entry take place in the first instance? This argument is easy to resolve since the apparent contradiction assumes that the entrant attempts to replicate the business model of the incumbent. This need not be the case and is often not the case since changes in technology and other factors usually have an impact on entry barriers and make entry possible in industries where there was previously little or no entry. The other way to look at the role of sunk costs is to argue that if the incumbent is able to evict the entrant, exit barriers must be low. If exit barriers are low then so are entry barriers, in which case the dominant firm is unlikely to be able to recoup losses. Predation without recoupment is not an antitrust concern.

Proponents of the post-Chicago view use game-theoretic models that rely on imperfect information, signalling, and reputation effects to show that predation could be a rational equilibrium strategy.[23] Consider first, Selten's chain store paradox, which can be interpreted as a game of multi-market entry or a game of sequential entry. Suppose a monopolist faces sequential entry for three time periods. The same game is repeated for three periods.[24] Does it make sense for the incumbent monopolist to fight and develop a reputation of toughness early on in the game so as to deter entry later on? This game is solved using backward induction, or the last period (or last market) is solved first. In the third (or last) market, accommodating entry is better for the incumbent than predation. This is because there is no one left to deter after the third period, so there is no benefit to developing a reputation for toughness. Moving back to the second period – does it make sense to deter entry in the second period given that the incumbent is going to accommodate in the third period? The answer again is – no, because the monopolist is going to accommodate in the third period; there is no value to developing a reputation for toughness to deter future entry. So the incumbent accommodates again and in this way the game unravels.[25] Enter-accommodate is a sub-game perfect Nash equilibrium for this game, or the threat of predation is not credible. Introducing asymmetric information can lead to entry deterrence.

[23] In a game theory context, the term dynamic is used for repeated games. The end period could be either finite or unknown (stochastic). If the end period is stochastic, players typically form beliefs about the likelihood that the game will continue to the next period. The probability beliefs affect the actions of the players.

[24] There are the usual informational assumptions: everyone knows the rules of the game, their own payoffs and those of other players and the history of the play. And everyone knows that everyone knows this … ad nauseum.

[25] A finite horizon repeated prisoner's dilemma unravels in a similar manner and cannot explain cooperation (or cartels). Cooperation is possible in an infinite horizon repeated game, or a repeated game with stochastic end period. See Spar (1994) for a more plausible treatment of cooperation. Spar's work is grounded in both good economics and business realism. She argues that from an economic theory viewpoint, the markets for diamonds, uranium, gold and silver have similar (structural) characteristics and should be 'similarly amenable' to the formation of cartels. Instead, the extent of cooperation varies and is explained by internal organization, which has an impact on the ability to threaten, punish and commit.

Consider the following game (Figure 2.1), which shows a 'rational' incumbent.[26] This game can be solved using backward induction, so the incumbent's decision is examined first. The incumbent accommodates since the payoff from accommodation (4) exceeds that from predation (0). Given this outcome, the entrant enters since the payoff from entry given the incumbent will accommodate (**4**) is greater than the payoff from staying out (**0**). The equilibrium is enter-accommodate. Next consider Figure 2.2, which shows an 'irrational' incumbent.

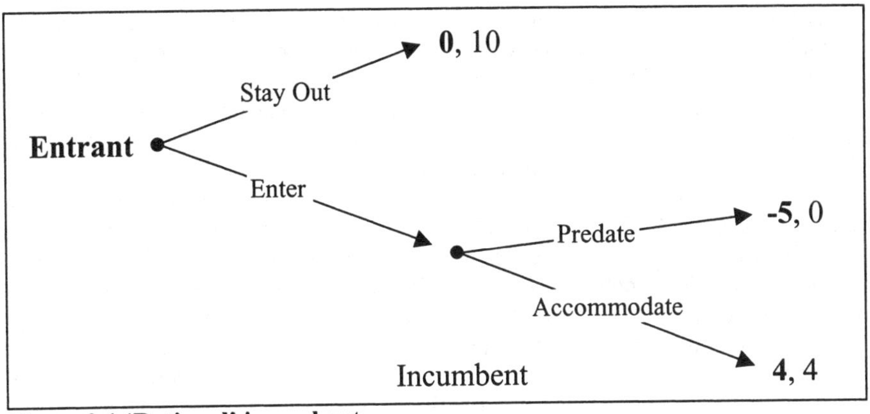

Figure 2.1 'Rational' incumbent

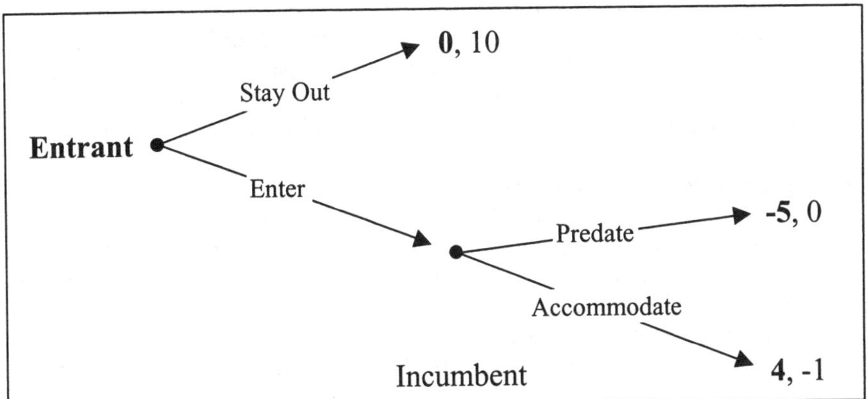

Figure 2.2 'Irrational' incumbent

Again, using backward induction, the incumbent predates since the payoff from predation (0) exceeds that from accommodation (–1). Given this outcome, the entrant stays out since the payoff from staying out (**0**) is greater than that from

[26] Payoffs for the entrant are shown in bold typeface. The two games are from Baird et al. (1994) at 180-182.

entering given that the incumbent will predate (–5). Entry does not occur in equilibrium.[27]

Now consider a situation where the entrant is uncertain about the type of incumbent it is facing, in other words, the entrant does not know which of the above two games they are playing. Let *p* be the entrant's subjective probability that it is facing a rational incumbent. If the incumbent is rational then it will accommodate entry and the entrant will get a payoff of **4**. If the incumbent is irrational, it will predate and the entrant will get a payoff of **–5** if it enters. The alternative, in either case, is for the entrant to chose staying out which yields a payoff of **0** in both cases. The entrant will enter if the expected value of entry exceeds that of staying out or if: $4p + (1 - p)$ g $(-5) > 0$ or $p > 5 / 9$. In other words, the entrant enters if it perceives that there is at least a 55% chance that the entrant is rational; otherwise it stays out. Beliefs about the possible responses of the other player are central to the outcome.

Consider another situation where entry occurs in two separate markets. In the first stage entrant 1 (*E1*) must decide whether to enter market 1 or stay out. The incumbent (*I*) must then decide whether to accommodate entry or fight (predate). In the second stage entrant 2 (*E2*) must decide whether to enter market 2 or stay out. *E2* of course has the benefit of observing the outcome of the previous stage where one of three things could happen: *E1* stays out; *E1* enters and *I* fights; *E1* enters and *I* accommodates. This is a representation of Selten's game and with the appropriate payoffs and perfect information assumptions; the outcome will be that the entrant accommodates entry in both markets. However, it may well be different if *E2* observes *I* fighting *E1* in market 1. *E1* may not enter, if *E1* believes that *I* believes that *E2* will believe that *I* is irrational after *E2* observes *I*'s aggressive response to entry in market 1. It may be in the interest of the incumbent firm to perpetuate such a belief. The previous game had perceptions; this one has perceptions of perceptions or products of probabilities, which can be very small fractions. Thus small numbers can affect the outcome of the game or the outcome of the game depends on weak conditions.[28]

In this vein, the work of Kreps and Wilson, and Milgrom and Roberts modifies the informational assumptions underlying Selten's model to resolve the paradox. Signalling models usually assume that the incumbent is better informed about the state of the industry and is able to manipulate and distort entrants' perceptions of reality.[29] Thus a high cost incumbent can pretend that it has low costs.[30] Similarly in models of test market predation an informed incumbent can

---

[27] In reality the incumbent is rational in both games because it always chooses the alternative with highest payoff. So predation is rational in the second game.

[28] This is Adam Brandenburger's example.

[29] In the American aviation sector, it is unlikely that firms can fool each other. There is a large amount of information in the public domain; moreover the typical incumbent has faced entry from airlines like Southwest in numerous markets. In general it is more probable that entrants have better information about incumbents. Incumbents are more likely to be publicly listed companies that routinely submit financial information to securities regulators.

[30] In nature fooling is usually not possible so all signals are honest and costly. A sickly peacock cannot pretend that it is healthy. This is also likely to be true in business. Many firms have competitive intelligence departments.

prevent the entrant from discovering true market conditions or demand by dropping prices in the entrant's test market.[31]

Game theoretic models of predation have not been as influential in courts as their proponents would like, primarily for two reasons. Firstly, the models do not produce robust results. They are sensitive to the assumptions and often produce multiple equilibria. They specify very stringent conditions under which predation *could* take place. Courts cannot reasonably be expected to base decisions on perceptions, or perceptions of perceptions, or the ability of one firm to bluff another.[32] Some observers have noted that the contribution of these models is to assert the obvious: 'bluffing may work when victims let themselves be fooled'.[33] The post-Chicago view is clearly interventionist and is more likely to favour false convictions.[34] Secondly, as the next section shows, empirical support for these models is limited at best.

The more recent debate on predation has centred on above-cost price cuts. Edlin (2002) argues that these could be harmful to consumers and should be considered predatory. The premise is based on a simple model of Bertrand price competition between an incumbent and entrant selling an identical commodity. The incumbent is a low cost monopolist while the potential entrant has higher costs. In this very restrictive framework prices equal the marginal cost of the entrant, but exceed the marginal cost of the incumbent. The incumbent makes positive profits and the entrant stays out because it is unable to recover the sunk costs of entry. Consumers are worse off because they are deprived of lower prices. Edlin views predation as an ex-ante issue and proposes a Williamson type rule that freezes the incumbent's prices for twelve to eighteen months if an entrant provides a 20% discount compared to prices charged by the incumbent monopolist.[35] If the incumbent does not observe the price freeze, the entrant can successfully sue for predation without the need to show either below-cost pricing or recoupment. An incumbent monopolist faced with this rule may use limit pricing to deter entry, which would benefit consumers because the monopolist will not charge monopoly prices. In other words, the entry deterring profit sacrifice occurs ex-ante or before entry. If entry takes place nonetheless, prices fall further.[36]

In markets where an incumbent monopoly enjoys significant advantages over potential entrants, but another firm enters and provides buyers with a substantial discount, the monopoly should be prevented from responding with substantial price cuts or significant product enhancements until the entrant has had a reasonable

---

[31] See Martin (1993) for these and other models, or the more accessible discussion in TEN Kate and Niels (2002), or Baird et al. (1994) at 178. More recently, Bolton et al. (2000 and 2001) provide a comprehensive discussion of the post-Chicago view.

[32] Arguments of a reputation for aggressiveness and bluffing (or outright misinformation) are not new. They were used about two decades ago in the U.S. Federal Trade Commission's capacity pre-emption case against DuPont. See Dobson et al. (1994) at 165.

[33] TEN Kate and Niels (2002) at 24.

[34] Elzinga and Mills (2001) at 2494.

[35] In the particular case of airlines where there are potentially many prices for the same seat, Edlin proposes applying 'a rough freeze' on a sales weighted average price (Edlin (2002) at 969). What constitutes a 'substantial' price discount and the duration of the price freeze would depend on the circumstances of each case.

[36] Edlin (2002) at 945.

period of time to recover its entry costs and become viable, or until the entrant's share grows enough so that the monopoly loses its dominance.

Edlin bases his argument on the premise that the primary goal of antitrust is to enhance consumer welfare.[37] Chicago scholars such as Bork (1978) hold a similar view and have identified promoting consumer welfare as the primary goal of competition and competition policy:[38]

> 'Competition' may be read as a shorthand expression ... designating any state of affairs in which consumer welfare cannot be increased by moving to an alternative state of affairs through judicial decree.
>
> Antitrust is about the effects of business behaviour on consumers.

Edlin refines his proposal with some qualifications. For example, the rule would be limited to situations where entrants don't face capacity constraints, or where entry takes place at a 'substantial' capacity since the rule is meant to encourage entrants to enter with 'gusto'.[39] Further, Edlin seeks only to tie the hands of the most efficient monopolies, these being 'incumbent monopolies with substantial proven advantages'.[40] Examples of advantages include cost reduction via learning, scale and/or scope economies, making a superior quality product, demand-side network externalities, a trustworthy brand. However, 'the very advantages that give a firm monopoly power' allow it to use above-cost predation to exclude or evict rivals from the market.[41] Since these advantages are not intrinsically exclusionary, Edlin proposes that reducing prices in response to entry be treated as exclusionary under the monopolization doctrine.

In Edlin's view, superior skill either in terms of product quality or low costs should not allow the monopolist 'to exploit customers by raising prices unreasonably'. Instead the monopolist can continue to be a monopolist and charge a price above its marginal cost, so long as that price does not exceed the marginal cost of a potential rival so as to invite entry. Edlin does not find such a 'limited profit-taking' monopoly unreasonable.[42] Implementation issues aside, Edlin's proposal is based on a restrictive model of price competition with homogenous goods and underlying goals of antitrust which are defined in an equally restrictive

---

[37] The total welfare effects are ambiguous, as benefits to consumers have to be weighed against losses due to inefficient pricing by the incumbent. A second goal of antitrust according to Edlin is to reduce or oppose concentration of economic power even if it entails lower costs. Edlin quotes Judge Brandeis and others to argue that one has to choose between democracy and concentration of power and wealth. See Edlin (2002) at 947, n. 24.

[38] Bork (1978) at 61 and 90.

[39] In the first instance it may seem that this rule does not protect low cost carriers since they usually enter on a small scale, often with a single aircraft operating on a single city-pair. However in his discussion of the American Airlines case, Edlin indicates that he would allow the incumbent to respond if the entrant is capacity constrained: 'American's licence to respond would be limited to passengers whom Vanguard is unable to serve.' Edlin does not indicate how potential demand for Vanguard's service should be measured. Edlin (2002) at 968.

[40] Edlin (2002) at 967 and 968.

[41] Edlin (2002) at 963.

[42] Edlin (2002) at 951.

manner. Posing predation as an 'ex-ante' problem is in effect an argument for regulation of prices from the outset, or one that mandates entry by less efficient firms by preventing the more efficient monopolist from responding to entry until such time that it ceases to be a dominant firm.[43]

Elhauge (2003a) argues against Edlin's proposal on both efficiency and administrative grounds. For example, entry of less efficient rivals may be temporary and they may be driven out over time or when the price-freeze restrictions expire. The restrictions may reduce incentives for efficient entry. In addition, restrictions on price may lead to inefficient changes to product quality and imposing restrictions on product enhancements may have a negative impact on innovation. Administrative problems include the inability to define the exact moment of entry and appropriately defining post-entry price and output ceilings, particularly if market conditions were to change during the proposed twelve to eighteen month period.

Most claims of predation rely on some type of dominance or market power argument, particularly those brought under the monopolization or abuse of dominant position provisions of statues. Building on the work of Levine (2002), Elhauge shows that price discrimination in aviation does not imply market power. Airlines in the U.S.,, at least at the network or system level, are not earning abnormal rates of return. Indeed average economic returns over the period 1978-96 were close to those in commodity industries such as steel.[44] Airlines have to incur common costs to produce output or to serve both business and leisure travellers. Thus the complex pricing structure of airlines represents an optimal revenue maximizing price discrimination schedule – it is a means of recovering common costs rather than an indicator of market power. Using a simple example of two competing airlines, Elhauge shows that uniform pricing leads to a situation where there is lower frequency, unmet business demand and all leisure travellers are not served because the uniform price exceeds their willingness to pay; in addition airlines have empty seats. Allowing for price discrimination rectifies all these problems. Thus price discrimination allows firms to recover common costs, provide more frequency and at the same time serve both customer types.

Levine makes the same argument at the level of the network. Infrastructure costs that are common at the network level include items such as station costs, maintenance and reservations labour and overhead. In addition there are other indivisibilities that are required to provide valuable frequent service to locations where demand is insufficient to support stand-alone point-to-point service. Deviations from the optimal price discrimination schedule are costly in the presence of competition. If business class customers are charged higher prices, revenues decline as they go elsewhere. If prices are lowered, revenues may be insufficient to cover the common costs of providing the service.

Entry usually occurs in one or more segments of the network carrier's market and this requires the network airline to re-optimize its pricing schedule subject to the low prices set by the entrant. The re-optimization rarely results in route abandonment because abandoning service between A and B affects service to all

[43] Edlin (2002) also proposes freezing quality (at 986-987) if it is measurable.
[44] Ghemawat (2001) Exhibit 2.1 at 20.

points that connect to B. Instead matching fares provides an 'incremental' revenue opportunity, and this usually results in the provision of more capacity in order to accommodate demand at the new fare.[45] The new capacity 'almost always pays' according to Levine because the common costs of serving the route are already in place; there may be economies of scale in functions such as marketing; capacity costs decline with aircraft size and there may be indivisibilities in other cost items. Thus the incumbent who matches (lower) fares of the entrant may not be better off compared to the situation prior to entry, but may be better off than they would have been had they not responded to entry. The new entrant typically discovers that the market cannot support both carriers and may exit, after which the incumbent re-optimizes, that is, reduces capacity and increases prices. There is no room for recoupment in this argument either because there is little difference in the competitive constraints before entry and after exit – there are no post-exit monopoly profits. Thus due to demand linkages and common costs, Elhauge suggests that the entire hub and spoke network should be considered as the relevant product market and that airlines may have to continue service in certain markets even if they lose money in the short term.[46] Continuing service on a route even if it does not make money may make sense if it increases demand on other routes in the network.[47] Above-cost price cuts may therefore be required in the airline industry because without them firms may not be able to cover their common costs – 'This reaction is involuntary'.[48]

4. Empirical Studies

Proponents of the game theoretic approach often cite the work of Burns (1986) and Morton (1997). Burns studied the Tobacco Trust between 1891 and 1906. During that time period American Tobacco and its affiliates acquired 43 competitors. Burns studied the impact of predation on asset values. In other words, the empirical study sought to show that American Tobacco used predation to depress the asset values of firms it subsequently acquired. Burns found support for this hypothesis and also for reputation effects, as American Tobacco was also able to acquire other competitors at discounted prices. Predatory behaviour led to savings of about 60% and reputation effects produced additional discounts of about 25%. Burns admits

---

[45] Levine (2002) at 33-34. See Dana (1998) for a theoretical model of price discrimination under competition when capacity is not storable.

[46] Edlin (2002) at 943, n. 12 also acknowledges that there are considerable efficiencies in the hub and spoke system that may result in network carriers having lower costs per passenger than low cost carriers. Edlin and Farrell (2002) indicate that one of the reasons why low cost carriers cannot always out-compete major network carriers despite the fact the former usually have lower costs per available seat mile because 'countervailing economics of scope and scale' resulting from hub operations are not captured in measures of cost per available seat mile.

[47] Levine (2002) states at 22: 'We know the cost of operating the whole network for a day a or a year … we simply do not know the marginal cost of carrying a passenger … between any two points in the network.'

[48] Elhauge (2003a) at 745.

however that the results were not conclusive and could also be consistent with 'intensified but lawful price competition'.[49]

Morton examined British shipping cartels and found that entrant characteristics, weakness in particular, helped predict the probability of a price war, or that weak firms were more likely to be preyed upon. Weak firms or young firms were defined as those that lacked experience, finances, an established customer base, or had little multi-market contact with incumbents. Due to the lack of cost data, Morton's work could not provide evidence on predation. Instead, Morton found that entry by small firms was more likely to invoke an aggressive response than entry by large firms because it was cheaper to fight smaller rather than larger firms. The results lend support to the long purse or deep pockets type arguments. Morton also used 'time since last war' and 'time since last entry' as explanatory variables in probit regressions, but both variables were statistically insignificant – indicating that the results do not support models that rely on reputation. Both Burns (1989) and Morton relied on qualitative evidence of intent such as conversations between executives to support their respective arguments of exclusionary conduct.

Isaac and Smith (1985) used experimental techniques to determine if they could observe predatory pricing in a laboratory setting. They specified a set of structural conditions, which they thought were favourable to the emergence of predatory pricing. Given their rules, they could not find any predatory pricing or support for reputation arguments. Their work is more a test of the structural rules under which predation is likely to be observed in a laboratory setting rather than a test of predatory pricing. Further, they examine the impact of imposing antitrust rules on a structural environment where predation is expected to emerge with the intent of testing for type II regulatory errors.[50] They use Williamson's rule, which prevents output expansion in response to entry and Baumol's rule, which prevents reversal of price cuts. They find the existence of type II regulatory error. In other words, adding rules against predation in an environment, which may be susceptible to predation, may lead to a less competitive and less efficient outcome.[51]

Lott (1999) suggests that the apparent richness of game theoretic models allows them to explain almost any set of results – like running many regression lines through one observation. Lott tested, but found no evidence to support models relying on signalling, reputation effects and the long purse theory.[52] In his view these models do not help to differentiate predatory behaviour from competitive behaviour. Instead, he found that government firms (non-profit maximizers and therefore 'irrational') were more likely to meet the assumptions of game-theoretic

---

[49] Burns (1986) at 290.

[50] A type II regulatory error would be banning a practice that would have been beneficial to the community.

[51] More recent work in this area emphasizes that outcomes are dependent on experiment design and structure. See for example Jung et al. (1994) who do find some support for Kreps and Wilson type models and are able to generate predation in an experimental setting.

[52] The long purse or deep pockets theory is clearly a stretch given the current aviation environment in the United States. It is low cost entrants that have larger market capitalizations and higher profits and not incumbent carriers; many of them are bankrupt. The same is true in Canada.

models than private firms and further that governments themselves viewed private firms as less of a predatory threat than their government counterparts.

The issue therefore is not that courts are summarily discarding the post-Chicago view and the 'new consensus' in industrial organization economics, but as Hovenkamp observes: proponents can 'only produce data that are minimally consistent with the theory, but often cannot rule out alternative explanations'.[53] Moreover, as the Canadian Competition Tribunal observed in the *Teledirect* case: they do not provide courts and antitrust tribunals with an objective judicial standard:[54]

> ... because of the absence of any criteria, the Tribunal is being asked by the Director to place itself in the shoes of a potential entrant with a view to assessing the credibility of the alleged 'threat' being issued by Tele-Direct by its responses to entry. The Tribunal must determine whether the response in the initial markets in which entry occurred was so 'overwhelmingly intense' that an entrant would be intimidated and future entry or expansion deterred. What may seem to be a response of 'overwhelming intensity' to one person may not to another. It is inevitably a highly subjective exercise. Decisions by the Tribunal restricting competitive action on the grounds that the action is of overwhelming intensity would send a chilling message about competition that is, in our view, not consistent with the purpose of the Act. ...

Recent empirical work on airlines in the United States casts considerable doubt on reputation type explanations of entry deterrence and indeed warns against setting blanket anti-predatory rules based on price reductions and capacity additions. The work of Ito and Lee (2004) is particularly compelling. They study the response of incumbents to entry by low cost carriers (LCCs) in the United States over the period 1991-2002. Over this period there were 370 unique episodes of LCC entry in hub markets, with only 89 cases of unsuccessful entry.[55] A number of interesting conclusions emerge from this study. Firstly, as would be expected, LCC entry leads to lower fares and stimulates traffic growth. The median fare reduction over the decade relative to incumbent's pre-entry prices was about 50% and this led to 60% growth in origin-destination passengers relative to the pre-entry passengers of the incumbent. Secondly, entrants are more aggressive than incumbents, both with regard to price reductions and capacity additions. Relative to pre-entry capacity and prices respectively, the median LCC introduced 32.7% additional seats and reduced prices by 49.5%. The median capacity response by incumbents was 4% (compared to pre-entry capacity) and average fares of incumbents dropped by 15.1%.(compared to pre-entry average fares).[56] In addition, 38% of incumbents either did not change capacity or reduced capacity in response to entry. In only 8% of the cases did incumbents add more capacity than LCCs. Lastly, the characteristics of unsuccessful entrants are different from those of successful entrants. Unsuccessful entrants typically add less capacity and use

---

[53] Hovenkamp (2001) at 271.

[54] *Director of Investigation and Research v. Tele-Direct Inc.*, CT-94/3 at 290-291. Available at: http://www.ct-tc.gc.ca/english/cases.html.

[55] The data they use in their regression analysis has 265 episodes of entry.

[56] This could partly be due to the fact that incumbents provide different classes of service and the study uses average fares rather than average economy class fares for incumbents.

deeper price cuts than successful entrants. The incumbent's response apparently has little to do with these events. In fact the incumbent's capacity response is slight lower in cases of unsuccessful entry than in cases of successful entry – 3.5% as opposed to 4.5% at the median.[57] The study does not find any 'discernable difference' in the incumbents' price response between the two cases of successful and unsuccessful entry. These results are supported by regressions, which show that incumbents' price and capacity responses have no statistically significant effect on the probability of exit by an LCC from a hub.

The more likely explanation therefore is that LCCs prefer to enter hub markets even though they are highly concentrated.[58] As Posner argues, entry is more likely to occur in more concentrated or more monopolized markets. Margins are high in hub markets; they have higher density and are more likely to be underserved by LCCs – they have a large number of potential customers. This is also supported by Lederman and Januszewski (2003), who find that LCCs are more likely to enter larger airports and routes 'which have a large number of established network carriers serving both endpoints'. Faced with LCC entry incumbents are eager to obtain a share of the benefits of the traffic stimulation effect of LCC entry. Thin markets or those that do not provide the entrant with a large base of origin-destination passengers explain LCC exit. This has nothing to do with the price or capacity response of incumbent carriers or their supposed reputation for toughness.

5. Conclusions

A number of broad conclusions emerge from this review. Firstly, the divergent views on aviation industry structure indicate that using a two-stage approach like the one suggested by Joskow and Klevorick may be preferable. Understanding the industry structure is essential to sensible antitrust. Secondly, there is no agreement on what the appropriate cost test should be. Courts in the United States favour a variable cost test and do not consider above-cost prices as predatory. This is a sensible approach and one that gives business the benefit of doubt. This is important, because price and capacity changes are also instruments of healthy competition. In Canada, the statue requires the use of the avoidable cost test in aviation cases. Baumol proposed this test as a multi-product generalization of the Areeda-Turner test. Problems are likely to show up in the application of the test. Examining alternatives, or asking what the firm could have done had it not re-directed capacity to the route in question, is speculative. It is likely to undermine the precision and reliability of the calculations. Including opportunity costs in avoidable costs will work against defendants, particularly if aircraft ownership costs are included in avoidable costs. The problem of the counterfactual also turns this test into a test for doing better or a test for profit maximization. Courts in the

---

[57] Goolsbee and Syverson (2004) examine the response of incumbent U.S. carriers to the *threat* of entry by Southwest. They do find a price response but only weak evidence of a capacity response. Thus lower prices lead to higher load factors. They interpret this as an attempt by incumbents to build loyalty before entry occurs.

[58] The regression results show that the Hirfindahl Index is negatively related to the probability of exit.

United States have rejected this approach – business cannot be penalized for failing to maximize profits. Thirdly, the game theoretic literature, though interesting, contributes to expanding the scope of the offence – from below cost pricing to strategic entry deterrence. Game theoretic models are capable of explaining almost any type of behaviour. More importantly, given that perceptions are often central to the outcome; they do not provide courts with an objective judicial standard. Courts cannot base their decisions on the ability of one firm to bluff another. Fourthly, Edlin's proposal that above-cost predation should form part of the doctrine is just a sophisticated argument for regulation. Edlin would penalize firms that are most efficient and most innovative – indeed he would prevent firms from responding at all to entry. Perhaps the best way to interpret Edlin's work is that it examines the implications of using a consumer welfare standard. If doing so results in tying the hands of business, then it may be better to adopt a total welfare standard. Lastly, empirical evidence on U.S. airlines supports the view that an aggressive response to entry is the exception rather than rule. An aggressive response is not necessarily a predatory response. In addition, the price and capacity responses of incumbents have no statistically discernable effect on the probability of exit. At least in the U.S. airline industry predation seems to be a rare occurrence.

Acknowledgements

I thank Hans-Martin Niemeier and William G. Morrison for their comments.

References

Baird, Douglas C., Robert H. Gertner and Randal C. Picker (1994), *Game Theory and the Law*, Harvard University Press, Cambridge, MA.

Baumol, William J. (1996), Predation and the Logic of the Average Variable Cost Test, *Journal of Law and Economics*, 39 (1) April, 49-72.

Bolton, Patrick, Joseph F. Brodley and Michael H. Riorden (2000), Predatory Pricing: Strategic Theory and Legal Policy, *Georgetown Law Journal* 88, 2239-2330.

Bolton, Patrick, Joseph F. Brodley and Michael H. Riorden (2001) Predatory Pricing: Response to Critique and Further Elaboration, *Georgetown Law Journal* 89, 2495-2529.

Bork, Robert H. (1978), *The Antitrust Paradox: A Policy at War with Itself*, Basic Books, New York.

Brock, James W. (2000), Industry Update: Airlines, *Review of Industrial Organization*, 16 (1) February, 41-51.

Burns, Malcolm R. (1986), Predatory Pricing and the Acquisition Cost of Competitors, *Journal of Political Economy*, 94 (2) April, 266-296.

Burns, Malcolm R. (1989), New Evidence on Predatory Price Cutting, *Managerial and Decision Economics* 10 (4) December, 327-330.

Dana, James D. (1998), Advance-Purchase Discounts and Price Discrimination in Competitive Markets, *Journal of Political Economy*, 106 (2) April, 395-422.

Dobson, Douglas C., William G. Shepherd and Robert D. Stoner (1994) 'Strategic Capacity Preemption: Dupont (Titanium Dioxide, 1980)', 157-188 in Kwoka Jr., John E. and Lawrence J. White (eds), *The Antitrust Revolution: The Role of Economics*, Second edition, Harper Collins, New York.

Easterbrook, Frank H. (1984), The Limits of Antitrust, *Texas Law Review*, 63, 1-40.

Easterbrook, Frank H. (1992), Ignorance and Antitrust, 119-136 in Jorde, Thomas M. and David J. Teece (eds), *Antitrust, Innovation and Competitiveness*, Oxford University Press, New York.

Edlin, Aaron S. (2002), Stopping Above-Cost Predatory Pricing, *Yale Law Journal*, 111 (4), 941-991.

Edlin, Aaron S. and Joseph Farrell (2002), The American Airlines Case: A Chance to Clarify Predation Policy, Working Paper, Competition Policy Center, University of California, Berkeley.

Elhauge, Einer (2003a), Why Above-Cost Price Cuts to Drive Out Entrants Are Not Predatory – and the Implications for Defining Costs and Market Power, *Yale Law Journal*, 112 (4), 681-827.

Elhauge, Einer (2003b), Defining Better Monopolization Standards, Regulatory Policy Program Working Paper RPP-2003-14. Cambridge, MA: Center for Business and Government, John F. Kennedy School of Government, Harvard University.

Elzinga, Kenneth G. and David E. Mills (2001), Predatory Pricing and Strategic Theory, *Georgetown Law Journal* 89, 2475-2494.

Ghemawat, Pankaj (1991), *Commitment: The Dynamic of Strategy*, The Free Press, New York.

Ghemawat, Pankaj (2001), *Strategy and the business landscape: Core concepts*, Prentice Hall Inc., N.J.

Goolsbee and Syverson (2004), *How do Incumbents Respond to the Threat of Entry? Evidence from Major Airlines*, Draft Version, July. The University of Chicago Graduate School of Business.

Hovenkamp, Herbert (2001), Post-Chicago Antitrust: A Review and Critique, *Columbia Business Law Review* 2001 (2), 257-337.

Isaac, Mark R. and Vernon L. Smith (1985), In Search of Predatory Pricing, *Journal of Political Economy* 93 (2) April, 320-345.

Ito, Harumi and Darin Lee (2004), Incumbent Responses to Lower Cost Entry: Evidence from the U.S. Airline Industry, Draft Version, June 10, 2004. Department of Economics, Brown University.

Jung, Yun Joo, John H. Kagel and Dan Levin (1994), On the existence of predatory pricing: an experimental study of reputation and deterrence in the chain-store game, *RAND Journal of Economics* 25 (1) Spring, 72-93.

Lederman, Mara and Silke Januszewski (2003), Entry Patterns of Low-Cost Airlines, Draft Version, August 8, 2003. Massachusetts Institute of Technology.

Lee, Darin (2003), An Assessment of Some Recent Criticisms of the U.S. Airline Industry, *Review of Network Economics*, 2 (1) March, 1-9.

Levine, Michael E. (2002), Price Discrimination Without Market Power, *Yale Journal on Regulation*, 19(1) Winter, 1-36.

Lott, Jr., John R. (1999), *Are Predatory Commitments Credible? Who should the courts believe?*, The University of Chicago Press, Chicago.

Martin, Stephen (1993), *Advanced Industrial Economics*, Blackwell Publishers, U.K.

Morton, Fiona Scott (1997), Entry and Predation: British Shipping Cartels 1879-1929, *Journal of Economics and Management Strategy* 6 (4) Winter, 679-724.

Niblett, Anthony, Joshua S. Gans and Stephen P. King (2004), Structural and Behavioural Market Power under the Trade Practices Act: An Application to Predatory Pricing, *Australian Business Law Review*, 32 (2) April, 83-110.

Niels, Gunner and Adriaan TEN Kate (2000), Predatory pricing standards: Is there a growing international consensus?, *Antitrust Bulletin*, 45 (3) Fall, 787-809.

Ordover, Janusz A. and Garth Saloner (1989), Predation, Monopolization, and Antitrust, 537-596 in Schmalensee, R. and R.D. Willig (eds), *Handbook of Industrial Organization, Vol. I*, Elsevier Science Publishers B.V.

Posner, Richard A. (2001), *Antitrust Law: An Economic Perspective*, 2nd edition, The University of Chicago Press, Chicago.

Scherer, F.M. and David Ross (1990), *Industrial Market Structure and Economic Performance*, Third edition, Houghton Mifflin Company, Boston.

Spar, Debora L. (1994), *The Cooperative Edge: The Internal Politics of International Cartels*, Cornell University Press, Ithaca, NY.

TEN Kate, Adriaan and Gunner Niels (2002),On the rationality of predatory pricing: the debate between Chicago and post-Chicago, *Antitrust Bulletin*, 47 (1) Spring, 1-24.

Chapter 3

Competition and Predation
Legal Aspects

Ruwantissa Abeyratne

1. Introduction

From an economic perspective, competition represents conditions that occur in markets allowing buyers and sellers to interact with a view to establishing prices and exchanging goods and services. Therefore, economic competition admits of a compelling synergy between buyers and sellers and their self interest to serve the needs of a society as well as those of individual market participants by producing a maximum number of goods at the lowest possible prices. However, from a legal perspective, competition is associated with the rights of the competitor as well as the consumer, whereby the former is precluded from exercising dominance over others in the overall competitive environment, by acting in a manner calculated to eliminate, restrict or deter competition.[1] Such action is termed 'predation' and is particularly associated in the field of air transport with pricing.[2] Unfair competition, which is believed to result in one competitor being a dishonest or fraudulent rival in trade and commerce, usually refers to misrepresentation of one's product through deceptive packaging, labelling, or other characteristics such as pricing of goods and services offered. In a broad sense, unfair competition refers to any activity that unfairly creates an advantage. Competition law *de lege ferenda*, which is law as it should be if the rules were changed to accord with good policy, would dictate that in a state of 'perfect competition', there would be a benchmark for evaluating performance in actual markets. Economists contend that under conditions of perfect competition, goods and services would be produced as efficiently as possible – that is, at the lowest possible price and cost – and consumers would get the maximum amount of the goods and services they need.

Unfortunately, the economic variants in air transport are far too complex to be treated on the basis of 'perfect competition'. Therefore, at best, the air transport industry can look to a 'workable competition' regime, such as the one introduced by American economist John M. Clark in 1940, through which he recognized that in most industries the number of business firms is not so great as to preclude an

---

[1] P. Hanlon, Discriminatory Fares: Identifying Predatory Behavior, *Journal of Air Transport Management* (1994) Volume 1/(2), pp. 89-102 at 91.

[2] Hugo B. Roos and Niels W. Sneek, Some Remarks on Predatory Pricing and Monopolistic Competition in Air Transport, *Air and Space Law*, Vol. XXII Number 3, 1997 pp. 154-157 at 154.

individual firm from having some power to influence market prices and conditions.[3] In addition, participants rarely have complete knowledge of market conditions. Clark theorized, however, that departures from the ideal of perfect competition are often not great enough to warrant government intervention into the market (through antitrust action or direct regulation) in order to improve the situation. He further asserted that competition may be workable in the sense that the results achieved could be approximately comparable to the outcome of the theoretical ideal of perfect competition in application. The main difficulty emerging from the workable-competition concept is its vagueness, in that no precise criteria have been developed to determine when workable competition actually exists. This is certainly true of the air transport industry.

The airline industry has always been in the throes of a dichotomy. On the one hand, while it has been prolifically international in terms of operations, the industry has been unobtrusively national with regard to matters of ownership and control of airlines and interests relating to market access. The latter, brought to bear by regulatory inhibition, which does not admit of an airline freely accessing markets by flying at will to anywhere in the world, and prevailing restrictions as to who owns and controls an airline which bears the nationality of a state, has been increasingly viewed as overtly restrictive in an expanding air transport market. This has led to a gradual liberalization of market access as well as ownership and control in many parts of the world, opening the door to increasing competition between carriers required to cater to a rapidly expanding demand for air transport services.

2. Competition

Current Trends in Competition

Competition in global air transport presents unique strategic issues. Global industries, such as multinational air carrier alliances, are characterized by the presence of competitors operating worldwide from home bases in different countries. The governments concerned may have deep rooted interests and objectives relating to employment and balance of payments including other factors that may not strictly be economic but nonetheless of critical importance to competition among airlines. Therefore, airlines will be increasingly called upon to conduct stringent competitor analysis through examination of the relationships between individual air carriers and their governments. The home country's industrial policy must be well comprehended, particularly in terms of political considerations which may be related to such issues as purchases of aircraft and the exchange of market rights.

A global industry, such as air transport, is one in which commercial entities offering their services view competition as global and build strategies accordingly. Therefore, it follows that competition involves a coordinated world wide pattern of market positions, facilities and investments. Factors to be taken into account are overlap between competitors, geographic location of carriers and defensive

[3] This model was the precursor to the modern oligopoly theories that are now used to represent the structure in airline markets.

investments in particular markets and locations so as not to let competitors gain advantages that can be factored into their overall global posture.[4]

The main consideration, leading up to efforts by the international aviation community to achieve a deregulated global airline industry, is involved with the question as to whether free market principles can be applied globally to air transport. What needs to be considered is whether we are ready to accept the throwbacks as consequences of free market competition in air transport, particularly in losing national prestige projected by flag carriers. One of the corollaries to industry deregulation is the introduction of free market competition when companies switch from operative performance to competitive performance. Competition therefore emphasizes the need to focus on a company's performance in relation to its competitors. This principle can be readily applied to various industries that have already been deregulated, such as the motor vehicle industry, chemical industry and information technology industry. The operative question is 'are these good analogies for application to the air transport industry?' Whatever be the answer to this question, if the deregulated domestic air transport industry of the United States were to be considered an analogy, one could say that a deregulated system in the United States, introduced in 1978, has led to a more efficient airline system in the country.[5] Whatever be the case, access to facilities in a competitive market is essential toward attaining fluidity of market forces. In the air transport industry, this can be translated to mean that if free markets do not exist in the supply of complementary facilities, there will be no positive impact of liberalization.[6] The complementary services in the supply of air transport are airport access, computer reservation systems and airport and air regulation services.

Any agreement to bring in an aspect of trade within a liberalized framework is generally a pro-active measure, which brings to bear the willingness and ability of the governments to face trading issues squarely in the eye. However, any agreement for trading benefits would be ineffective without the element of competition, both between enterprises and between States. The essential requisite for success in trading relations is competition, which in turn leads to national prosperity. A free trade agreement is merely the catalyst in the process.

Anti Competitive Conduct and Antitrust Issues

The regulation of competition within the European Community is governed by the EC Treaty.[7] Two provisions in particular, Articles 81 and 82, contain principles which outlaw anti competitive conduct. While the former prohibits the prevention, restriction or distortion of competition, the latter makes itself applicable against

---

[4] See Michael E. Porter, *Competitive Strategy*, The Free Press: New York 1980 at pp. 291-298.

[5] Pieter Bouw and Wendy Hall, Global Competition in the Airline Industry, *ICC World Business and Trade Review*, 1999, pp. 147-152 at p. 147.

[6] David Banister, Bjorn Andersen, Joseph Berechman and Sean Barrett, Access to Facilities in a Competitive Transport Market, *Transportation Planning and Technology*, 1993 Vol. 17, pp. 341-348 at p. 341.

[7] The EC Treaty, also called the Treaty of Rome, was concluded in 1957 to forge 'an even closer union among the people of Europe'. See Jeffrey Goh, *European Air Transport Law and Competition*, John Wiley & Sons: New York 1997 at p. 15.

abuse by one or more undertakings of a dominant position within the market. While the former essentially contains provisions for agreements, decisions or practices with anti competitive effects, the latter concerns itself with abuses of a dominant marketing position. The aim of these two provisions in particular are to preclude distortion of competition within the Common Market by supplementing the basic principles enshrined in Articles 81 and 82 with substance. The goals of the Treaty in general and Articles 85 and 86 in particular are to promote the free movement of services, goods, persons and capital whilst effectively obviating barriers to trade within the community. Both these provisions relate generally to all sectors of transport unless explicitly excluded by the Treaty provisions.[8]

Article 81 prohibits as incompatible such agreements as directly or indirectly fix purchase or selling prices or any other trading conditions; limit or control production, markets, technical development or investment; share markets or sources of supply; apply dissimilar conditions to equivalent transactions with other trading parties, thereby placing them at a competitive disadvantage; and make the conclusion of contracts subject to acceptance by other parties of supplementary obligations which, by their nature or according to commercial usage, have no connection with the subject of those contracts. These conditions are imposed on agreements between undertakings, which are defined as independent entities performing some economic or commercial activity.

Article 82 provides that any abuse by one or more undertakings of a dominant position within the Common Market or in a substantial part of it shall be prohibited as incompatible with the Common Market insofar as it may affect trade between member States. The Article prohibits: direct or indirect imposition of unfair purchase or selling prices or unfair trading conditions; limitation of production, markets or technical development to the prejudice of consumers; application of dissimilar conditions to equivalent transactions with other trading parties, thereby placing them at a competitive disadvantage; and conclusion of contracts subject to acceptance by the other parties of supplementary obligations which, by their nature or according to commercial usage, have no connection with the subject of such contracts.

In implementing these two provisions, air carriers have to exercise caution in not assuming that purely in view of a bloc exemption on air transport in the Treaty that may pertain to a particular issue, a related practice would be exempt from the prohibitions contained in Articles 81 and 82. In the air transport section of the Treaty, it is abundantly clear that block exemptions may apply only if abuse of dominant position is not evident in a given transaction.[9] Articles 81 and 82 are independent and complementary provisions and any exemption under Article 81 will not necessarily render the provisions of Article 82 nugatory.[10] 'Dominant position' was defined in the 1979 decision of *Hoffman-La Roche* v. *Commission*[11] as a position of economic strength enjoyed by an undertaking which enables it to prevent effective competition being maintained on the relevant market by affording

---

[8] Case 167/73 *Commission* v. *French Republic* [1974] ECR359 at 370.

[9] See Bernardine Adkins, *Air Transport and E.C. Competition Law*, European Competition Law Monologues, Sweet and Maxwell: London, 1994 at p. 81.

[10] Case T-51/89, *Tetra Pak Rausing SA*. v. *Commission* [1990] II E.C.R. 309, [1991] 4 C.M.L.R. 334 para 31.

[11] Case 85/76 *Hoffman-La Roche & Co. A.G.* v.*Commission* [1979] E.C.R. 461.

it the power to behave to an appreciable extent independently of its competitors, its customers and ultimately of its consumers. Such a position may necessarily preclude some competition except in monopoly or quasi monopoly situations. There is every indication, from existing jurisprudence and EC practice, that an assessment on an abuse of dominant position would not be predicated upon one factor alone or single characteristic but would rather be anchored on numerous factors such as market structure, barriers to entry and conduct of the business enterprise concerned.

In Europe, there is entrenched legislation of the European Union concerning predatory pricing, competition and fair trade. Article 82 of the *Treaty of Rome* prohibits abuse of dominant position by a carrier through a determination of the relevant market and market share enjoyed by the carrier under evaluation. This information is matched with the pricing practices of the carrier which must not be lower than average variable cost. The Competition Act of 1998 of the United Kingdom links predatory pricing with dominant position and uses a process similar to that of the European Union[12] in assessing price-cost relationships. Germany has similar legislation in the *Gesetz gegen Wettbewerbsbeschrankungen* (GWB) which is the Act Against the Restraint of Competition, which identifies predatory practices as an abuse of dominant position if the predator is dominant in the market; the conduct of predatory pricing is sustained and continuous and pricing is below average costs without objective justification.[13]

The Australian *Trade Practices Act* of 1974, which is administered through the Australian Competition and Consumer Commission, provides in Section 46 that, when a firm takes control of dominant market power, particularly with intent to lessen or eliminate competition, the onus is on the person holding the position of dominance to prove his actions are not tantamount to predatory practices. The criterion used is that recoupment through pricing at supra competitive levels was a *sine qua non* to prove predatory pricing.

In the United States, the term 'antitrust laws' encompasses federal and States legislation (statutes) which regulate competition with a view to wiping out unfair trade practices and preserving competition among sellers and buyers. Needless to say, antitrust laws apply equally to international air services, and are calculated to preclude both conduct and structural changes in business enterprises. A typical example of conduct coming under antitrust laws in the United States is a merger between competitors which would unduly limit competition. These laws are also meant to prevent producers or purchasers of goods from exercising a monopoly in imposing prices which significantly deviate from expected free market competition norms.

Antitrust legislation in the United States goes back to 1890 with the enactment of the *Sherman Act* which makes it criminally illegal for any contract, combination or conspiracy to be formed in restraint of trade. This all encompassing

---

[12] Both the EU and the United Kingdom uses the AKSO NV case as a benchmark where a Dutch chemical company, with a 65% market share of its flour bleach product was found to be abusing its dominant position. The European Court of Justice found that price below average variable cost by means of which dominant competitor seeks to eliminate its competition is regarded as an abusive practice. See *AKZO Chemie BV* v. *EC* (1991) ECR 1-3359 at para. 71-72.

[13] *Gesetz gegen Wettbewerbsbeschrankungen* (GWB) Section 20(4).

provision prohibits price fixing, anti discounting agreements, divisions of markets by pooling agreements and capacity agreements and exchanges of information that can be considered as competitively sensitive. The Act also prohibits monopolies and conspiracy to monopolize in Section 2.

In 1914, the United States Congress legislated the *Clayton Act*, primarily to supplement the *Sherman Act*. The *Clayton Act* outlaws certain types of 'exclusive dealing' and 'tied sales' and prescribes standards for determining the legality of mergers and acquisitions. Both the Acts admit of compensation to persons injured in their trade or business up to three times the amount of their loss plus attorney fees. Courts have also permitted consumer class actions on antitrust activity, allowing for significant recovery of damages.

Air transport in the United States, which was strictly regulated from 1938[14] to 1978, experienced strict control of air fares and rates.[15] The rigid fare structure which prevailed in the United States was criticized both by governmental authorities and commercial carriers for effectively precluding competition and thereby being less profitable and more inefficient. The restrictions placed on pricing flexibility compelled airlines to compete with other carriers in offering more frequencies in flights, more attractive in-flight amenities and higher levels of service on the ground. In 1976, the stultifying effect on competition brought about by the stranglehold on pricing in the United States, impelled the Civil Aeronautics Board to ease control on pricing in 1977,[16] leading the way to the Airline Deregulation Act of 1978. Prior to this legislation coming into effect in the United States, an air carrier was required to offer demonstrable and cogent evidence that its entry into a particular market was necessary in terms of public convenience and need. Market access within the United States has evolved since, particularly after 1982, when any carrier could offer services in any market in the United States provided it was found fit, willing and able, and it did not contravene applicable ownership and control rules and prevailing cabotage rules.[17] Thus, deregulation offered United States carriers ample opportunity to provide 'hub and spoke' operations in route structures which brought in passengers from spokes to hubs where they could be re-transported to their destinations.[18]

In its role as the sole international regulatory body in the field of air transport, the International Civil Aviation Organization has issued clear policy and guidance material on the avoidance or reduction of conflicts over the application of competition laws to international air transport. ICAO has issued these guidelines to address the conflicts that may arise between States which adopt policies, practices and laws relating to the promotion of competition and restraint of unfair competition within their territories. ICAO urges States to ensure that their competition laws, policies and practices, and any application thereof to

---

[14] *Civil Aeronautics Act of 1938*, 8 Stat 973.

[15] 49 U.S.C.S. 1373, 1482 (1970).

[16] Melvin A. Brenner, James O. Leet and Elihu Scott, *Airline Deregulation*, Connecticut: ENO Foundation for Transportation Inc, 1985, at 1-2.

[17] Steven Morrison and Clifford Winston, *The Economic Effects of Airline Deregulation*, Washington D.C.: The Brookings Institution, 1986 at 4.

[18] See Angela Cheng-Jui Lu, *International Airline Alliances: EC Competition Law/US Anti Trust Law and International Air Transport*, Kluwer: Netherlands, 1997 at 65.

international air transport are compatible with their obligations under relevant international agreements.[19] Within this guideline, there is a strong recommendation for close consultation between States and all interested parties in order that uniformity in practice be achieved across borders to the maximum extent possible. Accordingly, when a State is adopting laws pertaining to competition, it is expected to give full consideration to views expressed by any other State or States whose interests in international air transport may be affected. States are urged to have full regard to principles of international comity, moderation and restraint. The Guidelines also provide direction on dispute resolution and problem solving.

The regulation of air transport services lies within the purview of ICAO member States which maintains in its Legal Bureau a register of all bilateral air transport agreements. The bilateral air transport agreement usually includes a reciprocal agreement between States for their carriers to have fair and equal opportunity in operating air services between their territories without unduly affecting the air services operated by each other. Under a bilateral agreement, capacity offered by carriers must bear close relationship to the needs of the people using air transport.[20] These regulatory provisions have so far succeeded in protecting carriers of lesser developed States by obtaining for them fair and equal opportunity to operate air services in routes that are shared by more established carriers of wealthier nations.[21]

Airlines are faced with the imminent prospect of the future realm of commercial aviation being controlled by a group of air carriers which may serve whole global regions and operated by a network of commercial and trade agreements. Regional carriers will be predominant; easing out niche carriers and small national carriers whose economics would be inadequate to compare their costs with the lower unit costs and joint ventures of a larger carrier. It is arguable that a perceived justification for 'open skies' or unlimited liberalization exists even today in the bilateral air services agreement between two countries, where *fair and equal opportunity to operate* air services is a *sine qua non* for both national carriers concerned. This has been re-interpreted to mean *fair and equal opportunity to compete* and later still, *fair and equal opportunity to effectively participate* in the international air transportation as agreed.[22] Of course, there has been no universal acceptance of this evolution in interpretation and carriers and States whose nationality such carriers bear, have maintained their own positions tendentiously.

---

[19] *Policy and Guidance Material on the Economic Regulation of International Air Transport*, Doc 9587, Second Edition, 1999, Appendix 2 at A2-2. See also *Manual on the Regulation of International Air Transport*, Doc 9626 First Edition 1996, Appendix 5, Guidance Material for Users of Air Transport, at A5-1.

[20] These conditions are the result of an agreement reached on 11 February 1946 by the United States and the United Kingdom in Bermuda. For a clear analysis of the Bermuda Agreement see Ramon de Murias, *The Economic Regulation of International Air Transport*, McFarland: North Carolina 1989, 52-72.

[21] See generally, R.I.R. Abeyratne, The Air Traffic Rights Debate – A Legal Study, *Annals of Air and Space Law*, Vol. XVIII part 1: 1993, pp. 3-44 at p. 3.

[22] Henri Wassenbergh, De-Regulation of Competition in International Air Transport, *Air and Space Law*, Vol. XII, November 2, 1996 at p. 80.

ICAO has suggested the following preferential measures for the consideration and possible use of its member States who are at a competitive disadvantage when faced with the mega trends of commercial aviation and market access:

a) the asymmetric liberalization of market access in a bilateral air transport relationship to give an air carrier of a developing country: more cities to serve; fifth freedom traffic rights[23] on sectors which are otherwise not normally granted; flexibility to operate unilateral services on a given route for a certain period of time; and the right to serve greater capacity for an agreed period of time;
b) more flexibility for air carriers of developing countries (than their counterparts in developed countries) in changing capacity between routes in a bilateral agreement situation; code-sharing to markets of interest to them; and changing gauge (aircraft types) without restrictions;
c) the allowance of trial periods for carriers of developing countries to operate on liberal air service arrangements for an agreed time;
d) gradual introduction by developing countries (in order to ensure participation by their carriers) to more liberal market access agreements for longer periods of time than developed countries' air carriers;
e) use of liberalized arrangements at a quick pace by developing countries' carriers;
f) waiver of nationality requirement for ownership of carriers of developing countries on a subjective basis;
g) allowance for carriers of developing countries to use more modern aircraft through the use of liberal leasing agreements;
h) preferential treatment in regard to slot allocations at airports; and
i) more liberal forms for carriers of developing countries in arrangements for ground handling at airports, conversion of currency at their foreign offices and employment of foreign personnel with specialized skills.[24]

While the open skies policy sounds economically expedient, its implementation would undoubtedly phase out smaller carriers who are now offering competition in air transport and a larger spectrum of air transport to the consumer. Lower fares, different types of services and varied in-flight service profiles are some of the features of the present system. It is desirable that a higher level of competitiveness prevails in the air transport industry, and to achieve this objective, preferential measures for carriers of developing countries would play a major role.

---

[23] The right to uplift or discharge passengers, mail and cargo in a country other than the grantor State.

[24] See *Study on Preferential Measures for Developing Countries*, ICAO Doc AT-WP/1789, 22/8/96 at A-7 – A-9. For a more recent revision of guidelines, see, *Policy and Guidance Material on the Economic Regulation of International Air Transport*, ICAO Doc, 9587, Second Edition, 1999, Appendix 3 at A3-1 – A3-3.

Ownership and Control and Airline Competition

For a long and sustained period of time since the formal regulating of civil aviation, many countries have owned and controlled their national carriers, partly for national prestige and symbolism, and partly because of a traditional requirement in the standard bilateral air services agreement, that a designated carrier should be substantially owned and effectively controlled by nationals of a country which designated that carrier to operate air services under bilateral agreements entered into by that country. This requirement, although tolerable in the first decades of commercial aviation when demand for capacity was manageable, gradually evolved into being an inhibitor in the provision of air transport services. Many States were left with unprofitable State owned airlines that largely required subsidization. The circumscribing nature of an inflexible ownership and control requirement has prompted many States to permit privatization of air carriers, with a reduction in percentage of government held shares. For example, British Airways and Lufthansa have been completely privatized, while Air France, Alitalia, Sabena and Iberia have been partially privatized. Across the Atlantic, the United States deregulated its domestic carriers in 1978 in order to meet capacity demands efficiently and equitably, offering lower prices and improved services to customers.

Although liberalization of air transport is sweeping the globe with its various attractions, offering more capacity and competitive services, it is claimed that the bilateral air services agreement is still preventing pro active measures of airlines to merge with each other and enter into other strategic alliances, through antiquated requirements of national ownership.[25] The bilateral requirement of substantial ownership and effective control, which is based on the fundamental postulate that a majority ownership provision would effectively preclude foreign ownership from taking major control of a national carrier, has not been easy to enforce or put into practice in all situations. Although a blanket provision might require majority national ownership and control, airlines and States have had to contend in many instances with complex issues of nationality of members of a board of directors, the powers of a board and the powers of directors of such boards.[26] Often States have attempted to circumvent these difficulties by establishing a safeguard to ensure for the government concerned a 'golden share' which accords the owner government a greater voice in the decision making process on issues of importance and significance to the carrier concerned.[27]

Although airline alliances may offer a way round circumscriptions of the market access constraints that may be presented by bilateral air services agreements, such alliances are not usually effective against the inhibiting qualities of the traditional ownership and control provisions of the typical agreement, particularly in the context of facilitation of cross border investment, which is

---

[25] See Kirsten Bohmann, The Ownership and Control Requirement in U.S. and European Union Law and U.S. Maritime Law – Policy; Consideration; Comparison; *Journal of Airlaw and Commerce*, Vol. 66, 2001, at p. 690.

[26] See ownership and Control, Report of the Think Tank – World Aviation Regulatory Monitor, Geneva, 7 September 2000, prepared by Peter van Fenema, IATA, Government and Industry Affairs Department at p. 4.

[27] *Ibid.*

essentially regulated by the bilateral air services agreement. In order to find a practical and legitimate way out of this seemingly impossible situation, ICAO has devised a pro active approach based on making the 'principal place of business' and 'permanent residence' of the carrier the operative criteria for purposes of devolution of control.

Both the United States and member States of the European Union have protected their domestic markets from external operators by preserving these markets for their flag carriers or at least carriers that were owned by the State concerned or nationals of that State. In the European Union, in keeping with Article 4 of Court Regulation 2407/92, national authorities are vested with the power of granting operating licences based on the criteria that the principal place of business of the carrier applying for the licence must be located in the licencing member State; the carrier must be involved in air transportation as its main occupation; the holder of the licence must be under direct or majority ownership nationals of the European Union; and the licencee must be effectively controlled by such nationals. Effective control essentially means the power and ability to exercise a decisive influence on an air transport undertaking, including but not limited to the use, enjoyment and alienation of movable and immovable property of that undertaking. One of the reasons, at least from the perspective of the European Union, of retaining the ownership and control criteria within its territory, is to safeguard the interests of the member States of the Union and to preclude carriers of non EU States from capitalizing on a liberalized European Union Market.

United States law too contains explicit requirements pertaining to nationality in terms of management of airlines,[28] in some contrast to Regulation 2407/92 of the EU which does not expressly address issues regarding nationality of management. Arguably, the EU addresses external control by stockholders of a company, and not particularly, as envisaged by the United States law, management control of the administration and running of the air transport enterprise.

Be that as it may, both the United States and the European Union have shown, by their legislation, that the issue of ownership and control still remains to them a critical consideration in the overall picture of liberalization of and competition in air transport.

The airline industry is unique in terms of trade and competition. It is the only industry which gives States the prerogative to impose conditions on ownership and control of airlines. This discretion enables States to withhold permission for landing rights in their territories if they are not satisfied that substantial ownership and effective control of the airline seeking permission to land for commercial purposes are not vested in nationals of a Contracting State which lends its nationality to that airline. The International Air Services Transit Agreement (IASTA),[29] which is the only international agreement that provides for conditions upon which ownership and control of airlines may be restricted, in Article 1, Section 5, provides that:

---

[28] 49 U.S.C. 40102 (a) (15) 1994.

[29] International Air Services Transit Agreement, 7 December 1944, 84 U.N.T.S. 389, 394 ICAO Doc 7500, also reproduced in ICAO Doc 9587. This agreement has been ratified, as of 10 March 2002 by 118 States.

> Each Contracting State reserves the right to withhold or revoke a certificate or permit to an air transport enterprise of another State in any case where it is not satisfied that substantial ownership and effective control are vested in nationals of a Contracting State, or in case of failure of such air transport enterprise to comply with the laws of the State over which it operates, or to perform its obligations under this agreement.

Commencing in the early fifties, States responded, both through the Chicago Convention, and through IASTA, to the growth of civil aviation by increasingly availing themselves of the prerogative given by IASTA to withhold or revoke an airline's operational permit to enter into their territories for purposes of landing and takeoff on a commercial basis. By the same token, States used this right to restrict foreign investment in their own airlines or 'Flag carriers' by including a clause with specific conditions on nationality of their aircraft in the bilateral air services agreements which they signed with other States in conformity with Article 6 of the Chicago Convention.

There is no documented definition of, or agreed meaning to the term 'substantial ownership and effective control'.[30] This *in limine* creates a certain ambivalence in the field of trade and competition in the airline industry, particularly for an airline that is not fully owned and operated by a State or instrumentality of State. The real difficulty in arriving at a conclusive definition of the term arises when an airline is privatized and the government no longer holds the majority of the shares and there is no demonstrable evidence of national ownership. The international community has, as a practice, got used to identifying ownership of an airline with the voting shares of the company, often equating 'substantial ownership' to more than 50% of the voting shares. However, it can no longer be viewed in this simplistic manner, particularly owing to the wave of privatization experienced by the airline industry. For example, 45% of voting shares held by a private entity in a national airline may arguably be termed 'substantial' ownership even though 55% of such voting shares may be held by nationals of the State which designates the airline as its national carrier.

The issue of 'effective control' on the other hand is a more complex issue than ownership, as it is no longer a question of percentages but rather relates to who controls the airline concerned. In broad terms, this may mean who directs policy of the airline and hires and fires personnel. An indicative definition of 'control' can be found in the United States where legislation admits of control being inclusive of possession, direct or indirect, of the power to direct or cause the direction of the management and policies of a person, whether through the ownership of voting securities, by contract, or otherwise.[31]

---

[30] See *Ownership and Control, Report of the Think Tank* – World Aviation Regulatory Monitor 2000, IATA, Government and Industry Affairs Department: Geneva, 7 September 2000 at p. 4.

[31] The *Securities Exchange Act*, 17C.F.R. 240.12b-2 (1988 and Supp. 1995). The European Union, by Council Regulation No. 2407/92 of 23 July 1992, identifies 'effective control' as a relationship constituted by rights, contracts or other means which, *inter alia*, grants the right to exercise a decisive influence on an undertaking, in particular by the right to use all or part of the assets of an undertaking and determine the composition, voting or decisions relating to the running of the business or undertaking. See Article 2(g).

There is a distinct variance in characteristics and effect between 'substantial ownership' on the one hand, and 'effective control' on the other, in that the former may often be established through a presumption of nationality, while the latter may involve complications brought to bear by nationality issues, such as the nationality of members of a supervisory board and nationality and rights of directors of a board. Furthermore, the nationality criterion pertaining to ownership may even be obviated with some ease, as in the case of the European Union, which, in 1997, introduced legislation admitting of a 'community air carrier' which could operate air services anywhere within the fifteen member States of the Union and Norway, Iceland and Liechtenstein. The European Union effectively replaced the 'national carrier' with a 'community carrier by this legislation'.

Similarly in November 2000, the United States, New Zealand, Chile and Singapore, under the auspices of the Asia-Pacific Economic Cooperation Group (APEC) agreed on liberalizing air services between their territories on a multilateral 'open skies' basis. The agreement signed by the four States parties in pursuance whereof has done away with the standard term 'substantial ownership' but retained the 'effective control' portion of the equation.

The International Civil Aviation Organization (ICAO), at its 33rd Assembly held in September/October 2001, adopted Resolution A33-19, which recognizes that the strict application of the criterion of substantial ownership and effective control for the authorization of an airline to exercise route and other air transport rights could deny to many developing States a fair and equal opportunity to operate international air services and to optimize the benefits to be derived therefrom. In other words, the ICAO Assembly expresses its apprehension that continued insistence on the substantial ownership and effective control criterion may seriously jeopardize the opportunity for airlines of developing States to compete fairly and equally with airlines of developed States. Resolution A33-19, *inter alia*, urges States to recognize the concept of community of interest within regional and subregional economic groupings as a valid basis for designating airlines.

It has been traditionally accepted in the aviation industry that relaxation of the ownership and control restrictions, as they currently exist, may increase competition. It is also an incontrovertible fact that these restrictions came into being as a result of the stringent national interests which prevailed immediately after World War II and continued thereafter, until globalization and privatization became inevitable in the last decade. Both these premises give risc to the inexorable question as to whether States still wish to retain stringent national interests in their airlines for reasons of national pride, prestige and other considerations or relax the restrictions to allow foreign investment in airlines and thus encourage competition.

If indeed States were to retain current restrictive practices for the reasons given above, they would be faced with the inevitable prospect of such restrictions impeding the emergence of new entrants and competitors in the market. This in turn would result in monopolistic national carriers, who would have no reason or incentive to lower production costs that would pass on to the consumer in the nature of a lower cost product. As a result, such a monopoly would become inefficient, producing an increasingly high priced, low quality air transport product.[32]

Withdrawal of current restrictions and liberalizing investment in national airlines would also result in a distinct advantage for the air transport market's ability to attract capital. The infusion of foreign capital into the industry would in turn result in increased competition in the international market. A natural corollary to capital injection in a domestic market would result in increased coverage and give a competitive edge to a national carrier in the international market. Of course, it follows that the end result would be enhanced competition among airlines and also between States, enabling developing States in particular to actively participate in market competition and enable their carriers to effectively compete with dominant carriers.

On the issue of national competition being facilitated by the removal of ownership and control restrictions, it is relevant to note that air traffic rights which a State obtains for its airline through the process of bilateral air services negotiations that are made necessary by Article 6 of the Chicago Convention, are a national asset. The question that arises then is whether such traffic rights are jeopardized by the removal of ownership and control restrictions.

In the final analysis, the issue of ownership and control of airlines and its effect on competition has to be viewed in the context of the two operative principles contained in Article 6 of the Chicago Convention and Article 1 Section 5 of IASTA. If airlines are required to obtain permission of States before flying into their territories for commercial purposes, as indeed is the case, and if States continue to exercise the prerogative offered by IASTA, by insisting that airlines are substantially owned and effectively controlled by an identified category of person, such States should ask themselves the compelling question 'is this what we want for our airline industry'? and 'why'. If the answer to these questions is based on national pride and prestige, protectionism related to air traffic rights and, in some instances a certain reluctance towards implementing change, it is wholly understandable, although it may not be economically judicious. However, in such instances, States should realize that retaining the *status quo ante* would not contribute toward the need to provide needed services to the customer and promote air transport in their territories.

Although liberalization of air transport is sweeping the globe with its various attractions, offering more capacity and competitive services, the bilateral air services agreement is still preventing pro active measures of airlines to merge with each other and enter into other strategic alliances, through antiquated requirements of national ownership.[33] The bilateral requirement of substantial ownership and effective control, which is based on the fundamental postulate that a majority ownership provision would effectively preclude foreign ownership from taking major control of a national carrier, has not been easy to enforce or put into practice in all situations. Although a blanket provision might require majority national

---

[32] See C.W.L. Hill, International Business, competing in the Global Market Place (2 ed), Chicago: Richard D. Irwin, 1997 at p. 39.

[33] See Kirsten Bohmann, The Ownership and Control Requirement in U.S. and European Union Law and U.S. Maritime Law – Policy; Consideration; Comparison; *Journal of Airlaw and Commerce*, Vol. 66, 2001, at p. 690.

ownership and control, airlines and States have had to contend in many instances with complex issues of nationality of members of a board of directors, the powers of a board and the powers of directors of such boards.[34] Often States have attempted to circumvent these difficulties by establishing a safeguard to ensure for the government concerned a 'golden share' which accords the owner government a greater voice in the decision making process on issues of importance and significance to the carrier concerned.[35]

Although airline alliances may offer a way round circumscriptions in the nature of the market access constraints that may be presented by bilateral air services agreements, such alliances are not usually effective against the inhibiting qualities of the traditional ownership and control provisions of the typical agreement, particularly in the context of facilitation of cross border investment, which is essentially regulated by the bilateral air services agreement. In order to find a practical and legitimate way out of this seemingly impossible situation, ICAO has devised a pro active approach based on making the 'principal place of business' and 'permanent residence' of the carrier the operative criteria for purposes of devolution of control.

Both the United States and member States of the European Union have protected their domestic markets from external operators by preserving these markets for their flag carriers or at least carriers that were owned by the State concerned or nationals of that State. In the European Union, in keeping with Article 4 of Court Regulation 2407/92, national authorities are vested with the power of granting operating licences based on the criteria that the principal place of business of the carrier applying for the licence must be located in the licencing member State; the carrier must be involved in air transportation as its main occupation; the holder of the licence must be under direct or majority ownership nationals of the European Union; and the licencee must be effectively controlled by such nationals. Effective control essentially means the power and ability to exercise a decisive influence on an air transport undertaking, including but not limited to the use, enjoyment and alienation of movable and immovable property of that undertaking. One of the reasons, at least from the perspective of the European Union, of retaining the ownership and control criteria within its territory, is to safeguard the interests of the member States of the Union and to preclude carriers of non EU States from capitalizing on a liberalized European Union Market.

United States law too contains explicit requirements pertaining to nationality in terms of management of airlines,[36] in some contrast to Regulation 2407/92 of the EU which does not expressly address issues regarding nationality of management. Arguably, the EU addresses external control by stockholders of a company, and not particularly, as envisaged by the United States law, management control of the administration and running of the air transport enterprise.

---

[34] See ownership and Control, Report of the Think Tank – World Aviation Regulatory Monitor, Geneva, 7 September 2000, prepared by Peter van Fenema, IATA, Government and Industry Affairs Department at p. 4.

[35] *Ibid.*

[36] 49 U.S.C. 40102 (a) (15) 1994.

Be that as it may, both the United States and the European Union have shown, by their legislation, that the issue of ownership and control still remains to them a critical consideration in the overall picture of liberalization of and competition in air transport.

In order to face the exponential growth of the air transport industry, it is inevitable that competition and liberalization of the industry should be given serious consideration as a current and future trend in the aviation field. What is needed foremost, to improve international cooperation toward achieving a well meshed and overall competitive policy, is to consider the various possible options available. One of the options to promote competition and facilitate trade in air transport lies indisputably in combating and eliminating anti competitive practices. State responsibility toward achieving this goal is a key factor. One way of ensuring collective State action in this regard and enforcing the duties of States toward the international community is to establish an international entity charged with ensuring the implementation of a global competition code. States may also enter into understandings or agreements toward combating restrictive trade practices, either bilaterally or plurilaterally. Along with a plurilateral framework of competitive policy, there also should be a concomitant bilateral structure of individual agreement between States to stringently monitor anti competitive conduct.

The issue of ownership and control of airlines fits in well to the overall picture of liberalization and anti competitive practices in the air transport field. It is hoped that ICAO will be moving toward offering an acceptable formula in this regard at its upcoming 5th Worldwide Air Transport Conference, to be held in early 2003.

In order to face the exponential growth of the air transport industry, it is inevitable that competition and liberalization of the industry should be given serious consideration as a current and future trend in the aviation field. What is needed foremost, to improve international cooperation toward achieving a well meshed and overall competitive policy, is to consider the various possible options available. One of the options to promote competition and facilitate trade in air transport lies indisputably in combating and eliminating anti competitive practices. State responsibility toward achieving this goal is a key factor. One way of ensuring collective State action in this regard and enforcing the duties of States toward the international community is to establish an international entity charged with ensuring the implementation of a global competition code. States may also enter into understandings or agreements toward combating restrictive trade practices, either bilaterally or plurilaterally. Along with a plurilateral framework of competitive policy, there also should be a concomitant bilateral structure of individual agreement between States to stringently monitor anti competitive conduct.

As for liberalization of air transport, there has so far been no indication whatsoever that any State favours total liberalization calculated to open out its domestic market. Strategic alliances between airlines, be they through mergers or other arrangements, will be viewed with caution and objectivity by individual airlines and States so as to preclude the total overrunning of local interests. It is this consideration that would make liberalized ownership and control criteria less attractive to local entrepreneurs who would not encourage foreign ownership to encroach local control airlines have of their own markets.

Finally, although liberalization of air transport cannot be dismissed as a viable prospect for the future, particularly in trading terms, the players concerned must necessarily view air transport in its entirety, as a service composed of critical factors that are inherent in safe and efficient air transport.

Extra Territoriality Extraterritoriality is one concept which could affect more than one jurisdiction in the application of domestic trade policy. The United States, the European Community and Germany are proponents of extra territoriality where competition rules are applied to commercial conduct of foreign enterprises which conduct business in places other than those whose domestic markets are affected by such trade. In the seminal *Alcoa* case[37] of 1945, the United States courts established the 'effects' doctrine whereby commercial conduct carried out overseas but which is intended or calculated to affect the United States would be subject to United States antitrust laws. This doctrine has been followed by the courts in the United States with an unfailing consistency, culminating in recent guidelines on international commercial operations adopted by the United States Justice Department.[38] These guidelines contain principles that give the United States a wide scope of extra territorial jurisdiction in respect of anti competitive practices which foreign enterprises follow in countries outside the United States, provided such activities adversely affect the United States' market in that particular commercial activity. One of the most compelling features of this legislation is its emphasis on 'market access' relating to United States' businesses in foreign countries. A number of hypothetical examples incorporated in these guidelines reflect that the Department of Justice would challenge the conduct of foreign enterprises in foreign countries if such enterprises would hinder United States' enterprises from using opportunities of exportation of goods to a foreign country or investing in a foreign country.

In the famous *Woodpulp* case[39] the Court of Justice of the European Communities decided that the EC competition rules apply to agreements of foreign enterprises which are entered into outside the European Community as long as they are implemented within the common market.

One cannot deny that in this era of global economy, some degree of extraterritoriality in the enforcement of national competition rules is inevitable. A State would therefore be seen as being justified in applying its competition rules to the conduct of foreign enterprises abroad when conduct which occurs in a foreign country affects its economy adversely, particularly where the State in which such conduct occurs has no competition rules or has no intention to prohibit such conduct. This phenomenon is easily reflected by transnational business entities who may engage in restrictive business practices in a 'twilight zone' where no State can fully exercise jurisdiction and yet harmful effects of such restrictive business practices may be felt in one or more States. To say that there should be no extraterritoriality of any kind in the application of competition rules would mean that such transnational entities can engage in anti-competitive conducts with impunity.

---

37 *United States* v. *Aluminium Company of America*, 148 F. 2d 416 (2d Cir. 1945).

38 See *BNA, Antitrust and Trade Regulation Report*, Vol. 67, No. 1685 (20 October 1994) at p. 488 *et seq*.

39 *Ahalstrom Osakeyhtio* v. *Commission* (1988) ECR 5193.

There is of course the consideration that an extraterritorial application of competition rules is a costly business both for the enforcement agency and for the foreign defendants and is often a second-best solution to a problem which essentially inquires as to how to cope with transnational anti-competitive conduct. An extraterritorial application of competition rules is often not as effective as it would be if applied domestically. A State which attempts to apply its anti competitive laws extraterritorially to a defendant enterprise located abroad could always face difficulties of enforcement and considerations of forum and jurisdiction. There could also be disabling legislation in a foreign State which may effectively preclude extraterritoriality.

The *Watchmakers of Switzerland* case[40] of 1955 exemplifies the essential commercial law principle of the United States, that applicability of anti trust laws on foreign enterprises may often entail conflict with legislation of other States. The court in this case held that a watch repair enterprise, conducted in the United States by two Swiss corporations, could be subjected to the domestic laws of the United States. The court further held that in order for a foreign corporation to be present within the jurisdiction of a court for purpose of service of process, there must be proof of continuous local activities and a showing that under all circumstances of the case the forum is not unfairly inconvenient. Even though the two Swiss entities had no property in the United States and did not carry out their activities directly (the business activities of the Swiss corporations were carried out by an American corporation in the United States), since the Swiss corporations determined the prices and terms of the business enterprise, the court further held that the Swiss corporations could be subjected to anti trust statutes and tariff laws of the United States.[41]

In the watershed case of *Laker Airways Limited* v. *SABENA Belgian World Airlines*[42] it was held that territoriality-based jurisdiction allows states in the United States to regulate conduct or status of individuals or property physically situated within a territory even if effects of conduct are felt outside that territory, and conversely, conduct outside a territory, which is calculated to have a substantial effect on that territory, may also be similarly regulated. It was also held that a state has jurisdiction to prescribe law governing conduct of its nationals whether such conduct takes place inside or outside the territory of that state. Accordingly, the plaintiff Laker Airways Limited, a British corporation seeking remedy in the United States whose activities in question took place in countries other than the United States, was deemed to be subject to United States anti trust legislation on the basis that such activities gravely impaired United States' interests.[43] In deciding upon the contentious question whether the law of the United Kingdom should apply to the plaintiff, the court compared the diametrically opposed anti trust legislation of the United Kingdom and the United States and held:

---

[40] *United States* v. *The Watchmakers of Switzerland Information Center Inc. et al.* 133 F. Supp. 40.
[41] *Id.* at 41.
[42] 731 F. 2d 909 (1984).
[43] *Id.* 910.

> We find no indication in either the statutory scheme or prior judicial precedent that jurisdiction (by the United States) should not be exercised. Legitimate United States interests in protecting consumers, providing for vindicating creditors' rights, and regulating economic consequences of those doing substantial business in our country are all advanced under the congressionally prescribed scheme. These are more than sufficient jurisdictional contacts under *United States* v. *Aluminium Co. of America*[44] and subsequent case law to support the exercise of prescriptive jurisdiction in this case.[45]

In the United States, the scope of antitrust legislation and protection thereby extends to those persons who are either directly or indirectly affected adversely by antitrust violations by third parties. The adverse effect on the plaintiff must be one that the laws were written to guard against. An example of this principle can be seen in the *Uranium antitrust litigation* of 1979[46] where a business entity which indulged in a 'tying arrangement'[47] to sell its product was considered a violation of antitrust legislation. The tie-in resulted in a drop in demand for the product concerned, giving way to a drop in prices and adversely affecting other competitors of the product in the market.

3. Conclusion

There has been a marked surge of energetic and robust competition among carriers in the air transport industry in recent years, mostly due to the globalization of the industry and privatization of airlines and airports. Regionally, market deregulation in Europe and Asia added to the impetus of emerging trends, collectively resulting in the emergence of a number of new price-based carriers, as well as the restructuring of existing carriers to meet the competition. The business models that have been applied by some major network airlines have proved to be fundamentally flawed in the past few years, enabling emergent low cost carriers to establish themselves with robust business profiles.[48] These low cost carriers have adopted sustained pricing policies consistent with the recovery of costs and profit making. The end result has been the dwindling of market shares of legacy carriers. The main difficulty faced by legacy carriers was their inability to maintain a viable

---

44 148 F. 2d 416 (2d Cir. 1945), cited at n. 136 of this article.

45 731 F. 2d 909 (1984) at 945-946.

46 *In re Uranium Antitrust Litigation, Westinghouse Electric Corporation* v. *Rio Algom Limited et al.* 473 F. Supp. 393 (1979).

47 A tying arrangement is the sale of one item (the tying product) only on condition that the buyer would take the second item (the tied product) from the same source. Such arrangements are *per se* unreasonable and violative of antitrust laws if the tie-in involves two distinct products, and the party has sufficient economic power in the tying market to impose significant restraints in the tied product market.

48 Doganis observes that in 1999 a survey carried out on 19,000 leisure passengers had astonishing results where the majority had preferred low cost no-frills carriers to established scheduled airlines such as British Airways and other scheduled European air carriers. See Rigas Doganis, *The Airline Business in the 21st Century*, Routledge: London 2001 at 126.

business model that could drive a revenue base to cover a traditional cost base while allowing for an adequate return on invested capital.

Low cost carriers emerged primarily because of the globalization of the air transport industry and market deregulation in most parts of Asia and Europe, which encouraged new enterprises to approach the air transport market with vigour and energy.[49] The preeminent consideration in the low cost legacy carrier equation is that the low cost carrier is not a trend but rather a business model with imposing and permanent visibility in the market place. Faced with the prospect that low cost carriers are here to stay, legacy carriers (sometimes called full service network carriers) may serve a declining share of the market which could eventually stabilize at around 40% to 50%[50] although with a higher revenue share. One has to consider that the threat posed by low cost carriers to the legacy carriers was non-existent until the early 1970s, when charter carriers commenced encroaching on the market. This trend, together with the increasing awareness, particularly in Europe where controls on market access, monopoly of air services by legacy carriers who were receiving State aid, and restrictions on pricing and frequency of services were perceived to be overwhelmingly anti-competitive and thus detrimental to the interests of the travelling public.[51] In the United States, the *Airline Deregulation Act* of 1978 also paved the way for liberalization of the domestic air transport market within the United States.

The success of the low cost carrier lies mainly in much lower cost structures when compared to their larger competitors, efficient seat management policies and the absence of discrimination on price to the same extent as the legacy carriers. Particularly, the one way low air fare offered by the low cost carrier obviated the tedium of the Saturday stopover requirement for the traveller together with a drastically reduced air fare between two points, which more than compensated for the frills such as network connectivity and other value added services offered by the legacy carriers.

The surge of liberalization and the healthy growth rate of 4-6% of the aviation industry in the late 70s and thereafter for the next 20 years due to aggregate rises in gross domestic product, spurned an increased demand for travel, particularly by business travellers who had deeper pockets than did the average tourist. Major airlines contemporaneously launched a practice called 'network management' using sophisticated computer technology and optimized business models. This was the period of matching expected demand and offered capacity through advanced quantitative analyses enabling major carriers to build global networks using the famous 'hub and spoke' model. The trend encouraged network carriers to attract traffic to their designated hubs, even to the extent of creating multi-hub systems which were visible in the United States in the 1980s. However, intense competition between hub and spoke carriers created a standoff between major competitors who

---

[49] Thomas C. Lawton, *Cleared for Take Off, Structure and Strategy in the Low Fare Airline Business*, Ashgate Publishing Limited: Aldershot, 2002, p. 1.

[50] Michael W. Tretheway, Distortions of Airline Revenues: Why the Network Airline Business Model is Broken, *Journal of Air Transport Management*, Vol. 10, Issue 1, January 2004, pp. 3-14 at p. 4.

[51] R. Doganis, The Impact of Liberalization on European Airline Strategies and Operations, *Journal of Air Transport Management*, Vol. 1, No. 1, 1992 at p. 15.

had to match others in optimizing destination profiles even at the expense of productivity.[52] If carriers failed to match their competitors with their destination portfolios, they would be devoured by the competition through the computer reservation systems which penalized reductions in connectivity with reductions in bookings which in turn resulted in losses in revenue. This dilemma necessitated large network carriers to resort to other compensatory competition tools by forming partnerships and global alliances which were immensely popular in the mid 1990s and continue to flourish today in certain instances. These alliances would have been absolutely successful and thrived if existing regulatory restrictions on ownership and control air carriers were liberalized. However, since this did not happen globally, airline alliances failed to fully attain the desired cost reduction potential. The final nail in the coffin of the conventional legacy carrier was driven in late 2000 and throughout 2001 when the economic downturn which had started in an unprecedented manner grew worse and the fear factor spurned by the unfortunate events of 11 September 2001 saw a rapid decline in the demand for travel. More recently, the Iraq war of 2003 and the ominous doom of the spread of Severe Acute Respiratory Syndrome (SARS) threatened conventionally priced full air services, offering low cost carriers a niche market for price sensitive clients.

One distinct advantage of the low cost carrier service is its unwavering focus on the primary service, i.e. efficient and punctual carriage by air. The fact that the low cost carriers fly to secondary airports is proving attractive to passengers who increasingly prefer to go through airports which offer direct connections with less interactions than congested mega airport complexes. Low cost air travel also encourages the passenger – who would have chosen a cheaper alternate mode of transportation than the high priced fares of network carriers – to go by air instead of using other conventional modes of transport. Large full service network carriers are further confronted with the possibility, as demonstrated by Southwest which has operated low cost services for the past 30 years and has now established its own network of low cost destinations in the United States, that low cost services can drastically expand their own empires with their low cost based business models.

The point to point services offered by such low cost carriers as Ryan Air, Easy Jet and Southwest are offered with operational efficiency and simplicity of service. These carriers use simple services and processes which in turn result in a simple, lean organization. This approach has been called 'sustainable competitive advantage'.[53] Usually, this advantage is gained by offering one class of service with open seating, no meals. This strategy has the dual advantage of cutting costs and simplifying cabin services. Aircraft types used are standardized (such as all 737 fleet in Southwest) which lowers maintenance and crew training costs to the minimum. Subscribing to one aircraft type also endears the operator to the manufacturer where the latter can give special deals to the former. Arguably, from a cost base perspective, the most strategic focus of the low cost carrier is on the use

---

[52] Markus Franke, Competition Between Network Carriers and Low Cost Carriers – Retreat Battle or Breakthrough to a New Level of Efficiency? *Journal of Air Transport Management*, Vol. 10, Issue 1, January 2004, p. 17.

[53] David Gillen, Competitive Advantage of Low-cost Carriers: Some Implications for Airports, *Journal of Air Transport Management*, Vol. 10, Issue 1, January 2004, p. 43.

of secondary airports of small cities. These uncongested airports offer better facilitation of passengers and cargo by reducing circling time or in-air waiting to land or take off as well as ensure much lesser gate time than do congested airports of large cities. On a typical low cost flight, there is usually no interlining of baggage or passengers. All these factors result in the carrier's being able to lower its cost per available seat mile compared to its competition. Unfortunately, network carriers cannot attain this ultimate result because of the vicious cycle they are in – that they have to get rid of their networks to attain a sustainable competitive advantage of low cost but they are unable to, as their existence depends on their route networks.

The airline-airport relationship is an integral part of the low cost legacy carrier equation. In addition to the cost factor, there is also the psychological factor affecting the passenger. Shorter walkways at airports and the minimum of confusion at gates and check-in points are critical advantages offered to the low cost carrier's passenger. Furthermore, secondary airports do not entail the usually difficult transfer connections, delays in the delivery of baggage, and even boredom.[54]

There is no doubt that the low cost carrier business model has been accepted by the consumer and will continue to flourish. If legacy carriers were to be protected, States who have overall responsibility for policy would have to seriously consider eliminating needlessly complex business restrictions such as those applicable to ownership and control of air carriers, thus releasing the network carriers to forge new and innovative alliances. Furthermore, both low cost and legacy carriers have to understand the rules of the game in terms of working together to achieve a seamless and transparent process toward pricing and general airline practice. All parties concerned, including sovereign States, may need to adopt clear competition rules regarding cost bases of air carriers, particularly regarding available costs, product specific fixed costs and sunken costs. A key consideration for both types of carriers would be the manner in which future alliance models or cooperation arrangements including mergers could move towards optimal levels of efficiency. Airports play a critical role in this equation, and the first measure that airports can adopt is to align their business models to those of airlines and consider ways and means of revenue sharing. Airports could review their bargaining power and strategic approaches to non aviation revenue, particularly when factored in to the low cost carrier customer.

[54] The Mintel International 2000 Study, conducted on views of passengers, found that 33% of airport users found airports boring. The age group most affected was between 33 and 45. Of these, 29% found London Heathrow to be boring, while 32% and 33% had the same comment on Gatwick and Manchester respectively. See Sean B. Barrett, How do the Demands for Airport Services differ Between Full Service Carriers and Low Cost Carriers? *Journal of Air Transport Management*, Vol. 10, Issue 1, January 2004 at p. 36.

Part B:
Australia and New Zealand

Chapter 4

Predatory Behaviour in Australian Aviation Markets

Peter Forsyth

Introduction

In Australia, there has not been an extensive history of competition regulators pursuing allegations of predatory behaviour. There have been a few cases, though many fewer than in other areas of antitrust, such as price-fixing. This is possibly because the ideal conditions for predation, a dominant firm with relatively weak new entrants, have not often been present. With the deregulation of markets such as the air transport and telecommunications markets, this is changing.

Prior to airline deregulation in 1990, there was little scope for predatory behaviour in the domestic market, since there were two dominant carriers, and entry on to trunk routes was prohibited. However, after deregulation, new entrants sought to compete with incumbents. There were two periods of intense competition – in 1990-1991 and in 2000-2001, and during both of these, allegations of predation were frequent. In 1992, the competition regulator of the day, the Trade Practices Commission (TPC) investigated the allegations, and dismissed them. More recently, the regulator, the Australian Competition and Consumer Commission (ACCC), which succeeded the TPC in 1995, has responded to complaints by Virgin Blue, a new entrant, and took Qantas to court, alleging predatory behaviour on the Brisbane–Adelaide route. The case was dropped in November 2003 (ACCC, 2003). Thus, so far, there has not been much of a case law established in Australia which gives guidance on predation in aviation. As will be argued, the TPC report did not provide very helpful guidelines.

In this chapter, the Australian experience with predation in aviation is discussed, and particular attention is given to the specific Brisbane–Adelaide case. It begins with a brief review of how predation has been handled in general in Australia, and then the 1991 episode, as examined in the TPC Report, is discussed. Next, the Brisbane–Adelaide case is considered in detail; attention is focused on a range of key issues including: the argument for predation having taken place; the cost conditions and the relationship of prices to costs; cost disadvantages of incumbents; market definition and market power; whether the behaviour could have been profitable; and the test for predation. Finally, the scope for predation in the post 2001 environment is considered.

Australian Experience with Predatory Behaviour

Australian markets often include firms with significant market power, however this is likely to occur within an oligopolistic setting, where there are a small number of fairly equal competitors. This is especially true of the manufacturing industry. Australia has had relatively few cases of industries comprising of a dominant firm, with a fringe of smaller competitors, some of whom are new entrants. This is partly because the type of industry which is likely to generate this market structure has typically been government owned in Australia. In the utility and transport industries (with the exception of road freight), most firms have been government owned. Furthermore, until recently, competition for these firms had been prohibited. The situation has changed significantly over the last two decades, with the privatization of utility and transport firms, and with the extensive deregulation of markets. Thus now there are several markets which are characterized by large, dominant incumbents, faced with much smaller competitors, many of which are new entrants to the market (such as telecommunications markets). In short, the conditions for more widespread predatory behaviour have now been created.

Predatory behaviour is illegal in Australia – predation is covered by section 46 of the Trade Practices Act. This is a general section covering abuse of market power, but it has been established that the actions it prohibits include predation. For predation to be established, it must be shown that the alleged predator *possesses* market power in the relevant market – thus a definition of market will be needed. Secondly, it is necessary to establish that the behaviour had the *purpose* of damaging a competitor, preventing entry or preventing competitive conduct. The requirement to prove intent has been controversial, though courts have been willing to infer intent from conduct (Hurley and Wiffin, 1999, pp. 134-34).

There is a small but developing case law on predation (see Hurley and Wiffin, 1999, Clarke and Corones, 1999). An early successful prosecution of predatory behaviour occurred with the *Victorian Egg Producers* v. *Parkwood Eggs* case. There have been other cases which have been settled out of court – a recent one of these involved the dominant telecommunications carrier, Telstra, which was alleged to have behaved in a predatory manner towards a new entrant which sought access to its monopoly facilities (Brenchley, 1999). This new entrant was a subsidiary of British Telecom, which itself is much larger than Telstra.

The most important case, which was resolved in 2003, was the *Boral* case. In this, a large established brick producer was alleged to have behaved in a predatory manner when a small regional brick producer entered the Melbourne market. After losing this case at a lower court, the ACCC won an appeal at the Federal Court level (ACCC, 2001). This, however, was overturned on appeal to the High Court. In the light of this loss, the ACCC considers that it would be difficult for it to win any predatory case (Fels, 2003). This loss was undoubtedly a factor in its dropping of the Qantas case.

In their approaches to predation, the competition regulators have paid much attention to US cases and analysis. For example, the ACCC is sympathetic to the Areeda-Turner rule in determining whether predation has taken place. The courts likewise are strongly influenced by the US.

Predation in Australian Aviation

Prior to 1990, there was little real scope for airlines to indulge in predatory behaviour. Airlines were tightly regulated, and only two airlines, Ansett and Australian Airlines (since merged into Qantas) were permitted to operate on trunk routes. Towards the end of the regulatory period, it would have been feasible for one of the airlines to have behaved in a predatory manner with respect to the other, though it would have had little or no incentive to do so.

In 1990, the airline market was deregulated and new airlines were permitted to enter all markets. Deregulation created an environment in which predation could take place, and one in which the conditions for predation were about as conducive as they could be. There were two large well established airlines which had been financially secure for many years. Any competitors would be, at least for a time, much smaller. As it turned out, the entrants which appeared were, in most cases, very weak financially. In the early post deregulation days, the incumbent airlines may have had some incentive to establish for themselves a reputation as tough competitors, to discourage further entrants – thus they might react strongly towards new entrants.

There have been two periods of strong competition in the domestic market during which predation became an issue. Firstly, there was the period 1990-1991 when Compass Mark I entered and exited the market. A successor to Compass Mark I, known as Compass Mark II entered later, but collapsed soon after due to financial problems. From this period, until 2000, there were no entrants, and the two incumbent airlines had the market to themselves. While these airlines gradually differentiated themselves, they did not enter any sustained and damaging price wars. In 2000, two new airlines entered; first Impulse, followed by Virgin Blue. By early 2001, Impulse was failing, and it merged into Qantas. Virgin Blue survives; it has been helped in this by the collapse of Ansett in 2001/2002. For the two or three years following the failure of Ansett, there was inadequate capacity, which resulted in little concern over predatory issues. Qantas and Virgin Blue have been adding capacity to make up the shortfall caused by the exit of Ansett. Over time, competition may become more intense again, especially if any new independent carriers enter.

Allegations of predatory behaviour have been made frequently during the periods of intense competition. Compass Mark I alleged predation by the incumbents, and claimed that it was a key factor in its failure. The competition authority published a report into the failure of Compass, and examined the claims of predation, though after it had failed (TPC, 1992). This report represents the most substantial discussion so far of predation in the airline context in Australia. It was a quickly written report, and did not go into great detail, but it does give a good summary of the way a competition authority would tackle the issue. It concluded that the major airlines had not indulged in predatory behaviour.

The context of the entry attempt, and the strategies of the entrant, were very relevant factors in the authority's assessment. Compass had entered the market in late 1990, just when the Australian economy was entering a deep recession, and at a time when the airline market would be expected to be soft. This also was a time of worldwide recession, when the market for airliners was weak. Compass had relatively low costs, and it was an aggressive price cutter. The established airlines responded by matching Compass's fares, but they used yield management to ensure that their yields did not fall too much. Compass kept on cutting fares, and the others kept offering discounts to match them.

The TPC concluded that the two major airlines did not behave in a predatory manner, and that there was no evidence of predatory purpose. Rather, it accepted the airlines' argument that they were responding to price competition, and were simply attempting to minimize losses in the face of it. They did not reduce capacity, partly because they were concerned about the long term market share implications of doing so, and partly because the opportunity cost of the airliners was very low at the time. It was thus sensible to keep the capacity scheduled, but to attempt to fill it by discounting, while maintaining yields as far as possible on higher fare categories. The TPC concluded that yields were most likely above average variable cost, which was low at this time (it also noted that the yields of Compass were below those of Australian Airlines on comparable routes).

The TPC also looked at other possible indicators of predatory behaviour. It considered the issue of price leadership, and concluded that if there were any price leader, it would have been Compass, not the incumbent airlines. It also considered non-price competition, e.g. through loyalty programmes, and found no evidence of predation. It noted that the airlines had not been very helpful in providing Compass with terminal facilities, but that this had not been very harmful, except perhaps towards the end when Compass was targeting business travellers.

Most commentators are agreed that Compass's own strategies were an important factor in its downfall – it scheduled too much capacity too quickly, and dropped fares to unsustainably low levels (Nyathi et al., 1993). In doing this, it made it difficult for it to maintain an allegation of predation with much credence. By simply selectively matching Compass fares, something which can be defended as maximizing profits or minimising losses by meeting the competition, the incumbents made it very difficult to show that they were behaving in a predatory manner. The incumbents did not need to indulge in predatory behaviour to help Compass on its way out.

Three aspects of this case are worth noting. One is that the TPC undertook its investigation after Compass had failed. Even if predation had been proven, it would have made little difference to competition in the market, even though it would have had some legal implications. In the longer term, however, a successful prosecution would probably have influenced future pricing behaviour by the remaining airlines. Essentially, the Compass case did not lead to the setting of any guidelines or precedents.

A second aspect concerns the TPC's interpretation of predation. At various points, it adopts the Areeda-Turner rule; i.e. that prices above average variable costs are not predatory. At other points, it seems to adopt a much more broad definition of predation – it states that if the incumbent sets a price which is below the *profit maximizing* price, given the entrant's price, it would be behaving in a predatory manner (TPC, 1992, p23). It creates a simple link between 'profit maximizing' and 'predatory' behaviour. Thus it would claim predation even if price were still above average variable cost. In other words, in places it adopts a much more inclusive test of predation than the Areeda-Turner rule would imply.

A third aspect of interest is that the TPC only considered predatory behaviour in general; it did not take a market by market approach. It did not, for example, examine behaviour of the airlines on selected route markets or groups of related markets. Predation could take place on one route market even though it was not taking place on any other markets.

The Brisbane–Adelaide Route: Was Qantas Predatory?

When the second period of intense competition began, there were several allegations of predatory behaviour. Virgin Blue is a subsidiary of the Virgin Group, which is quite experienced in litigation (vide the case brought against British Airlines). Several specific instances of possible predatory conduct were brought to the ACCC. Eventually, in June 2002, the ACCC decided to test one of these in court. The ACCC brought a case against Qantas for predatory conduct on the Brisbane–Adelaide route.

Brisbane and Adelaide are both cities of over 1 million inhabitants, and Brisbane is a gateway for the tourist areas of the Gold Coast and Sunshine Coast. For many years, the route had been served mainly indirectly, via Melbourne or Sydney. There are frequent services between Adelaide and Melbourne and Sydney, and between Brisbane and Melbourne and Sydney. In 2000, there were some three daily direct services, two by Qantas and one by Ansett. Then Virgin Blue entered, with two services daily. Virgin Blue also offered lower fares than the incumbents. Qantas responded by matching Virgin Blue's fares, at least for some fare categories. But in addition, it added another flight. The ACCC alleged that Qantas was capacity dumping, and that this overall response was predatory. In September 2001, the situation changed with the demise of Ansett.

The actions of Qantas almost certainly hurt Virgin Blue. But were they predatory? It is necessary to show more than hurt to a competitor for an action to be predatory. Whether the Qantas action was predatory depends on several factors; the impact of additional capacity, market definition, the impact on indirect as well as direct services, cost conditions and the criteria for predation (e.g. whether the Areeda-Turner rule, or some other rule, is accepted). The actions of Qantas could have been the profit maximizing response to Virgin Blue's entry. Even if it were not a profit maximizing response, it need not have been predatory.

The Case for Predation

There is probably no problem with the incumbent matching the entrant's price (though see below), as Qantas did. If it did not match price, it might not have survived in the market. In the case of airlines offering differentiated products, it is a little difficult to be exact as to what 'matching' a price means. Qantas may have higher costs than Virgin Blue, because it offers a superior quality of services (such as better connections) which entail higher costs. This is an aspect which has been discussed in other predation cases – see the chapters on the Lufthansa-Germania case by Ewald and Hueschelrath in this volume). Through yield management, the full service carrier may be able to achieve a higher average yield than the low cost carrier, though it may offer discount fares at the same level. There may be some difference between the average cost levels of the two carriers, but some of this is due to different types of operation; for present purposes, assume that the quality adjusted costs are not very different.

Consider the addition of capacity. The addition of another flight is similar to, though not exactly the same as, a further reduction in price. In most, though not all, circumstances, the addition of a flight will lower the effective price received by all of the airlines serving the route. An additional flight will add more to capacity than it adds to patronage, and thus load factors will fall. At any given price, this leads to a reduction in revenue per flight. It is not impossible that the additional flight will improve convenience so much that it adds more to patronage than capacity – however if this were so, the market would not have been in equilibrium.

This possibility aside, there remains the issue of whether the revenue from the flight is below its costs – if so the addition of the flight could be predatory. It can be assumed that the new price is sufficiently high to cover costs (average total costs-ATC) with the passenger load spread over five flights (unless the entrant set the price and capacity at a loss making level). If price were above costs, and costs for the incumbent are the same as for the entrant (an assumption relaxed below) it may be high enough to cover costs when there are six flights. Alternatively, the price may not cover ATC, but still cover average variable costs (AVC) – if so, the action would not be predatory according to the Areeda-Turner rule. However, if fixed costs are a small proportion of the total, it is possible for AVC to be close to ATC. If this is so, the price may be below AVC, and the revenue from the additional flight may be below the variable cost of scheduling it. If this were so, it could be maintained that the action was predatory according to the Areeda-Turner rule (which is not the only criterion which could be used).

Cost Conditions

The case for predation can be seen as depending on the relationship of price or revenue to specified cost measures, especially marginal and average variable cost. How high these costs are for an airline depends critically on the relationship of demand to capacity. If capacity is ample, and an airline has aircraft which it cannot easily dispose of, average variable cost could be relatively low. However, when capacity is tight, the opportunity cost of aircraft is high. To make aircraft available to be used on a route, they must be rescheduled from other routes on which they are profitable, or additional aircraft must be leased (or purchased). In this situation, the average variable cost and short run marginal cost will be high.

As the basis for analysis, it is probably best to consider the flight rather than the passenger. Thus we should look at the revenue from the flight and how this compares to the total and incremental costs of operating that flight. The decision which was allegedly predatory was one of scheduling an additional flight, and it is the costs and revenue consequences of this which are relevant. This also avoids any fruitless arguments of whether, if the flight is not full, the marginal cost of a passenger is nearly zero.

There will be some costs of operating a route which do not depend on the number of flights very closely; for example some terminal costs. However, most of the costs are likely to depend on passenger and flight numbers. Adding another 20% more flights will not add 20% to costs, but it will add something fairly close to this.

Another issue is whether there is likely to be a large difference between average total costs, and average variable costs, or incremental costs, of adding a flight. The difference between short run average total costs and average variable costs consists of average fixed costs. The fixed costs of operating an airline are substantial, so it might be argued that AVC would be well below ATC. However, the relationship of AVC to *long run* ATC is not systematic. If there is excess capacity, AVC can be well below LR ATC (the situation in the 1990s). If capacity is scarce, AVC can exceed LR ATC – the opportunity cost of capacity is quite high. In 2000/2001, Qantas was adding new capacity. In short, this implies that AVC was close to LR ATC.

The upshot of this is that the incremental cost of adding a flight in 2000/2001 to a route was quite high, probably close to the LR ATC. If revenues per flight were, after the addition of the sixth flight, below the LR total costs associated with that flight, there is some chance that they would also be below the variable costs of the flight, and that if an Areeda-Turner rule were used, the addition of the flight could be construed as being predatory. In short, the 'AVC is very low and thus is lower than average revenue' defence, which was raised in the Compase case, is not strong.

Incumbent versus Entrant Cost Levels

In the discussion above, it was tacitly assumed that the costs of the incumbent and entrant are the same, on a quality adjusted basis. In fact, this may not be the case, and the incumbent's costs could be above those of the entrant. This is a likely situation when a low cost carrier LCC enters competition with an established incumbent which has not faced strong competition for some time. The incumbent does not face the pressure to keep costs at a minimum, and it becomes locked into contracts and work practices which keep costs high. The new entrant can achieve lower costs because it has greater flexibility in sourcing labour and other inputs, and it can achieve higher productivity. Furthermore, the incumbent is not able to match the lower costs of the entrant for a long time because it takes several years for contracts to expire. This is the case with Qantas. It is recognized that its costs are well above those of Virgin Blue, and even though it is actively seeking to reduce them, it is not able to match them because of its lack of flexibility in sourcing and labour arrangements (this is a major reason why it has established its LCC subsidiary, Jetstar – see below).

This poses a problem for determining whether an incumbent's response has been predatory. If the incumbent merely matches the fare of the entrant (or matches it on a quality adjusted basis as in the Lufthansa-Germania case), it could be accused of predation. If the entrant sets its fare at or close to its own costs,, the incumbent may have to set fares below its own costs, including below its average variable cost, to match the fare of the lower cost entrant. If it does not match the fare, its own survival could be put at risk, since it will rapidly lose its market, unless it has other advantages (which it might have). The incumbent could be setting prices below cost in the short run to enable its presence in the market in the long run, not to drive its competitors out of the market. Given Qantas's higher costs, this could have been a real issue in the Brisbane–Adelaide case.

In this situation, it would be inadvisable for the competition regulator or court to take too rigid a definition of predation. Cost based tests of predation, such as the Areeda-Turner test, are, ultimately, arbitrary, and alternative tests may be appropriate in specific circumstances. In this context, both the intent to behave in a predatory manner, and whether a predatory effect has been achieved, will be difficult to determine. The cost disadvantages of the incumbent need to be recognized, but equally, they should not be used to grant the incumbent a blank cheque to price as it wishes.

Market Definition and Market Power

For predatory behaviour to be shown, it is necessary that the alleged predator have 'a substantial degree of market power' in a market. Thus it is necessary to determine what the boundaries of the market are and whether the firm did indeed possess substantial market power.

A narrow or broad definition of the market could be used. The narrow definition would encompass only direct Brisbane–Adelaide flights and traffic. Alternatively, the market might consist of direct and indirect services, and traffic. (In Australia, the 'market' is sometimes taken as the entire domestic market, though this very broad definition will not be used here). Direct and indirect services are likely to be good substitutes for one another. The direct service would normally be preferred by most passengers, some of whom would be prepared to pay a premium for convenience. Indirect services could cheaper for the airline to operate, in that they make use of marginal capacity scheduled on other routes (or enable economies of density to be reaped on these routes through scheduling of lower per unit cost, larger aircraft). The density of the Brisbane–Adelaide market may not be high enough to warrant direct services all day, and indirect services during the off peak may be cost effective.

The more difficult issue to decide is whether Qantas possessed substantial market power, in either the broadly or narrowly defined market. In determining whether market power exists, Australian practice has looked at whether there are substantial barriers to entry, at market shares, and at whether the firm has the ability to raise its price. Barriers to entry in the Brisbane–Adelaide market can be taken as not being substantial. As to market shares, prior to the entry of Virgin Blue, Qantas would have had about two thirds of the direct traffic, and something more than half of the indirect traffic. Does this consist of substantial market power? There are no clear guidelines in Australian practice on what market shares are critical for market power to exist. As to the ability to raise prices, if competition between Qantas and Ansett was not intense, they would have had the ability to raise prices above costs, and in this sense, Qantas would have possessed market power (shared jointly with Ansett). It would have been difficult for Qantas to have raised prices unilaterally, however. It might have been able to do this if it were the recognized price leader, and Ansett simply followed the lead set by Qantas. After the entry of Virgin Blue, Qantas would have possessed less market power, though it might still have possessed considerable market power, jointly with Ansett, in the high fare markets which Virgin Blue did not serve.

Could the Additional Flight have been Profitable?

Central to the ACCC's case for predation is that it would not have been rational or profit maximizing for Qantas to have added capacity when prices on the route fell due to the entry of Virgin Blue. At first sight, it does seem that adding capacity would dilute profits. This need not be the case – it could even be that higher capacity would be more profitable than lower capacity.

Adding capacity could be profitable if the elasticity of demand for direct flights was very high. This is possible. Suppose most travellers fly on indirect routes, but would prefer direct flights, if they were less expensive. When the price of direct flights falls on entry of Virgin Blue, many travellers would wish to switch to direct flights – since the direct and indirect services are very good substitutes, own price and cross price demand elasticities are very high. It is possible that the reduction in price could result in demand for more than five full, and even six full flights. If it did not add the extra flight, demand for direct flights would have to rationed by waiting lists, to the benefit of none of the airlines. It would only be a matter of time before one of Qantas' competitors scheduled an additional flight. In short, it is perfectly possible that scheduling an additional flight could have been profit maximizing. Whether it was so depends on the demand elasticity.

Alternatively it is also possible that not adding a flight would have been the more profitable option for Qantas. Adding the flight may have diluted load factors, resulting in lower revenues per flight. This could have been an aggressive response to Virgin Blue's entry, though it need not have been predatory. If an Areeda-Turner rule is adopted, this response would not have been predatory so long as the revenues from the additional flight covered the costs of scheduling it. Qantas may have considered that it would be in its long run interest to schedule more than adequate capacity, and ensure that its higher paying customers gained access to preferred direct flights, rather than gain temporary profits from operating high, but inconvenient, load factors.

In interpreting what has happened, it is important to remember that all the competitors were operating in a context of uncertainty, and were not able to forecast the response of passengers to price reductions accurately. Virgin Blue may have erred in setting too low a price when it entered the market. Qantas, with its greater experience in the market, may have realized that passenger response to the price fall would be greater than Virgin Blue had anticipated. It is also possible that, in the past, Qantas may have underestimated the attractiveness and viability of direct flights, and relied too heavily on indirect flights prior to Virgin Blue's entry.

Another possible interpretation of what happened lies in price discrimination. Direct flights may have been relatively cheap to operate, but Qantas and Ansett may have been using them to segregate the market. They may have set high average fares for the more convenient direct flight, and offered lower fares on indirect routes. Passengers who valued convenience and were prepared to pay for it would have been able to access direct flights at a price. However, such price discrimination can only be supported when competition is absent. When Virgin

Blue entered, it became unsustainable. The most profitable option for Qantas might then have been to adapt to the new, lower direct fare environment, and make the most of it by making sure that it had plenty of capacity.

Ex post evidence is not very conclusive on this issue. Qantas now schedules two or three direct flights on the routes, and Virgin Blue schedules two. Thus the number of flights is down from its peak of six, when Ansett also offered a flight. Six flights may have been too many, and Qantas' actions could either have been a mistake or could have been predatory. However, the airlines are only now emerging from a situation of tight capacity, and this may have been constraining how many flights they offered.

The Test for Predation

Whether behaviour is determined to be predatory depends on the test for predation. The Areeda-Turner test has been mentioned above. An alternative test is the Willamson test (Ordover and Saloner, 1989). This test looks at the output response of the incumbent; if its output, adjusted for the increase in market size as a result of the lower prices, has increased, this would be taken as evidence of predation. A further alternative would be to define predatory in terms of whether the incumbent's response was the profit maximizing response to the new circumstances. This is not often argued, but the ACCC seems to have it in mind as the appropriate criterion.

The Williamson test can be applied, though it is not likely to be very decisive. If the narrow definition of the market is accepted, it is necessary to examine Qantas' market share in direct traffic. Most likely its share fell, since it operated only half the flights whereas before it had operated two thirds. The picture is less clear if the broader, and preferable, definition of the market is used. It is not very meaningful to examine the issue in terms of capacity offered, because capacity is offered for several markets, not just single city pair markets (Melbourne Adelaide flight serve Melbourne Adelaide traffic as well as Brisbane–Adelaide traffic). Qantas' market share of the Brisbane–Adelaide traffic could have risen or fallen. It would have gained from the increase in traffic, and to an extent, at the expense of Ansett's indirect traffic. It would have lost some of its own indirect traffic. Its share could have fallen even if it were indulging in capacity dumping on the direct route to harm its rivals.

The ACCC appears to be indicating that it is using a profit maximization criterion as a test for predation – behaviour which results in less profit than the profit maximizing response is taken as predatory. This is a different criterion from the Areeda-Turner type of criterion, and it is much more likely to conclude that behaviour has been predatory. It has been noted that the TPC used this criterion in its discussion of the Compass case, though it did not seem to note the difference between the two criteria.

The difference between the two can be shown in the Brisbane–Adelaide example. Suppose that demand is moderately, but not highly, price elastic for the direct option. The reduction in price coming about after the Virgin Blue will increase demand, but suppose that the new demand warrants about five, not six, flights. If Qantas schedules an additional flight, it will dilute load factors, and it will add little to its revenues, and the addition to revenue will be less than the cost of operating the extra flight. It will also take some traffic away from Qantas' indirect flights, further lessening the addition to revenue. In this situation, it is very likely that the additional flight will result in a reduction in profit. According to the 'profit maximization' criterion, this would be predatory behaviour.

If the more commonly used Areeda-Turner criterion is used, this response is much less likely to be deemed predatory. Adding the extra flight may have reduced revenue per flight, but it may still be well above the variable cost per flight (and possibly total cost per flight). Equivalently, average fares may still be above average variable costs of passengers. The Areeda-Turner criterion is set out in terms of the relationship of price to average variable cost (or marginal cost, depending which variant is used). If the test is to be one of whether the firm is setting prices or average revenues below the profit maximizing level, the criterion would be in terms of the relationship of marginal revenue to average variable or marginal cost. In contexts such as this, price or average revenue is typically less than marginal revenue. The average revenue of a flight will be above the marginal revenue from operating it, because the latter takes into account the effect of the flight on reducing revenues on other flights which the airline operates.

Summary

It is clear that many issues would need to be resolved if the ACCC were to establish its case that Qantas indulged in predatory behaviour on the Brisbane–Adelaide route. Certainly, the behaviour could have been predatory. However, it could even have been a profit maximizing response to the new low fare environment. Whether this was so depends on demand elasticities – it could have been so if they were very high. Even if they were not very high, they could have been sufficiently high to result in the additional flight being viable, though not so if the revenue dilution effects on other flights are taken into account. In other words, prices may still have been above marginal and average variable costs, the most commonly used criterion for predation. There can be some good reasons why Qantas responded as it did. In contrast to the earlier period of intense competition, however, it would not be convincing for the incumbent to argue that variable costs were very low because of excess capacity – during the period under consideration, capacity was far from excessive, and variable costs would have been close to average total costs.

Predation in the Post 2001 Environment

Since 2001, predation has not been an issue in the Australian domestic aviation market. The exit of Ansett took over 40% of capacity from the market, and there has been a situation of demand exceeding supply until 2004. Virgin Blue grew rapidly, and it achieved around 30% market share in 2004. With the downturn in international traffic, Qantas was able to switch some aircraft from international to domestic routes, though it took time to increase its fleet. Both airlines took up the slack left by Ansett. Virgin Blue was able to add flights and extend its network without having to enter any price wars. Discount fares were available, but yields were not low relative to costs, and both airlines have been profitable.

In this environment, predation is unlikely. Virgin Blue is relatively well funded and successful, and it is clearly in the market to stay. It competes on a head to head basis with Qantas on most of the main trunk routes. Qantas would have little incentive to indulge in predatory behaviour, and its ability to capture market share from Virgin Blue has been limited by its lack of capacity.

The excess demand period is coming to an end, and both carriers have added to their fleets. Thus, a period of more active competition is more likely. There will be some price wars on individual routes, and while Virgin Blue has been exiting routes which have not been successful (e.g. the business route from Sydney to Canberra), this is not a response to predation. The pattern of competition which emerges will depend on the objectives of the carriers. If Qantas is determined to preserve its market share, and Virgin Blue is equally determined to expand its, then strong competition will result – however, there would be no predation in this. Qantas and Virgin Blue may well decide to tacitly accommodate each other.

The most significant recent development has been the formation of a LCC subsidiary of Qantas, Jetstar. This airline has been set up as a separate operation from Qantas, though some facilities are shared. Jetstar commenced flying in May 2004 on some major trunk routes, and on some routes from secondary airports. It operates with more restrictive conditions (e.g. very early check in) than Virgin Blue, and it does not offer connections. Its costs are well below those of Qantas, and probably slightly below those of Virgin Blue. It is primarily oriented to leisure travellers, unlike Virgin Blue which is also making a strong pitch for price sensitive business traffic. While it is difficult to measure fares, its fares seem to be lower than those of Virgin Blue. It has been quite successful in its first few months.

One possibility is that setting up a LCC could be a predatory response by a full service carrier to the entry of a LCC. The LCC subsidiary could be a 'fighting brand', designed to undercut the entrant and force it from the market. The incumbent might need to subsidise the subsidiary for it to be able to offer lower prices. If this were so, the action in setting up the LCC could be construed as predatory.

In the Qantas-Jetstar case, there is no evidence that this is so. Jetstar has low fares, but its costs are low – there is no reason to believe that fares are below costs. In several cases, it has chosen to avoid head to head competition on routes with Virgin Blue and its parent, Qantas, by flying from a secondary airport. Establishing Jetstar would seem a sensible strategy for Qantas. By doing so, it has been able to achieve significantly lower costs than those which Qantas itself is able to achieve. Qantas is cutting its costs, but it can only go so far given its current contracts. By setting up Jetstar it has been able to get around contracts which the parent airline is locked into. Qantas has done this before, by setting up a low cost international airline, Australian Airlines (not to be confused with the old domestic airline of the same name). Jetstar will enable Qantas to compete successfully in the low fare market, while not diluting its higher fare markets.

The presence of Jetstar has the effect of making the market rather crowded. With one full service carrier, and two LCCs, the market is quite well served. Since both Jetstar and Virgin Blue are both serving the low fare market, there is no market niche which a new entrant can easily serve, even if it could achieve slightly lower costs and offer slightly lower fares. Thus potential entrants will think twice about entering the market as it is. To an extent, the establishment of Jetstar may have been an effective entry forestalling strategy on the part of Qantas. The environment in which predatory conduct is most likely is one of strong incumbents with a weak new entrant. The establishment of Jetstar has lowered the likelihood of this situation coming about for the near future in the Australian domestic market.

Observations and Conclusions

In the Australian domestic airline market, there have been periods of intense competition, as well as periods of stability and quiet competition. During each of the periods of intense competition, which occurred in the early 1990s and in 2000-2001, there were several allegations by new entrant airlines that the incumbents were behaving in a predatory manner. Since then, with Qantas and Virgin Blue adding to capacity to make up for the demise of Ansett, the situation has been one of restrained competition, and predation has not been an issue. No doubt, during the next phase of intense competition, predation will again become an issue.

The history of Australian competition policy, as applied to predation in the airline market, has not been particularly satisfactory. There was an investigation of predation after the Compass episode, but even if it had discovered overt predation, it would have been too late to have had much effect. The TPC report gave little guidance for the longer term. The opportunity for clearer principles to be established has been lost with the dropping of the Brisbane–Adelaide case.

As has been evident overseas, predation in the airline context is not a simple matter to establish. There are particular problems associated with airline markets, coming about because of networks of direct and indirect routes, subtleties in the definition of costs, problems of determining the unit of output (passengers, flights), handling variations in quality associated with different load factors and the complexities of airline price schedules. Predation is never easy to define or police, but there are additional difficulties with airlines. This issue is important, especially in markets like those of Australia, because effective competition depends on small new entrants taking on large and well established incumbents.

So far, Australia has relied on the occasional testing of predation allegations, and slow development of precedent. The problem is that these tests give little guidance for behaviour in the future. At the moment, entrant airlines have little guidance as to what protections they can expect to receive, and incumbent airlines have little guidance as to what behaviour will lead to challenge.

Acknowledgements

I am grateful to Bill Morrison for helpful comments on an earlier version. Any errors are my own.

References

Australian Competition and Consumer Commission (2001), *Media Release: Court Upholds Predatory Pricing Appeal Against Boral Besser Masonry Ltd – Important Case About Misuse of Market Power*, ACCC February 28.

Australian Competition and Consumer Commission (2002), *Media Release: ACCC Alleges QANTAS Misused its Market Power in Anti-Competitive Manner*, ACCC May 14.

Australian Competition and Consumer Commission (2003), Enforcement: Qantas Airways Limited, *ACCC Journal*, No 49 November-December, p. 21.

Brenchley, F. (1999), 'Telstra Exposed', *The Bulletin*, April 6.

Clarke, P. and S. Corones (1999), *Competition Law and Policy: Cases and Materials*, Melbourne, Oxford University Press.

Fels, A. (2003), 'Competition policy: A report card for the last 12 years and an agenda for the future', *ACCC Journal*, 46, May-June, pp. 7-12.

Hurley, A. and G. Wiffin (1999), *Outline of Trade Practices and Consumer Protection Law*, Second Edition, Sydney, Butterworths.

Nyathi, M., P. Hooper and D. Hensher (1993), 'Compass Airlines: 1 December 1990 to 20 December 1991: What went wrong?', Parts I and II, *Transport Reviews*, 13, 119-149 and 185-206.

Ordover, J. and G. Saloner (1989), 'Predation, Monopolization and Antitrust', in R. Schmalensee and R. Willig, *Handbook of Industrial Organization*, *Volume I*, Elsevier Science.

Trade Practices Commission (1992), *The Failure of Compass Airlines Report by the Trade Practices Commission – Main Report*, February.

Chapter 5

When does Airline Competition become Predation?

David Greig

Introduction

2001 saw the demise of two Australian domestic airlines: Impulse (an efficient airline with unsound fare policies) and Ansett (a larger, inefficient airline). The dominant airline, Qantas, then briefly accounted for around 90% of the domestic aviation market, and still has 60-70%. From time to time there have been accusations that Qantas has been overly aggressive in its competitive tactics. For example the main new entrant, Virgin Blue, has accused it of dumping capacity on new Virgin routes. Impulse earlier made the same accusation over its Melbourne–Newcastle route. Investors who had been looking at reviving Ansett were reported as asking for the government to restrain Qantas. The Chairman of the ACCC (the Australian regulator) expressed concerns about competition in this market.[1]

It is normal for firms to react to their competition – to modify the goods or services they offer or the way they offer them or their prices, in response to what a competitor does. But when does aggressive commercial behaviour go too far and become predation? – that is, pricing below marginal cost[2] in order to drive rivals away and then raising prices again. This article looks at:

- what is meant by predation
- evidence of it among US airlines
- how established airlines can use their computer reservations systems as tools for responding to competition
- implications for Australia.

---

[1] ACCC *Airports and Aviation – Regulatory and Competition Issues*, speech to Airports and Aviation Outlook Conference 12 November 2001, on www.accc.gov.au.

[2] The appropriate definition of cost is the centre of much debate in the literature. The general concept is that of a fare and traffic mix which will produce revenue for a particular flight that at least covers the marginal costs of operating that flight.

What is Predation?

Predation involves a company deliberately pricing below marginal cost on certain routes and carrying a loss until it has driven a rival on those routes out of business, after which it again raises prices to a monopoly level (as Plutarch said, a dead man can't bite). Besides the effect on the rival, the action is intended to deter others from considering entry – this is important, as otherwise the predator might have to incur losses again and again. In other words, the predator needs to develop a reputation as a fierce competitor.

Predation is only rational when the predator differs from the prey – for example it may be bigger or have 'deeper pockets'. However predation is difficult to distinguish from normal commercial conduct. There are many good reasons why a firm might cut its prices – as a promotion, as a response to variations in demand, as an investment in 'learning by doing' (so that future costs decline) or as a legitimate response to competition (where it is still pricing above marginal cost). With airlines it is difficult in practice to determine what the relevant costs are, both theoretically and from an accounting point of view – especially if allowance is made for network effects and the costs of competitive tools such as frequent flyer schemes, extra frequency/capacity, fancy terminals and lounges etc. The only way of being sure about predation may be when it is too late – when the rival has gone and it is possible to observe whether or not prices have been put up again. A predator's action in one market can be taken as a signal of how it would behave in other markets. If an airline can develop a reputation as a fierce competitor, so that a potential prey believes that a predatory threat will be carried out, there is no need for actual predation. 'Thus ... the most successful use of predatory threats...is difficult for outsiders to observe.'[3] Similarly, new airlines are unlikely to have access to as much capital as established airlines, and a predatory threat may further reduce newcomers' access to capital.

In the case of airlines predation is made easier by:

- the ease of obtaining information. A newcomer's services, fares and other details are known in advance
- the flexibility of key assets. Aircraft can be quickly redeployed from one route to another, and there is a liquid market in good second hand and leased aircraft
- the power of modern reservations and revenue management systems (see below).

US Aviation Evidence

There is evidence of possible predatory practice by large US airlines. These airlines dominate certain hub cities – e.g. American Airlines dominates Dallas and Northwest dominates Minneapolis. There is evidence of high fares being charged,

[3] Comanor W.S. and French H.E., Predatory Pricing and the Meaning of Intent, *Antitrust Bulletin*, Summer 1993.

then reduced when a low-fare competitor enters one of the hub routes, then increased again when the newcomer has been seen off. Some examples are set out in Table 5.1.[4]

Table 5.1 Examples of apparent predation

		Average fare US$		
Route	**Carrier**	**Quarter before entry**	**Second quarter after entry**	**Eighth quarter after entry**
Detroit – Boston	Incumbent New entrant	257 -	99 70	232 -
Atlanta – Cleveland	Incumbent New entrant	190 -	57 55	153 -
Atlanta – Pittsburgh	Incumbent New entrant	168 -	93 79	217 -

In some other cases the new entrant had survived but was looking precarious. Where low fare carriers survive and share the markets with traditional airlines, one study found that average fares were only 53% of fares in monopoly markets (for flights up to 1600km) and about two thirds the monopoly level for longer flights.[5] At hubs where one airline dominates, average fares were 40-60% higher than in competitive markets.[6] The table below gives some examples.[7]

Table 5.2 Low cost airlines' impact on fares

Origin	**Destination**	**Nonstop Distance (miles)**	**Average One-Way Fare**	**Low-Fare Carrier**
Atlanta, GA	Dayton, OH	432	$126	AirTran
	Indianapolis, IN	432	$242	-
St. Louis, MO	Detroit, MI	440	$83	Southwest
	Minneapolis, MN	449	$259	-
Cincinnati, OH	Kansas City, MO	539	$156	Vanguard[8]
	Philadelphia, PA	507	$278	-

---

[4] From *Predatory Practices in the US Airline Industry*, Clinton V. Oster and John S. Strong, January 15 2001, at www.dot.gov.
[5] US Department of Transportation, Office of Aviation and International Economics
[6] Severin Borenstein, presentation to US Transportation Research Board study on airline competition, 1999.
[7] From Dominated Hub Fares, US Department of Transportation, January 2001.
[8] No longer operating.

In one study, in 5 out of 6 cases where the new entrant exited the route, 'total incumbent revenues not only were higher, but had increased sufficiently to offset any revenue losses that came from additional low-fare traffic during the period that the new entrant was in the market'.[9] In other words, predation may be rational as temporary revenue losses may be recouped.

The US Department of Transportation concluded:

> ... that incumbent airlines at times have responded to new competition with fare cuts, capacity increases and other practices that are apparently designed to eliminate or reduce competition.[10]

However, the Department has failed to secure convictions. For example in a case involving American Airlines the court concluded in effect that it could not, with adequate legal robustness, distinguish between normal commercial behaviour and predatory behaviour.[11]

Revenue Management Systems: a Competitive (and Predatory?) Tool

Good airline practice now includes sophisticated revenue management systems, earlier called yield management systems, linked to computer reservations systems. The aim of revenue management is to maximize revenue per flight by achieving two objectives which sometimes conflict:

- to get the highest average fare ('yield') possible
- to fill a high percentage of the seats.

If all fares are high, many seats would be empty on off peak flights. If all fares are low, planes would be full but potential passengers will be turned away at peak periods and the airline will not maximize profits and possibly make a loss. Revenue management systems are designed to exploit the differing characteristics of different types of air traveller:

- many business travellers are prepared to pay higher fares in order to travel at peak periods and have the flexibility to change the reservation at short notice
- many leisure travellers and some business travellers are prepared to travel at less busy times and observe other conditions in exchange for a lower fare. For some such travellers the alternative is not to pay the higher fare but instead to travel by land or not travel at all.

---

[9] Oster and Strong op cit.

[10] Enforcement Policy Regarding Unfair Exclusionary Conduct in the Air Transportation Industry, US Department of Transportation, 17 January 2001 on www.dot.gov.

[11] *United States* v. *AMR Corp*, 2001-1, cited in Thomas B. Leary, The Need for Objective and Predictable Standards in the Law of Presentation, May 10 2001 on www.ftc.gov/speeches.

The difference between these two categories is explained by price elasticity of demand. The business category has low price elasticity (i.e. demand is not very responsive to price changes). The leisure category has high price elasticity of demand (if prices are high demand will be low and vice versa).

Airlines use their revenue management systems to exploit these characteristics in order to segment the markets, or in other words to offer a number of different 'products'. Some of the familiar types of market segment are:

- first class
- business class
- full economy fares
- off-peak fares
- lower fares with some or all of
  - restrictions on ability to change reservations or a penalty for doing so
  - 'use it or lose it' provisions
  - a requirement to book well in advance
  - maximum or minimum periods of stay, or a
  - requirement to stay over a Saturday night (which excludes most business travellers).

It is not sensible for airlines simply to offer these different products and different fares with certain numbers of seats set aside for different categories on different flights. Airline booking patterns are complex and it is profitable for airlines to take account of these complexities. For example:

- a peak business flight may have little demand at holiday times of the year
- a normally less popular flight may be full if there is a major convention or sports fixture at that time
- longer term underlying changes in demand affect airline bookings, for example different rates of economic growth of different regions in the country.

Airlines have designed their revenue management systems to take account of these complexities and uncertainties. They start with the historic booking pattern e.g. the previous year's booking for each flight on each route on each day and adjust for known changes such as special events. This then determines how many seats on each flight will be offered at cheap fares e.g. none on a normally busy business flight and many on an off-peak or weekend flight. The system is designed so that all seats on the plane are kept available for demand at higher fares – it is not an either/or situation. This concept is known as 'nesting'.

This initial allocation is varied, usually daily, in the light of actual bookings and continues to be varied until the day of departure. For example, if it appears that the level of higher fare business bookings is greater than would have been expected from historical experience, the number of seats made available at lower fares is reduced. If bookings turn out lighter than expected the number of cheap fares is increased.

Up to this point revenue management appears commercially sensible and benign. Business travellers get the flexibility they need and leisure travellers can obtain cheaper seats, sometimes for fares which are below the airline's marginal cost. However, there are some characteristics to note:

- on the basis of its high-fare business traffic, an established airline can offer very low leisure fares and still make a profit.
- if the airline were to increase its capacity on a route, the revenue management system would automatically increase the number of cheap fares on offer.
- if a competitor increased its capacity on the route, there would be an impact on the established airline's bookings which would automatically lead to an increase of cheap fares on offer.

It follows that revenue management systems can be used as an aggressive competitive device.

The response to a new entrant can be to increase the number of cheap seats on offer, especially when combined with a capacity increase, while still protecting the lucrative business traffic. Furthermore, this can be done without changing any of the published fares. The range of fares can stay the same as before but the proportions on offer can be changed. Thus the established airline need not draw attention to its responsive behaviour by announcing a new low fare.

This ability of established airlines to behave aggressively at the cheap end of the market while protecting the expensive end is difficult for new entrants to cope with. At first, new entrants are unlikely to have the flight frequency, feeder networks, reputation, business lounges, etc. to attract much business travel. They are therefore likely initially to rely on low fare leisure travel, which up to a point they can do given their low cost base. (The cost base of a new entrant is usually much lower than that of an established airline which typically has a more complex set of products, higher wages and lower productivity accumulated over many years of protected existence prior to deregulation.)

The new entrant faces a situation where the established airline can offer a significant number of seats at fares which are below its average costs and yet still make some profit overall – or has the financial resources to carry a loss for a while.

The result, as seen with Impulse, may be the demise of the newcomer even though it is more efficient than the established airline – and even though, had it been able to get better established, it might have been able to offer a frequent low fare profitable service, which would have forced the incumbent to improve its own efficiency or lose business.

Thus astute use of revenue management systems, combined with financial resources, can be used to achieve a predatory effect. However, in practice predation is very hard to prove. Computer reservation data are changing constantly and it is difficult to disentangle normal daily fluctuations and normal 'unaggressive' competitive behaviour from the more predatory type.

Implications for Australia

Has airline predation happened in Australia? In particular, has Qantas exhibited predatory behaviour in response to Virgin Blue and Impulse? Some of Qantas' responses to the new competition have been:

- to offer substantial quantities of cheap seats on key routes
- to divert aircraft capacity from the less profitable international routes (though this was mainly to fill the gap left by Ansett's demise)
- put more flights onto some of the routes chosen by the new entrants
- more recently to establish a low cost subsidiary, Jet Star.

All of these developments *could* be explained by commercial factors other than competition. For example, it is sensible to offer cheap seats on off-peak flights and Qantas, being the major business airline, has a lot of such flights. The diversion of aircraft capacity was sensible given the higher profitability of domestic routes compared with certain international routes.

Thus it would be hard to prove, on the basis of what Qantas has done, that it has been predatory. Indeed, the ACCC dropped a case against Qantas of alleged predation against Virgin Blue on the Brisbane-Adelaide route. The level of proof required in Australia (Section 46 of the Trade Practices Act) is even more difficult, as the ACCC would have to prove *intent*. It has suggested that the law needs to be changed to cover behaviour which has the *effect* of changing competition (even if the intent is absent or unable to be proved).[12] The ACCC also sought a 'cease and desist' power where the Commission suspects a breach of the Act, with the onus of proof switched to the company involved it argued that under present law competition may have been burnt off during the time it takes to get to court. However, the ACCC proposals, since watered down, could give it powers of widespread intervention in normal commercial activity. The lesson for Virgin Blue or any other startup airline is:

- expect an aggressive response
- don't rely for too long on the low fare leisure market alone; rapidly build up frequency so as to attract business passengers (as the most successful low cost airline in the US, Southwest, always does)
- don't count on the ACCC being able to help
- have deep pockets to be able to weather competitive storms.

New entrants will then be able to exploit Qantas' Achilles's heel which, like established airlines in other parts of the world, is an inability to offer cheap business fares other than for short periods or on particular routes. Established airlines can offer cheap leisure fares with restrictive conditions but they cannot,

[12] Alan Fels, 12 November 2001, ACCC Airports and Aviation – Regulatory and Competition issues on www.accc.gov.au.

with their current costs and inefficiencies, afford to offer substantially cheaper business fares.

The evidence suggests that Virgin Blue understands this and that Impulse did not. Impulse started with an unsustainable flat fare policy (the same fare on all seats on all flights) which had the predicable result: planes at off peak times had very low loadings. It then lurched to excessive reliance on very cheap fares. Virgin is now succeeding in getting business travel, as evidenced by the difficulty in finding cheap fares on many flights unless booked well in advance.

For the public, governments and regulators, the possibility of predatory practice is of concern because it could stop aviation developments that confer widespread benefits. If cheap airfares are easy to obtain, there is a large gain in consumer welfare. People on more modest incomes can (unlike before) afford to travel; others who would have travelled anyway can spend less on it and have money left over for other things.

Current Situation

Australia has now passed the period when there was a high risk of predatory behaviour. Virgin Blue, like some low cost airlines elsewhere, is now well enough established for it to be too difficult (and expensive) to pick off. It is now operating on all the main business routes and making good profits. It is one thing for a US airline to act as predator on a particular route, relying on its other routes to maintain cash flow, but another for Qantas to act aggressively across the bulk of its domestic system. Qantas will continue to be aggressive at the leisure end of the market and has set up a low-cost subsidiary, Jetstar, for leisure routes, but it cannot afford to undermine its high-fare business traffic (e.g. its unrestricted A$357 Melbourne-Sydney fare, compared with Virgin's A$239[13]).

Attention in the South Pacific has now swung to New Zealand domestic and international routes. There have been suggestions that Air New Zealand will be exposed to the risk of predatory behaviour by the much larger Qantas, especially granted that a proposed alliance between the two was not approved. Once again, it will be difficult to prove.

[13] At 10 December 2004.

Part C:
North America

Chapter 6

Predation in Aviation: The North-American Divide

David W. Gillen and Ashish Lall

1. Introduction

The aviation industry continues to be in a state of upheaval as emerging business models are buffeted by external shocks. Largely, this has been due to factors such as the events of September 11, 2001 and the subsequent wars in Afghanistan and Iraq. Heightened security concerns and more recently, SARS have led to declining profitability. Changes in the industry landscape however, preceded these events. In the domestic United States prices of air travel have declined by more than 50% since deregulation in 1978. Low cost carriers continue to successfully challenge full-service incumbents even in Canada where a single carrier dominates domestic air travel. In both countries, the increased competition has been accompanied by allegations of predation and other exclusionary conduct by dominant network carriers. In the United States, the Department of Justice brought charges against American Airlines and in Canada the Commissioner of Competition took Air Canada to task. The allegations were broadly similar, with both full-service carriers being accused of increasing capacity on routes where they faced entry from low cost competitors. Both governments alleged that the capacity additions were predatory because revenues from the incremental capacity did not cover associated costs. However, the outcomes were quite different, as were the tests for predation, their application and the underlying approaches of Courts in the United States and the Competition Tribunal in Canada.

The differences in the historical development of antitrust in the United States and Canada have perhaps been as stark as the differences in the market structure of their domestic aviation sectors. The Sherman Act was passed in 1890; one year after Canada's single provision (conspiracy) legislation, and remains the same to this day. Canada on the other hand, has always pursued a policy of continual amendment – adding offences and changing the law to suit economic circumstances. To its merit however, unlike the United States, Canada continues to have one federal antitrust statue – The Competition Act. The incremental approach and weaker enforcement have prevented the development of a body of law in many areas of antitrust in Canada. As a result, Canadian law is heavily influenced by American jurisprudence and until recently the predation doctrine was very similar in both countries. In the year 2000 the Canadian government amended the law yet again, to include aviation specific provisions in the Competition Act. These are based on a set of administrative rules which list aviation specific exclusionary practices; the tests to be applied; and how they should be applied. In addition, the amended statute provides the Commissioner of Competition with sweeping powers to issue temporary cease and desist orders if there is suspicion of misconduct by the dominant domestic carrier. Two months after the amendments, the Commissioner exercised these powers against Air Canada.

In the United States the doctrine has not changed and was most recently enunciated by the Supreme Court in *Brooke Group Ltd.* v. *Brown & Williamson Tobacco Corp.*[1] The most recent application in an aviation context was the American Airlines case. This paper compares both the doctrine of predation and its application to aviation in the United States and Canada. There is now a major divide between the two countries on both counts. Section 2 summarizes what has come to be known as the 'Brooke standard' in the United States. The Canadian approach and the recent aviation related amendments are discussed in section 3. Section 4 discusses the American Airlines case in the United States and section 5 is devoted to Phase I of the Air Canada inquiry. This is followed by concluding remarks.

2. The Brooke Standard in the United States

The predation doctrine in the United States is enunciated in the *Brooke* case, which provides an example of alleged predation in an oligopoly setting.[2] The case related to the cigarette industry in the United States, which was highly concentrated and profitable since the 1920s.[3] Firms spent heavily on advertising and instruments of competition included product differentiation and brand proliferation.[4] In the early 1980s the cigarette industry was facing an unfavourable market environment. Demand was declining partly due to health concerns but also because firms had excess capacity. Such situations are ideal for disruption of conscious parallelism in oligopolies as firms are tempted to play market share games. Brooke was on the verge of bankruptcy when it introduced a generic line of 'black and white' cigarettes. The product brought price competition to the market and competitors responded in kind, introducing their own low-priced economy brands. A discount or rebate war ensued at the wholesale level and Brooke alleged that Brown & Williamson was offering to sell its generics below cost. In addition, it filed a suit under § 2(a) of the *Clayton Act* alleging that Brown & Williamson practised illegal price discrimination between its full-priced branded product and low-priced

---

1 *Brooke Group Ltd.* v. *Brown & Williamson Tobacco Corp. 509 U.S. 209 (1993)* (henceforth *Brooke*).

2 One peculiarity of *Brooke* was that neither firm involved in the case was dominant. Brown & Williamson was the third largest firm with a market share of approximately 12% during the time frame relevant to the case. Brooke was the smallest of the six firms, with a market share of 2.3% in 1980 and 5.7% in 1984 (Table 10-1 in Burnett (1999) at 240). Predatory pricing complaints usually involve dominant firms and economic models of predation generally assume that the incumbent or predator is a monopolist.

3 According to Table 10-1 in Burnett (1999), the four-firm concentration ratio was 88.0 in 1980 and increased to 90.2 in 1988.

4 Scherer and Ross (1990) provide a brief history of the industry at 250-251.

generics.[5] Brooke contended that the predatory pricing scheme was a means of coercing it to reduce the price gap between generic and branded cigarettes, which would facilitate the return to oligopoly pricing.

The Supreme Court ruled that regardless of whether a complaint of predation was brought under § 2 of the *Sherman Act* or § 13(a) of the *Robinson-Patman Act*, to establish injury, the plaintiff must first show that:[6] '... a rival's low prices ... are below an appropriate measure of its rival's costs.'

The difference between the two statutes arises in the second element, which is recoupment. In this regard, the Court ruled:[7] '... we interpret § 2 of the Sherman Act to condemn predatory pricing when it poses "a dangerous probability of actual monopolization" ... whereas the Robinson-Patman Act requires only that there be "a reasonable possibility" of substantial injury to competition before its protections are triggered ...'

The Court did not deal with the issue of the appropriate cost test for predation because Brooke did not meet the second hurdle.[8] It was unable to show that its competitor Brown & Williamson had a 'reasonable prospect' of recouping its investment in predatory pricing. Citing *Matsushita* the Court observed:[9] 'Recoupment is the ultimate objective of an unlawful predatory pricing scheme; it is a means by which a predator profits from predation. Without it ... consumer welfare is enhanced ...'

In *Matsushita*, not only did the Court view predation as an investment, but also as an 'inherently uncertain scheme' because the short run losses or costs of predation were certain, but the benefits in terms of future monopoly profits were uncertain. The Court required evidence of both, the ability to attain *and* maintain monopoly power to allow the predator to recoup losses and 'harvest some additional gain'.[10]

In both *Brooke* and *Matsushita* the Court was not convinced that there was a possibility of recoupment, without which there can be no predation. In both cases, the predator was not a monopolist; in *Matsushita* the alleged vehicle for recoupment was explicit collusion or a price fixing conspiracy between major Japanese electronic goods producers. In *Brooke*, the supposed vehicle was tacit collusion, which in the opinion of the court was 'the least likely' method of recouping losses from predation.[11]

Brooke set a particularly high barrier for plaintiffs and the Court identified three essential elements for a conviction. Firstly, the plaintiff must show either a 'dangerous probability' or a 'reasonable prospect' of recoupment via subsequent

---

[5] Sections 2(a) through 2(f) of the *Clayton Act* were amended as sections 13(a) through (f) of the *Robinson-Patman Act* 15 U.S.C.A.

[6] *Brooke*, p. 222.

[7] *Brooke* at 222. With respect to § 2 of the *Sherman Act*, the Court cited *Spectrum Sports* 506 U.S. 447 (1993). In *Spectrum Sports* (at 459) the Court required both, a dangerous probability of monopolization *and* specific intent to monopolize. Intent alone is insufficient to establish a dangerous probability of success (at 448).

[8] However the Court ruled out above cost predation (*Brooke* at 222, n. 1 and 223).

[9] *Brooke* at 224. *Matsushita Electric Industrial Co., Ltd., et al.* v. *Zenith Radio Corp. et al.*, 475 U.S. 574 (1986) (henceforth *Matsushita*). Elzinga (1999) provides a good discussion.

[10] *Matsushita* at 589.

[11] *Brooke* at 227-228.

monopolization.[12] Secondly, to prove that prices were below costs and lastly, that the predatory pricing scheme would likely injure competition in the relevant market. The *Brooke* standard conforms to the Chicago view of antitrust and critics such as Edlin (2002) contend that it represents a very narrow interpretation of the law. Nonetheless, the doctrine is clear.

3. Sector Specific Administrative Law in Canada

Unlike the United States, Canada has a single Federal antitrust statute. The *Competition Act* (henceforth the *Act*) has both criminal and civil provisions. Predatory pricing is mentioned explicitly as an 'offence in relation to competition' under the criminal provisions of the *Act*, but a complaint may also be brought under the civil provisions as an exclusionary practice by a dominant firm.[13] Under the criminal provisions, the offence of predation is described as a 'policy' of selling products at 'unreasonably' low prices, leading to a substantial lessening of competition or the elimination of a competitor. Conviction may result in imprisonment of up to two years.[14] Under the civil provisions, predation is listed as one of many exclusionary practices that may be used by a dominant firm. Here, predation is described as the selling of articles below acquisition cost for the purpose of disciplining or eliminating a competitor.[15]

The Commissioner of Competition used the civil provisions in *Nutrasweet*, which was the first case brought before the Competition Tribunal under the abuse of dominance provisions of the *Act*. In *Nutrasweet* the Tribunal ruled that this provision did not lend itself to manufacturing situations. Instead the Tribunal interpreted it as being applicable to distribution or where goods are purchased for the purpose of resale. The Tribunal did rule however that the term 'anti-competitive act' in § 78 could be interpreted broadly enough to encompass other types of predatory conduct.[16] Though the Commissioner did not present any evidence on costs, the Tribunal was of the view that the Areeda-Turner standard or comparing price with marginal cost would be appropriate in predation cases. Average variable cost would be an acceptable proxy for marginal cost if *Nutrasweet* were producing to the left of the minimum point of the average [total] cost curve. At the minimum point the appropriate proxy would be average [total] cost since this would equal marginal cost. Further, the Tribunal observed that showing recoupment was an essential element to support a claim of predation:[17] 'Even if NSC was pricing below cost after 1988, it is highly unlikely that NSC would be able to recoup from Canadian consumers the foregone profits resulting from below-cost pricing.'

---

[12] The former applies to § 2 of the *Sherman Act* and the latter to § 13 (a) of the *Robinson-Patman Act*.

[13] No cases have been brought under the criminal provision of the *Act*. The most recent case *R* v. *Hoffman-La Roche Ltd. (Nos. 1 and 2) 58 C.P.R. (2d) 1; 1981* was brought under the criminal provisions of the previous legislation, the *Combines Investigations Act.*

[14] See § 50 (1) (c) of the *Act*.

[15] See § 78 (1) (i) of the *Act*.

[16] *Director of Investigation and Research* v. *The Nutrasweet Company*, CT-1989/002, 74-76.

[17] *Nutrasweet* at 77.

In a more recent case, the Tribunal reiterated its earlier position:[18]

> The essence of an allegation of predatory pricing is that the firm foregoes short-run revenues by cutting prices, driving out rivals and thus providing itself with the opportunity to recoup more than its short-term losses through higher profits earned in the longer term in the absence of competition. A predatory pricing allegation is difficult because, at least in the short-run, consumers apparently benefit from lower prices. In addition, predation can only succeed if the predator has greater staying power than its rivals and a reasonable prospect of recouping its losses. In order to distinguish competitive pricing action from predation, therefore, the 'Areeda-Turner test' for predatory pricing was developed and has been adopted by the courts.

Until the year 2000, it was clear that Canada and the United States had very similar approaches to predation. In both countries, predation was interpreted as pricing below cost with an Areeda-Turner type test being generally acceptable to courts. Both jurisdictions accepted that the initial period of price-cutting was beneficial to consumers, and that recoupment through subsequent dominance or the exercise of market power was essential for successful predation. In late 1999/early 2000 Air Canada acquired Canadian Airlines International making the former a virtual monopoly in the domestic market. The Government amended the abuse of dominance provisions of the *Act* in August 2000 to include aviation specific provisions so as to dissuade Air Canada from using exclusionary conduct toward potential entrants.[19] These were expected to be Canadian low cost carriers. The amendments allowed Government to make regulations that specify anti-competitive acts and conduct; and identify facilities that were considered essential to the provision of aviation services.[20] In addition, the amendments granted the Commissioner broad powers to issue temporary prohibition or cease and desist orders. This was because the government viewed capacity in the aviation industry as being very mobile. It could be deployed quickly to discipline or exclude competitors.

The regulations list a gamut of specific anti-competitive acts and tests, including operating or increasing capacity on routes at fares below 'avoidable cost'; using a 'second-brand' low cost carrier; commissions; incentives; loyalty programmes and a reputation for predation as a means of disciplining competitors; or pre-empting and/or denying competitors access to essential facilities such as landing slots. The Government was anticipating entry by low cost carriers therefore the draft enforcement guidelines issued by the Competition Bureau refer to a real or quality adjusted prices. They suggest that it may be considered anti-competitive for a full service carrier to match the prices of a low cost carrier. At identical nominal prices, a full service carrier offers higher quality either in terms of on-board service or through the availability of frequent flyer programmes.[21] Thus in

---

[18] See *Director of Investigation and Research* v. *Tele-Direct Inc.*, CT-94/3 at 290-291.

[19] § 78 (2) of the *Competition Act*.

[20] *Regulations Respecting Anti-Competitive Acts of Persons Operating a Domestic Service* SOR/2000-324, available at: http://laws.justice.gc.ca/en/C-34/SOR-2000-324/74279.html. Henceforth referred to as *Regulations*. These are very similar to the (now withdrawn) guidelines issued in the United States by the Department of Transportation in 1998.

[21] The February 2001 draft of the enforcement guidelines is available at: http://cb-bc.gc.ca/epic/internet/incb-bc.nsf/vwGeneratedInterE/h_ct02126e.html.

order to compete effectively, the low cost carrier may be compelled to charge lower prices than the full service incumbent so as to compensate consumers for the lack of frills. The guidelines define the relevant geographic market as an origin-destination city-pair and indicate that the avoidable cost test would be applied to a flight, on a daily basis for a period of at least one month.[22] Avoidable cost is defined as the cost that would have been avoided had the dominant firm chosen not to provide the service in question. For illustrative purposes, the guidelines defined four categories of costs. These include costs that are avoidable outright such as fuel costs and landing and navigation charges; costs that are avoidable through redeployment such as crew and flight costs; potentially avoidable costs which include primarily labour costs associated with maintenance, ticketing, baggage handling and reservations; and unavoidable costs such as general overhead, executive salaries and building expenses.

Canada now has a dual track approach to predation. The first track is based on statute and is similar to the approach in the United States, as it has always been – it applies to all sectors other than airlines. For the dominant airline there is another track – a very specific set of administrative rules which list a variety of anti-competitive acts including predatory pricing. They specify that the avoidable cost test must be applied to determine if the airline is in violation of the law. They also list the type of costs that would be included in the computation of the avoidable costs. The benefit, as many have argued, is transparency – at least the dominant airline knows what it is up against. The downside, as the discussion of the Air Canada case in section 5 (Air Canada) shows, is that transparency alone is of no value. The questions are: Are the transparent rules the right set of rules? Do they allow competition authorities to distinguish legitimate business conduct, no matter how aggressive, from predation? And can competition authorites and expert economists apply the transparent rules correctly?

4. American Airlines

On May 13, 1999, the U.S. Department of Justice brought a complaint under the monopolization provisions of § 2 of the *Sherman Act* alleging that American Airlines (AA) engaged in predatory conduct from 1995 to 1997 and intended, by monopolizing or attempting to monopolize seven routes, to recoup the losses from below cost pricing. These routes were centered at AA's hub at Dallas-Fort Worth Airport (DFW), and linked DFW with airports in Kansas City, Wichita, Colorado Springs, Long Beach, Phoenix and Oakland. The alleged anticompetitive conduct was AA's response to attempted entry by various low cost carriers (LCCs) such as Vanguard, Western Pacific and Sunjet. AA responded to entry by reducing prices and increasing capacity. These changes were reversed when the low cost entrant either moved operations or left the market.[23] In addition, AA allegedly sought to

---

[22] The Bureau proposed using average daily revenues calculated over a one-month period on an available seat mile basis.

[23] AA sought and was granted summary judgement by the U.S. District Court of Kansas, which was upheld by the Tenth Circuit U.S. Court of Appeals, No. 01-3202, July 3, 2003 (henceforth *AMR Appeal*).

develop a reputation for predation, which would help extend its monopoly power to forty other routes. Lastly, the effects of AA's conduct were allegedly 'felt' on five other routes even though it did not engage in predatory behaviour or attempt to monopolize these routes. The action related primarily to the seven routes mentioned earlier, which District Judge Martens refers to as 'core' routes because they were the focus of the government's case against AA:[24] 'The remaining routes are frequently treated ... as afterthoughts tacked onto the underlying claims involving LCC competition on the core routes.'

In assessing the allegations of the government, the District Court made some general observations on the nature of the airline business. Major carriers made large and sunk investments in setting up hub and spoke operations which provided them with a number of advantages.[25] Hubs provide significant economies of scale, scope and density leading to a lower cost per passenger therefore potential entrants to hub routes typically expect to lose money during the initial periods of operation.[26] Dominating a hub has other benefits. Providing more 'frequency and scope of service' allows the dominant carrier to obtain a 'disproportionate' share of traffic and revenues. Concentrated hubs allow airlines to charge higher prices; otherwise known as hub premiums.[27] Market share at hubs and yields are correlated therefore AA's price-variable cost margins were higher for flights originating at and departing from DFW compared to other flights in its system.

Low cost carriers were entering markets at many hubs and they brought lower fares because they had lower costs. For example, AA estimated Southwest's costs to be about 30% lower than its own. In 1994 AA calculated ValueJet's stage length adjusted cost per available seat mile (CASM) to be 4.32 cents compared to AA's 8.54 cents. Competition from LCCs reduced yields and generally AA's yields and revenues were higher on routes where it did not face any competition from LCCs. In determining its response to this serious competitive threat, AA conducted a variety of scenario planning exercises, conducted ramp counts at the gates of competitors and studied competitors' balance sheets and break-even load factors.[28] Further, the court observed that LCCs did not follow generic strategies and selected their markets carefully. For example Vanguard stayed away from Southwest routes because they would have nothing to bring to markets where

---

[24] *United States of America* v. *AMR Corporation, et al.* (henceforth *AMR*) at 92, n. 4. All page numbers cited here refer to the version of the Memorandum and Opinion (dated 24/07/2001) posted on website of the Antitrust Division of the U.S. Department of Justice: http://www.usdoj.gov/atr/cases.html. The correct citation for this case is 140 F. Supp. 2d 1141 (D. Kan. 2001).

[25] The business strategy literature (Ghemawat (2001)) also considers investment in hubs as 'commitment' or an investment in firm specific resources. Commitment is key to firm performance. In addition to providing scale and scope advantages hubs provide a barrier to imitation due to physical/locational uniqueness and long-term contracts for gates and slots.

[26] In *AMR* at 43, Judge Martens observed that losses accompany entry into any new route, including network expansion by incumbents. Plaintiff's expert agreed that this was also true for established carriers.

[27] *AMR* at 5 and 11. Scale advantages at hubs reduce marginal costs while product differentiation advantages allow hub carriers to charge higher prices.

[28] The government interpreted these activities as proof of intent of exclusionary conduct, however the court characterized them as 'generally monitoring competitors'.

prices were already low. Instead, Vanguard entered AA's routes. Others such as Access Air (based in Iowa) did not want to attract the attention of major carriers and so chose to serve large destinations which had not been turned into hubs. Access Air also ensured that its fares were above the variable costs of major carriers. Access Air observed the following rules:[29] '... stay off of elephant paths ... don't eat the elephant's food ... and keep the elephants more worried about each other than they are about you ...' The court observed that price matching was routine in the airline industry and entrants expect incumbents to respond in this way: [30]

> It is uncontroverted that new entrant airlines with low fare strategies, including Vanguard, Western Pacific, Frontier, National and Jet Blue, expect existing competitors to match those fares. Officers of these airlines do not believe matching another carrier's fare is anti-competitive conduct, so long as pricing is not below cost. Further, an airline that does not match fares is likely to lose business to its low priced rivals.

AA had a typical pattern of responding to entry, which was to reduce prices and increase capacity. As a result, the entrant either reduced or discontinued service. For example, Vanguard started three daily non-stop flights on the DFW-Kansas City route in December 1994. AA had eight daily non-stop flights on the route and Delta had six. In early 1995 AA matched fares, though it had penalties for refunds and it added six new flights in mid-1995. Eventually Delta stopped serving the route and in December 1995 Vanguard discontinued direct service but continued two one-stop (via Wichita) flights. AA then reduced its flights to ten. During this one-year period Vanguard's share of traffic on the route varied between 16 and 27%. AA responded to Midway's entry on the DFW-Chicago (MDW) route in a similar manner. AA responded with inventory parity in May 1994 and matched prices in September 1995. Midway stopped serving the route in 1995 and by May 1995 AA had gained, at the expense of Delta and Southwest, more than the initial market share it lost to Midway.

AA responded to entry aggressively because in the mid-1990s it had observed ValuJet's success in establishing a hub operation in Atlanta. Over the two-year period 1994-96, ValuJet had forty-eight aircraft and was serving twenty-eight cities. This included the Atlanta hub, which had twenty-two spokes. AA attributed ValuJet's success to the lack of an aggressive pricing response from Delta. It estimated that Delta lost $232 million in annual revenue due to the success of ValuJet. AA concluded that giving up part of the market to an entrant was not the appropriate response to entry. Instead, it sought to match prices even at the cost of lower profits in the short-term. The government used evidence such as notes taken by AA officials at internal meetings to show predatory intent or to buttress its claim that AA was developing a reputation for predation so as to exclude or deter rivals in other markets. The government sought to demonstrate sacrifice and recoupment; or that the loss in short-term profits was considered by AA as an investment, which would pay-off when the entrant left the market.[31] 'Ms. Block

---

[29] *AMR* at 73. Access Air appears to follow the precepts of judo strategy. Yoffie and Kwak (2001) advise small entrants not to 'moon the giant'.

[30] *AMR* at 15.

[31] *AMR* at 18.

recorded a statement made by American's then-CEO, Robert Crandall to the effect that: "If you are not going to get them out then no point to diminish profit."'

The court accepted neither the predatory intent nor the reputation argument. It insisted that the standard had to be an objective one. To meet the *Brooke* standard the government had to show below cost pricing and a dangerous probability of recoupment. Recoupment based on a reputation for predation in other markets, and more generally, strategic entry deterrence arguments were also rejected as subjective, speculative, and providing no limiting principle:[32] 'The government's theory offers no principled basis for the court to distinguish between a general reputation for aggressive but lawful conduct on the one hand, and illegal predatory conduct.'

More importantly, these arguments were at odds with *Brooke* which required objective evidence of recoupment in the relevant antitrust market, or 'in market' recoupment. The government had stated in its allegations that each city-pair route was a separate market.[33] Thus complaints by 'other competitors' in 'other markets' were found to be too broad based and speculative because there was no way to distinguish 'reputation' arguments from 'vigorous competition'. The court noted that the government's expert Professor Stiglitz had relied on 'industry folklore' to infer that AA had a reputation for predation:[34] 'Professor Stiglitz has admitted that an airline can acquire a reputation for aggressive behaviour without engaging in predatory pricing, and that such a reputation for aggressiveness ... is "by definition" not anti-competitive.'

The government used AA's four internal 'decision measures' to make its case on objective grounds. These earnings measures were developed and used by AA for assessing flight and route performance rather than for financial reporting purposes. FAUDNC and FAUDNS used fully allocated costs whereas VAUDNC and VAUDNS used variable costs.[35] FAUDNC and FAUDNS included between 97% and 99% of AA's costs. The other two earnings measures – VAUDNC and VAUDNS were based on costs that were variable over an eighteen-month planning horizon and accounted for about 72% of the total costs in AA's decision accounting system. FAUDNC and VAUDNC include net upline and downline revenues from connecting passengers – net of variable costs associated with those

---

[32] *AMR* at 133.

[33] The government identified the relevant antitrust market as city-pair non-stop airline service.

[34] *AMR* at 131, n. 23.

[35] FAUDNC (VAUDNC) refers to Fully allocated (Variable) earnings plus upline/downline contribution net of costs. FAUDNS (VAUDNS) refers to Fully allocated (Variable) earnings plus upline/downline contribution net of spill. Upline/downline contribution refers to beyond revenue net of incremental costs. Inclusion of upline/downline net contribution to profit recognizes hub complementarities. Despite the double counting in these measures, Edlin and Farrell (2002) suggest that this is a sensible way to do route level decision-making. As an example, consider a flight from point A to a hub (H) and a subsequent leg from H to point B. Some passengers on the A to H flight may take that flight because they want to travel between those two points; others may take the A to H flight because they can take the connecting flight from H to B. Passengers who travel from A to B via H would have taken an alternative carrier to travel from A to B if the A to H flight did not exist. Thus the A to H flight generates business on the H to B flight. Similarly, if the H to B flight were cancelled, this would affect business on the A to H flight.

passengers as well as any incremental flight costs assigned to connecting passengers. FAUDNS and VAUDNS incorporate the impact of 'spill' or that accommodating an additional passenger on an upline or downline flight may result in some other passenger being lost to a competitor. The government proposed its own measure VAUDNC-AC as a proxy for short-run average variable costs. This measure added aircraft ownership costs to VAUDNC and accounted for about 79% of AA's costs. The court observed that aircraft ownership costs were treated as fixed costs in the airline industry and should therefore not be included in the calculation of avoidable costs.

FAUDNC and FAUDNS, according to the government, were proxies for long run costs and included costs that were avoidable over an eighteen-month planning horizon. The court noted that none of the government experts had identified avoidable costs either in general or with respect to the core routes. The court interpreted these measures of route level performance as long-term break-even benchmarks, which had been negative on a persistent basis for several domestic routes. Further, AA had endured periods longer than 18 months when the system wide average FUADNC was negative. In June 1994, decision FAUDNC was negative for 55% of AA's routes. VUDNAC and VUADNS were interpreted by the court as measures of average avoidable costs of a route and were used to evaluate flight and route performance. The court observed that these could also be negative for a couple of months if an airline was entering a new route.

The government's experts suggested four tests for predation. The first asks: Did AA forgo better profit opportunities elsewhere on its system when it increased capacity on the core routes? In other words, do VAUDNC, VAUDNC-AC and FAUDNC decline as a result of the capacity changes on the route in question? If they do, then this would imply that the incremental cost of redeployment was higher than the incremental revenue which would provide a measure of sacrifice. The second test examines the level of FAUDNC on the route in question. If the effect of the redeployment was to make this measure negative then one could conclude that average revenue or price was below long run average variable cost. The third test looks for persistence of negative FAUDNC at the route level. In particular, the government's experts looked for instances where FAUDNC was negative for more than one year. Three percent of AA's domestic routes had negative FAUNDC for one year and 1.2% for eighteen months. This test essentially shows that that the route violated AA's internal measure for its planning horizon of eighteen months, suggesting perhaps that AA should not add capacity to routes if doing so violates their own route-level decision making parameters. The fourth test is similar to the first, but government's experts attempted to directly calculate incremental costs and revenues of capacity additions on four of seven core markets using VAUDNC-AC and revenues from 'incremental' passengers.[36]

---

[36] By evaluating short run costs and persistence, the government was in effect evaluating both the initial and the subsequent responses to entry. So long as the initial response covered avoidable cost at the route level, it would be allowed. However if the subsequent more aggressive response was less profitable than the initial response it may be deemed predatory. We thank Professor Donald G. McFetridge for pointing this out. See *AMR* at 71 for an example of AAs response to ValuJet.

The court rejected tests one and four because they were not tests for predation but of failure to maximize short-run profits. Using such tests would prevent consumers from benefiting from price reductions and would also condemn capacity additions for which average revenues exceeded average variable costs. In addition, the court rejected the tests because they examined incremental capacity, which the court interpreted as representing a fraction of the relevant antitrust market. The second and third tests were rejected because they used a fully allocated cost measure – FAUDNC, which the court interpreted as being equivalent to applying an average total cost test, or, the wrong test. Fully allocated costs included costs of city ticket offices, some station expenses, sales and advertising costs, and flight simulator expenses. These were allocated arbitrarily over the entire fleet and could not be avoided by not operating a particular flight or route. The court noted that the government had only provided evidence on four core routes and on these routes AA's price exceed all measures of variable cost: VAUDNC, VAUDNS and the government's proposed VAUDNC-AC. Thus there was no predation, and since the claims relating to other routes were dependent on a finding of predation on the core routes, they were also dismissed.[37] The court re-affirmed the Areeda-Turner rule as the appropriate test for predation:[38]

> Average variable cost, as a measure of predatory pricing, enjoys not only the weight of authority, it is also most congruent with the goal of the Sherman Act: prohibiting unfair competitive practices while simultaneously encouraging open, indeed vigorous price competition.

Two other issues deserve mention. The first relates to recoupment, which under *Brooke* requires evidence of a dangerous probability of recoupment via monopolization. The second is the role of price matching. The court examined 'structure' at DFW to deal with the first issue. In other words, it sought to determine the extent of competition at DFW. The government painted a picture of dominance by AA. Both Delta and AA use DFW as a hub and in the early 1990s Delta attempted to increase its operations at DFW but was unable to, partly due to the aggressive response of AA. As a result Delta suffered operating losses over the period 1992-94 and downsized its operations at DFW. Over the period July 1993-96 Delta's share in terms of passengers boarded at DFW decreased from 28.4% to 19.2% whereas AAs share increased from 64.7 to 71.8%. In May 2000 AA's share of passengers boarded was 70.2% where as that of LCCs was 2.4%. The government noted that compared to other hubs, the share of LCCs was lower at DFW. For example, in the third quarter of 2000, LCCs had a market share of 15.8% in Denver and 16.8% in Atlanta.

The court rejected the government's view and indicated that in mid-2000 there were seven new entrant low cost carriers serving DFW and this hub had more low fare airlines than any other hub airport. LCCs served at least thirty-one of the top fifty destinations from DFW. Further, airport officials stated that they had

---

[37] The court did not consider the remaining routes because the government did not provide any evidence on these. This suggests that a few routes cannot be used for illustrative purposes and that evidence is required for every market. See *AMR* at 105.

[38] *AMR* at 103.

successfully attracted foreign carriers, which had contributed to a decline in AA's market share. The airport facilitated entry of new carriers through advertising support programmes and access to common use gates. Even though LCCs had low shares, the growth in shares had been quite substantial – they had a 25% increase in passenger share over the one-year period May 1999 to May 2000. Over the decade 1990-99 there were a total of forty-four instances of entry at DFW – on average 4.7% of DFW routes were being entered per year. Based on such evidence the court concluded:[39]

> ... there are no structural barriers to entry at DFW, which can accommodate any domestic carrier that seeks to establish or expand service. Not only do the uncontroverted facts fail to show any strategic barriers to entry by new entrant carriers, supra-competitive pricing on DFW routes is also disproven by the active presence of other strong competitors in the Dallas-Fort-Worth market.

On meeting the competition, the court indicated that the statutory defence available under the price discrimination provisions of § 13(b) the *Robinson-Patman Act* might be applicable in the case at hand. The court emphasized that AA had matched, rather than undercut, prices charged by entrants. Antitrust laws were designed to encourage this kind of activity:[40]

> Nor has the plaintiff identified any instances in which American undercut the published DFW-MCI fare of Vanguard with a published American fare during the relevant time periods. American's average fare throughout the period of Vanguard's DFW-MCI service was higher than Vanguard's average fare.

Judge Martens went so far as to quote Areeda and Hovenkamp who argue that the meeting competition defence would apply even if price matching were predatory, so long as the price matched that already being charged by the entrant and was maintained for the same or lesser duration. Further, the court viewed this defence as having particular application in the case at hand because the alleged predator's revenue exceeded its variable costs.[41] Since price reductions stimulate demand; price matching 'implicitly but necessarily requires the ability to increase sales capacity'.[42] The court did not agree with the government's argument that if quality differences between AA and competing low cost carriers were taken into account, then allowing AA to charge the same dollar price as competitors would be allowing it to 'effectively undercut' competitors. This:[43] '... would require courts to engage in a series of subjective price comparisons based on intangible values.'

The government appealed to the Tenth Circuit Court of Appeals.[44] The Appeals Court declined to rule on a definite cost measure, but noted that courts may need flexibility to examine various proxies to marginal cost and sole reliance

---

[39] *AMR* at 122.
[40] *AMR* at 24.
[41] *AMR* at 116.
[42] *AMR* at 119.
[43] *AMR* at 119.
[44] *U.S.* v. *AMR Corporation et al. United States Court of Appeals, Tenth Circuit, No.013202.* Henceforth *AMR Appeal.*

on average variable costs may in some instances 'obscure the nature of a particular predatory scheme'.[45] While conceding that the average variable cost measure was a good proxy in most cases, any alternative proxy that courts may consider 'must be accurate and reliable in the specific circumstances of the case at bar'.[46] Tests two and three were rejected because they relied on FAUDNC, or fully allocated costs, which were not representative of either marginal or incremental costs. Test one was rejected because it treated forgone profits as costs. 'Rather than determining whether the capacity itself was priced below an appropriate measure of cost' this test compared profits before and after capacity additions.[47] That is, it sought to determine if profits were lower after adding capacity. Test four compared the (directly computed) cost of incremental capacity with the incremental revenue.[48] The Court did not view this as test for short-run profit maximization; instead it found this to be the correct interpretation of Baumol's avoidable costs test. The only costs that should belong in test four were those that could have been avoided by not adding capacity.[49] However this test was also rejected because it was not implemented correctly. As AA had argued, VAUDNC-AC included variable non-proportional common costs such as the costs of airport ticket agents, arrival agents, ramp workers and security. Arbitrarily allocated variable costs should not be included in avoidable costs. In this way, the Appeals Court rejected all four measures of cost and affirmed the judgement of the district court. Thus while there may still not be a single cost standard in predation cases, it is clear that average variable cost or average avoidable cost, if computed correctly would be acceptable proxies for marginal cost; and avoidable costs do not include allocated or attributed costs. Opportunity cost tests, or those that include forgone profits as costs; tests for short run profit maximization or for higher profit would likely be rejected.

5. Air Canada

After 1999 Air Canada took over the failing Canadian Airlines International and became the dominant domestic carrier in Canada with a market share in excess of 80%. There were a few smaller airlines in the market such as Canada 2000, WestJet, CanJet and Royal Air. CanJet was a low cost carrier that commenced service in September 2000 on six routes in Eastern Canada. Ticket sales had started in end-July 2000 and even prior to commencement of service CanJet lowered its announced fares in response to competition from another low cost carrier – Royal Air. CanJet claimed that it would only have to make a few seats available at fares lower than its 'announced fares' since Royal Air provided limited service and few seats at discounted prices on the two routes which CanJet also served. Thus a price war of sorts was in progress when Air Canada started offering its L14EASTS fare.

---

[45] *AMR Appeal* at 14.

[46] *AMR Appeal* at 14.

[47] *AMR Appeal* at 19.

[48] This is at odds with *AMR*, where the District Court rejected tests one and four emphasizing that they looked at incremental costs and revenues '*only from the added capacity*'. *AMR* at 109.

[49] *AMR Appeal* at 21.

L14EASTS matched CanJet's announced fare but was between $10 and $60 higher than the latter's discounted fare. Unlike CanJets's fare, L14EASTS came with advance purchase, maximum stay and other restrictions and penalties. In response to a complaint by CanJet the Commissioner of Competition exercised the sweeping powers provided to him under the aviation specific amendments to the *Act* and issued a temporary order preventing Air Canada from offering its L14EASTS or any similar fare on five city-pair routes in Eastern Canada.[50]

Air Canada applied to the Tribunal to have the order set aside, and in the event that it was not, to modify it to remove reference to 'any similar fares'.[51] The Tribunal agreed with Air Canada that the Commissioner's order prevented it from competing with entrants for the duration of the order or until the Commissioner had assessed the conduct of Air Canada under the abuse of dominance provisions of the *Act*. Without the order, the risk was that CanJet might not survive. The Tribunal noted that one of the objectives of the *Competition Act* was to ensure that small and medium-sized enterprises had an equitable opportunity to participate in the Canadian economy.[52] The Tribunal also interpreted § 104.1 to imply that when issuing a temporary prohibition order, the Commissioner need not be certain that an anti-competitive act had taken place.

In March 2001, the Commissioner commenced a civil action against Air Canada under the abuse of dominant position provisions of the *Act* alleging that Air Canada had engaged in a policy of 'fare matching' with low cost carriers without regard to its own profitability or to the 'additional benefits associated' with its service offering and had operated and increased capacity that did not cover the avoidable costs of providing air services on eight domestic routes.[53] The Tribunal divided the inquiry in two phases. Phase I focused on the application of the avoidable cost test in the context of the *Regulations* on two sample routes for the

---

[50] The order was issued in October 2000. The routes were Halifax–Ottawa, Halifax–Montreal, Halifax–St. John's, Toronto–Windsor and Ottawa–Windsor. A few weeks later the Commissioner extended the duration of his first prohibition order on the first three routes.

[51] *Air Canada* v. *Commissioner of Competition*, 2000 Comp. Trib. 26. The Tribunal did order deletion of reference to 'any similar fares' on grounds that it was imprecise and therefore unenforceable.

[52] See § 1.1 of the *Act*.

[53] The routes were St. John's (Nfld.)–Halifax, Montreal–Halifax, Ottawa–Halifax, Toronto–Moncton, Toronto–Fredericton, Toronto–St. John (NB), Toronto–Charlottetown and Hamilton–Moncton. In the statement of grounds and material facts (see Notice of Application to the Competition Tribunal (*Notice*) available at: http://www.ct-tc.gc.ca/english/cases/ct-2001-002/air-canada.html) the Commissioner lists barriers to entry at [79] and [80]. These include a 'reputation for predation' barrier, which is essentially an issue of perception rather than an objective one. In addition the list includes lack or traffic feed, business class lounges and frequent flyer programmes. This is rather curious because all competitors listed by the Commissioner are low cost carriers of one form or another – these carriers choose to provide point-to-point service and no connections (even though the Commissioner identifies WestJet as a network carrier); no frills; and to target the price sensitive leisure segment. Further, the list includes the cost of purchasing aircraft and hiring cabin crew. Air Canada's competitors are in the airline business by choice, and therefore it would make sense for them or anyone else seeking to enter this industry, to purchase aircraft and hire cabin crew.

period April 1, 2000 to March 5, 2001. Failing the avoidable cost test however would not lead to the conclusion that Air Canada abused its dominant position. The reminder of the application, or the elements required under § 79 of the *Act* would be assessed in Phase II.[54]

The Commissioner argued that failing the avoidable cost test would imply that Air Canada committed an anti-competitive act. The Tribunal noted that in previous jurisprudence and prior to the implementation of the *Regulations* the offences listed under §78 required an:[55] '... "object", "design" or "intent" to engage in an exclusionary conduct that is having the effect of augmenting, entrenching or extending market power. The presence of such wording made relevant the concept of legitimate business justification ...'

The *Regulations* however do not require the consideration of legitimate business justification for revenues below avoidable cost therefore such justifications could only be applied to the period before the regulations came into effect.[56] The Tribunal reserved the right to consider such justifications in the Phase II hearing. In aviation cases the Tribunal's role is to evaluate the conduct of the dominant carrier in accordance with the *Regulations* and not necessarily with the other provisions of the *Act*, which would apply in non-aviation cases.

In his *Notice* the Commissioner identified the relevant geographic market as all domestic city-pair markets, and all affected routes in the Halifax, Moncton and Toronto areas.[57] The areas are catchment areas; so Toronto includes Hamilton, which is WestJet's base of operations in the Toronto area.[58] The relevant product market was identified as 'airline passenger service' being separate from other modes of transportation such as bus and rail.[59]

The test as stated in the *Regulations* and as agreed to by all parties was Baumol's avoidable cost test. This is a multi-product generalization of the Areeda-Turner test. The test was applied to a hypothetical unidirectional flight between two cities. Implicitly therefore, an air carrier was viewed as a multi-product firm where each unidirectional flight is a different product. Air Canada argued for the test to be applied at the route level, but the Commissioner contended that this might hide predatory conduct because a particular flight may not cover avoidable costs but the route as a whole may. The Tribunal's interpretation of the test was as follows:[60]

---

[54] At the time of writing only the Reasons and Findings for Phase I were available: *Commissioner of Competition* v. *Air Canada*, 2003 Comp. Trib. 13. Henceforth, this case is referred to as *AC* and all citations in square parentheses indicate paragraph numbers.

[55] *AC* at [54] and [55].

[56] The application covers the period April 1, 2000 to March 5, 2001. The Tribunal noted that legitimate business justifications could be considered for the period April 1, 2000 to August 23, 2000. The latter being the date the Airline Regulations came into effect.

[57] *Notice* at [64].

[58] It is usual for low cost carriers such as WestJet to operate from secondary airports such as Hamilton.

[59] *Notice* at [51].

[60] *AC* at [80], emphasis added.

> As discussed by Dr. Baumol, the multiproduct enterprise will rationally continue to produce a given product as long as the revenue from the sale exceeds the product's avoidable costs. Since no enterprise will rationally set the price per unit of any of its products below their respective (average) avoidable costs in the pursuit of maximum short-term profits, deviations from rational pricing *may* be predatory.

The next issue before the Tribunal was the determination of avoidable costs. At a conceptual level all parties generally agreed that product specific fixed and variable costs were avoidable and common costs and sunk costs were not. The application of these principles was more controversial. The Tribunal adopted a broad interpretation of avoidable costs. One interpretation is that costs that do not have to be incurred when a flight is not operated are avoidable. The Tribunal called these costs 'outright' avoidable, or avoidable through 'outright shedding'. In addition it noted that costs could be avoided through redeployment, passenger recapture and via disposal of assets such as aircraft in secondary markets. Resources that are released by cancelling one flight can be used to offer new services elsewhere in the network – this is redeployment. Cancelling a flight does not imply that the airline loses all passengers that would have taken that flight; some of them may use other flights on the network and therefore be recaptured.

The cost items under consideration were those listed in Air Canada's 328 Report. The report assesses flight profitability by comparing revenues with fully allocated costs. The operating cost items relate to activities and cost allocation or attribution is done using cost drivers. There was no dispute over seventeen cost items, which were classified as variable. All parties agreed that overhead costs were not avoidable. The twenty-six disputed items were grouped into the following five categories: system labour costs, station labour costs, aircraft labour costs, non-labour system and sunk costs, aircraft ownership and insurance costs. The Commissioner's cost accounting expert Mr. Vettese referred to many of these costs as 'step variable' which could not be shed outright. Step variable costs, as the name suggests, change in steps as output changes and not continuously. Thus if one baggage handler can service up to five flights, reducing the number of flights to four does not allow the firm to avoid the wage cost of the baggage handler. However the Tribunal treated these costs as avoidable via redeployment or passenger recapture. The Tribunal also clarified that step variable costs could not be treated as common costs. So for example, a pilot *simultaneously* services passengers and cargo and so the pilot's remuneration is a common cost of providing passenger and cargo service. However a baggage handler performs tasks *sequentially*, handling baggage from one flight at a time. Suppose a baggage handler is paid on a daily basis and can handle five flights per day; this does not imply that the daily wage is a common cost of servicing five flights. The fact that wages are paid on a daily basis is an administrative issue; all one has to do is calculate the wage on a per-flight basis. Since all costs that are not common can be avoided either through recapture, redeployment or disposal, the baggage handler's wage in this example would be considered avoidable. In this way, 90% of Air Canada's fully allocated operating costs for the hypothetical cancelled flight were classified as being avoidable. The Tribunal did not require the Commissioner to identify specific profitable redeployment opportunities or to provide evidence on the proportion of passengers recaptured even though it noted that the Commissioner's expert economist Dr. West 'relied heavily' on recapture; indeed as discussed later, Dr. West assumed full recapture in his calculations.[61] The Commissioner acknowledged that in practice it would be difficult to analyse the ability of Air Canada to avoid costs through redeployment. Fortunately, the Tribunal did not want to 'burden' the Commissioner with the requirement to identify specific redeployment opportunities.[62] Instead the Commissioner was only required to show that these opportunities were 'generally and realistically' available. The Commissioner relied on press releases and media articles to show that Air Canada had started new services and also increased frequencies on many routes during the time period under consideration.[63] The Tribunal did not accept Air Canada's argument that it always tried to do better, so any profitable redeployment opportunities would already have been exploited.[64] Aircraft ownership costs are treated in a similar manner, that is, as being avoidable via redeployment or through other avenues such as resale or sublease. The Tribunal used Air Canada's annual report to infer that it had adjusted its fleet during the period under consideration by taking delivery of new aircraft and parking others either for potential sale or for return to lessors.[65]

Next, the Tribunal had to determine when costs become avoidable and over what time period to apply the avoidable cost test. The Commissioner argued that 'substantially all of Air Canada's costs except overhead are avoidable within three months' and that the avoidable cost test should be conducted for a three-month period on each route at issue.[66] The Commissioner referred to three successive months so in fact, the calculations were done on a monthly basis for three months – costs that were identified as being avoidable in three months were also treated as avoidable in each of the three monthly calculations. It would seem however that in fact the Commissioner was arguing that all of Air Canada's fully attributed operating costs with the exception of overhead become avoidable instantaneously – the moment a flight is cancelled. The Tribunal observes:[67] 'In the Tribunal's understanding, Commissioner conducts the avoidable cost test ... regardless of when avoidabilty commences.'

In the same paragraph the Tribunal quotes counsel for the Commissioner:[68] 'It didn't strike me appropriate to say a cost is not avoidable, is not avoidable, is not avoidable and all of a sudden it is.'

In other words, adjustment is instantaneous in the airline industry. The fact is, that the one-month period is based on Air Canada's reporting system, or when

---

[61] *AC* at [331].

[62] See [189] in Commissioner's Final Argument (*Final*) available at: http://www.ct-tc.gc.ca/english/cases/ct-2000-004/air-canada.html and *AC* at [134].

[63] *Final* at [199].

[64] Air Canada also argued (*AC* at [128]) that since the Commissioner's expert found that 42% of its flights failed to cover their avoidable costs this implied that opportunities for recapture and profitable redeployment were generally not available.

[65] *AC* at [280]. A careful reading shows that Air Canada actually did not get rid of any aircraft during the period of predation even though thirty were taken out of service.

[66] *AC* at [168].

[67] *AC* at [174].

[68] *AC* at [174].

information becomes available. The three-month period merely implies that within three to four months Air Canada should know that its revenues are below its fully allocated operating costs.[69] Air Canada argued for a one-year period and its expert economist Dr. Baumol suggested that the appropriate period is the period of the alleged predatory conduct. In addition, Air Canada argued that allowance should be made for seasonality so as not to confuse low demand with predatory conduct. The Commissioner's riposte was that seasonality is predictable.[70] The Tribunal's position was that that the *Regulations* 'do not specify a time period for the anti-competitive act'. Thus the dominant carrier violates the *Regulations* 'any time' it operates or increases capacity at prices that do not cover avoidable costs.[71] The Tribunal did not rule on this issue suggesting that since the Regulations did not specify any time period, it was the Commissioner's obligation to indicate the time period – it therefore accepted the Commissioner's position with regard to timing.

The last issue was whether beyond contribution should be given any recognition.[72] Beyond contribution reflects the network nature of the airline business. Suppose a passenger travels from A to B and then from B to C. Part of the contribution of this passenger on the B to C flight is pro-rated to the A to B flight. This recognizes the fact that the A to B flight contributed to the B to C journey, or if the initial flight were cancelled then the passenger could not have connected to the subsequent leg. Air Canada wanted beyond contribution to be counted as revenue in the implementation of the avoidable costs test. The Commissioner argued that beyond contribution should not be given any recognition as it would imply double counting and overstate revenues thereby concealing possible predatory conduct. The Tribunal ruled that beyond contribution should not be given any recognition but observed that it did not have enough evidence on recapture and revenue displacement to determine what the beyond contribution would be and that the burden was on Air Canada to provide this information. It stated that this could be considered in Phase II under the legitimate business rationale for operating a flight below avoidable cost.[73]

The test was applied to two of the eight routes identified in the Commissioner's complaint: Toronto–Moncton and Halifax–Montreal. Seventy-three flights were operated on the first route over the period April 2000 to February 2001. Forty-three of these flights had monthly revenues below avoidable costs. On the second route seventy-two of the one hundred and eleven monthly scheduled flights had revenues below avoidable costs over the period August 2000 to February 2001. Thus Air Canada failed the avoidable cost test, however whether this constitutes abuse of dominant position will be determined in Phase II.

[69] AC at [172] indicates that the Commissioner's two economic experts Drs. West and Tretheway thought that three months should be enough where as the Commissioners accounting expert Mr. Vettese said that this may take as much as four months.

[70] The Tribunal also noted at [194] that under the *Regulations* there was no exemption for operating below avoidable costs due to random factors.

[71] *AC* at [186].

[72] In the American Airlines case beyond contribution was referred to as upline and downline net revenues.

[73] *AC* at [301].

Regardless of the eventual outcome, this case is likely to remain controversial for a variety of reasons. For example, there are many alternative plausible characterizations of the airline industry. Consumers do not buy flights and air carriers do not sell them. Consumers perceive a network or a full service carrier as selling three qualities of service, or types of output – these being economy, business and first class. Once an airline decides to offer service in the form of a flight from point A to B there are very few costs it can avoid. Indeed most costs become fixed and common. Thus after this decision is made, all the airline can do is to maximize revenues or contribution to the costs of providing that flight. This is done through price discrimination using sophisticated yield or revenue management systems where seats within the same class of service may be sold to individual consumers at different prices. This is one way to characterize the airline industry – in other words, network carriers are multi-product firms offering a differentiated product and entry by low cost carriers affects primarily one segment, which is the low quality segment. The low cost carrier can then be thought of as a single product or specialized firm providing one low-quality class of service to price sensitive customers. Indeed this is the way low cost entry has occurred, not just in aviation but also in other industries such as steel. The (low cost) mini-mills entered the low-end market of vertically integrated steel producers in the United States and gradually moved up the quality ladder. The Commissioner's view that an airline is a multi-product firm producing unidirectional flights fails to recognize the network characteristics of the industry and the substantial scope and scale economies resulting from hub and spoke operations. It could be reasonably argued that the analysis should have been conducted either at the route level or even at the network level.[74]

The Commissioner contended that 'substantially all' or 90% of Air Canada's fully attributed operating costs with the exception of overhead were avoidable, and that Air Canada was able to avoid these costs the moment a flight was cancelled. Together these two observations point toward contestability, in the sense of easy entry and exit, or at least suggest that exit costs are very low. If this is the case then entrants cannot be hurt by predation.

The Commissioner's version of the avoidable costs test includes costs that are avoidable, or those that can be shed outright; and increased profits from (full) recapture and profitable redeployment. This is an opportunity costs test. Failing this test implies that the firm could have done better in the short-run but did not. Such behaviour is deemed 'irrational' but the goal of antitrust is not to impose economic notions of rationality on business.[75] As Elhauge (2003a) observes:[76]

---

[74] In the Commissioner's view 'a route is a fiction; it is a mere "collection of flights" between two places'. See *Notice* at [75] and *Final* at [245].

[75] For example, failure to take advantage of better alternatives could also be due to managerial incompetence.

[76] Elhauge (2003a) at 694. In his original work on the avoidable cost test, Baumol (1996) indicates (at 68 and 69) that short run profit sacrifice may be viewed as predatory if it can be made up through future monopoly profits after exit of rivals.

> ... it is vital ... to avoid using cost measures that effectively include forgone profits. Otherwise, one cannot keep predatory theories based on a failure to maximize short-run profits analytically distinct from theories based on pricing below costs.

As is discussed below, the opportunity cost test was applied incorrectly because the Commissioner's economics expert Dr. West, assumed full recapture. If recapture were minimal, cancelling a flight would lead to losses and not profits.[77] In closing arguments, counsel for Air Canada argued that it was contradictory to assume both full recapture and profitable redeployment: [78]

> The recapture theory assumes that Air Canada would lose no revenue if it cancelled a scheduled flight, which stands in direct contrast to the assumption that Air Canada can always generate incremental revenues by redeploying assets to routes with existing service. How can one simultaneously assume incremental revenues from increased service, but no revenue decline from service reduction?

McFetridge (2003) uses some simple numerical examples to show that opportunity costs could either exceed, or be less than inherently avoidable costs. In addition, opportunity costs could either exceed, be less than, or equal to fully allocated costs. What is crucial to these calculations is the extent of recapture and/or redeployment. One of McFetridge's examples is reproduced here for illustrative purposes. Suppose a flight can accommodate 50 passengers at a fare of $100 per passenger. Variable costs are $20 per passenger and flight specific fixed costs that are inherently avoidable are $2000. The fully allocated cost of the flight is assumed to be $6000. In this example, revenues ($5000) are below fully allocated costs but exceed inherently avoidable costs ($3000). If the flight is cancelled and all 50 passengers are recaptured, their contribution to revenue on other flights is $5000 and their contribution to variable costs is $1000. Thus with full recapture, their contribution to profit elsewhere in the network is $4000. The opportunity cost of continuing to offer the initial flight is $7000, which includes $3000 in inherently avoidable costs and $4000 in lost profits. This amount exceeds the fully allocated cost of $6000 and cancelling the flight would increase profits by $2000. The extent of recapture does not affect the inherently avoidable cost of $3000 or the fully allocated cost of $6000, but it does affect the opportunity cost because lower recapture results in lower additional profits and may be even losses. If only 20 of 50 passengers are recaptured, the additional contribution to profits on other flights due to recapture is $1600 (revenues of $2000 less additional variable costs of $400) and the opportunity cost is now $4600 (the additional profits due to recapture $1600 added to the inherently avoidable cost of $3000). The airline would lose $400 by cancelling the flight.

This example shows that the extent to which opportunity costs exceed inherently avoidable costs depends on the extent of recapture and not on the fully allocated costs and there is no reason to expect fully allocated costs to equal

[77] The Tribunal notes in *AC* at [136] 'Dr. Baumol insists that redeployment be profitable, that opportunities be available, and that only the truly avoidable portion of a cost be included in the test.'

[78] Air Canada's Phase I Closing Argument at 8 and 9, available at: http://www.ct-tc.gc.ca/english/cases/ct-2000-004/air-canada.html.

opportunity costs as the Tribunal's decision implies. Though the Tribunal acknowledged that the extent of recapture could depend upon historical load factors and the number of flights offered by competing airlines, it nevertheless accepted calculations based on full recapture.[79]

6. Conclusions

The American Airlines and Air Canada cases provide an interesting contrast in the application of each country's respective law reflecting their approach to controlling predatory behaviour. In both instances, full service carriers faced entry by low cost carriers and responded by increasing capacity and cutting prices. In the American Airlines case the Department of Justice (DOJ) constructed four measures of cost, three of which included beyond revenues. All tests were conducted at the route level. Two of the four measures were based on fully allocated costs, which included about 97% of total costs. These were put forward as proxies for long run costs. The court rejected fully allocated costs as being the correct proxy of variable or avoidable costs. The other two measures were based on costs that were variable over an eighteen-month period. They included about 74% of total costs and did not include aircraft ownership costs. One of these tests was rejected because it included forgone profits. Thus the court rejected opportunity costs tests, tests for foregone profits and tests for short-run profit maximization. The fourth test compared the costs and revenues of incremental capacity and would probably be acceptable to the court as a correct interpretation of the avoidable cost test. It was rejected because it was implemented incorrectly and included arbitrarily allocated common variable costs.

The DOJ also set out to show with specific calculations that American recouped on the four routes where it attempted to show below-cost pricing. Professor Berry, the government's expert used two benchmarks. The first was American's FAUDNC margin on those routes prior to the period of alleged predation; this was argued to be representative of margins under legitimate competitive conditions. The second benchmark was American's FAUDNC margin on routes where it competed with Southwest Airlines; this was put forth as a good benchmark for competitive returns. Comparing American's actual margin to these benchmarks during and after the period of alleged predation the government's expert estimated predatory sacrifice and in-market recoupment respectively. The court rejected the government's claim on a number of grounds. Firstly, the estimates were based on FAUDNC margins calculated using fully allocated costs and not variable costs. Thus the court found the calculations to be invalid as a matter of law. Secondly, the government did not prove that American had monopoly power on the routes in question, thus ruling out the feasibility of supra-competitive pricing in the post-predation period. Lastly, even after years of supposed supra-competitive pricing, the calculations failed to show the recovery of losses incurred during the alleged period of predation. The court viewed the

[79] *AC* at [118], [331] and [333]. McFetridge (2003) also provides examples that include both recapture and redeployment.

calculations as demonstrating that no substantial recoupment had occurred – there was instead a net sacrifice of $30 million.

To summarize, in the United States the average variable cost test is generally acceptable but the avoidable cost test or a test based on incremental cost may also be acceptable if it is implemented correctly. The cost measure should not include attributed or common costs or count forgone profits as costs – in other words, an opportunity costs test will not do. Prices must be below the appropriate measure of cost. Failure to maximize profits does not constitute predatory sacrifice and neither does failure to take advantage of more profitable opportunities. It is essential to show a reasonable likelihood of recoupment through use of monopoly power after the entrant has left the market. The court also rejected reputation-based arguments for predation.

In Canada, the avoidable cost test is applicable to dominant airlines. The Tribunal accepted that 90% of Air Canada's fully allocated costs were avoidable and no recognition was given to beyond revenue. In addition, the Tribunal noted that the *Regulations* did not specify a time period for the anticompetitive act and neither did they provide any exemptions for random factors. In essence it accepted the Commissioner's argument that the time period over which cost become avoidable was irrelevant – either costs were avoidable or they were not and Air Canada would contravene the *Regulations* the moment revenues did not cover the avoidable costs of a unidirectional flight. The Commissioner and the Tribunal failed to acknowledge the network aspects of the airline business. The Commissioner's version of the avoidable cost test, which was accepted by the Tribunal, included costs that could be shed outright and increased profits from [full] recapture and profitable redeployment. This is an opportunity cost test and one that was applied incorrectly because it assumed full recapture. This test and its application will probably be the most controversial aspects of this decision.

In Canada, there are two standards – administrative law for aviation and statute for everyone else. Which is better – precise administrative law or vague statute? To borrow from an American legal scholar:[80]

> We may not be able to define precisely how many hairs one needs to lose before one turns bald, but we all understand the general concept of baldness and what moves you closer or further from that state. Vague standards might be uncertain around the edges as applied to tough facts, but at least offer genuinely guiding normative principles.

The Canadian situation is curious. Foreign carriers are not allowed to enter domestic Canadian routes. The collapse of Canadian Airlines and its subsequent merger with Air Canada made Air Canada a virtual monopoly in the domestic market. During the merger, the Commissioner sought some undertakings from Air Canada to alleviate the anti-competitive effects of the merger, but there were others as noted by the Tribunal; such as the inability to lay off employees until March 2002; to continue to serve small communities that had been served prior to the merger, until January 2003; and the requirement that any aircraft that Air Canada wished to sell would have to be offered to Canadian carriers first.[81] Air Canada

---

[80] Elhauge (2003b) at 1.

[81] *AC* at [15].

filed for bankruptcy in April 2003. Would it not make more sense to apply notions of economic efficiency and market based tests of profit maximization to firms that are allowed to operate on market principles in the first place? Is this about antitrust or about the misuse of antitrust to rectify structural problems in Canadian aviation?

Acknowledgements

We thank Donald G. McFetridge and Hans-Martin Niemeier for their comments.

References

Baumol, William J. (1996), Predation and the Logic of the Average Variable Cost Test, *Journal of Law and Economics*, 39 (1) April, 49-72.

Burnett, William B. (1999), Predation by a Nondominant Firm: The Liggett Case (1993), 239-263 in Kwoka Jr., John E. and Lawrence J. White (eds), *The Antitrust Revolution: Economics, Competition, and Policy*, Third edition, Oxford University Press, Inc. New York.

Edlin, Aaron S. (2002), Stopping Above-Cost Predatory Pricing, *Yale Law Journal*, 111 (4), 941-991.

Edlin, Aaron S. and Joseph Farrell (2002), The American Airlines Case: A Chance to Clarify Predation Policy, Working Paper, Competition Policy Center, University of California, Berkeley.

Elhauge, Einer (2003a), Why Above-Cost Price Cuts to Drive Out Entrants Are Not Predatory – and the Implications for Defining Costs and Market Power, *Yale Law Journal*, 112 (4), 681-827.

Elhauge, Einer (2003b), Defining Better Monopolization Standards Regulatory Policy, Program Working Paper RPP-2003-14. Cambridge, MA: Center for Business and Government, John F. Kennedy School of Government, Harvard University.

Elzinga, Kenneth G. (1999), Collusive Predation: *Matsushita* v. *Zenith* (1986), 220-238 in Kwoka Jr., John E. and Lawrence J. White (eds), *The Antitrust Revolution: Economics, Competition, and Policy*, Third edition, Oxford University Press, Inc. New York.

Ghemawat, Pankaj (2001), *Strategy and the business landscape: Core concepts*, Prentice Hall Inc., N.J.

McFetridge, Donald G. (2003), *The Commissioner of Competition* v. *Air Canada*: Phase I, http://www.carleton.ca/~dmcfet/personal/images/some_of_my_recent_papers.htm.

Scherer, F.M. and David Ross (1990), *Industrial Market Structure and Economic Performance*, Third edition, Houghton Mifflin Company, Boston.

Yoffie, David B. and Mary Kwak (2001), *Judo Strategy*, Harvard Business School Press, M.A.

Chapter 7

Airline Predation in Canada

Scott M. Butler

Introduction[1]

Aviation history is littered with failed airlines. Some have suggested that there is no natural equilibrium in a de-regulated airline industry,[2] only a constant state of flux. This constant rise and fall of new airlines has led to many accusations of predatory behaviour, few of which have proved to withstand the scrutiny of the legal system. The most notorious recent example is the American Airlines case, where Judge J. Thomas Marten wrote in his judgement:

> There is no doubt that American may be a difficult, vigorous, even brutal competitor but here it engaged only in bare, not brass knuckle competition.

Indeed, to understand airline predation, one might ask Vanguard, Sun Jet International and Western Pacific (all named in the case as suffering from American's actions) whether they felt the sting of that bare knuckle competition. Unfortunately, they have all ceased to operate; perhaps brass knuckles weren't required.

In Canada, WestJet's entry into new markets from February 1996 up until April 2000 was accompanied by across the board price matching, capacity increases, travel agency commission over-rides, and the doubling and tripling of frequent flyer points on competitive routes by Canadian and Air Canada. Yet none of these tactics were enough to slow WestJet's progress. By the year 2000, after taking over Canadian, Air Canada controlled in excess of 90%[3] of the domestic market as measured by revenue. WestJet was expanding eastward in early 2000, and with WestJet's entry into the Hamilton–Moncton market on April 9, 2000, Air Canada undercut WestJet's walk-up fare and added more capacity than ever before as a competitive response.

---

[1] It should be noted that this collection of ideas and gathering of facts on predation is influenced through the author's work at WestJet, a consistently profitable low-cost airline that is currently involved in an 'abuse of dominance' case involving one of their routes in eastern Canada. Despite the working relationship with WestJet, any ideas and observations should be considered the author's own, in no way reflecting the corporate view of WestJet.

[2] Attributed to Ken Button via 'The Empty Core'.

[3] Konrad von Finkenstein, Q.C., in a letter regarding the 'Proposed Acquisition by Air Canada and 853350 Alberta Ltd. (the "Offeror") of Canadian Airlines Corporation' dated December 21, 1999 and addressed to Lawson Hunter, Q.C., Air Canada's legal counsel.

In Canada, the fists were out in April 2000, whether the knuckles were bare or brass is yet to be determined.

This paper explores predation[4] in the airline industry, with specific references to the Canadian experience. An attempt is made to understand the true origin of predatory behaviour, not to simply describe the anti-competitive acts in the industry. Airline predation proves to have many possible causes including: government policy, industry structure, airline strategy, and weak enforcement of competition legislation.

Aviation markets worldwide have been facing deregulation, leading to industry disruption in the form of many failures and start-ups. One could easily surmise that the presence of many start-ups and failures could indicate the presence of predation, making the airline industry a ripe study for regulators and academics. Lumping deregulation into the broad category of 'government policy', the history of predation and related government policy in Canada is explored.

Industry structure is in part affected by government policy and has an effect on the existence or lack of predatory behaviour. An industry characterized by one dominant player is more likely to contain predatory conduct than one with many competitors of equal size, simply because it would be easier to extract monopoly profits through the elimination of competitors if a position of dominance already exists. The ability to eliminate small competitors is also a component of industry structure that would affect the degree to which predation is used as a tool. The development of the Canadian airline industry since 1999 is reviewed in this context.

Airline strategy plays a vital part in understanding predatory behaviour. The strategy of network, legacy airlines is contrasted with the strategy of so called low cost airlines, and the results demonstrate a propensity of legacy airlines to engage in 'predatory' behaviour, particularly in Canada.

Finally, the current Canadian competition legislation for airlines is reviewed. Provisions in the Competition Act in Canada related to the airline industry are generally viewed as very proscriptive when compared to other jurisdictions. Both the rules and the mechanism for enforcement are examined in detail in search of causes for predation to occur despite, or perhaps, in spite of the Competition Act.

Understanding the origin of predatory behaviour in the airline industry should prove instructive in predicting future predation in any industry and should assist in developing future regulations to deal with predatory conduct.

Government Policy

In Canada, government policy has had an effect on both the development of the current airline industry structure and the strategies of the airlines.

Government policy has a long history of intervention in normal competitive interactions in Canada, culminating in competition legislation. The introduction of the National Policy tariff in 1879 led to the protection and encouragement of

[4] Predation in the context of this paper is meant to describe aggressive acts of a dominant competitor on a smaller competitor, and is not meant to reflect any legal definition.

various industries, including textiles, footwear, and refined sugar. Simultaneously, the development of a cross-country train system, the Canadian Pacific Railway, led to an easier exchange of goods between regions and allowed for larger quantities of production in a centralized location.

In a somewhat ironic fashion, the companies in protected industries had difficulty competing with one another. High profits early on led to many new entrants, and the businesses subsequently attempted to form combinations, or trade associations, to fix prices and limit production. The tariff and subsequent combinations allowed inefficient companies to proliferate. The price increases experienced by consumers and consequent outcry raised the ire of politicians, resulting in the passing of 'An Act for the Prevention and Suppression of Combinations formed in restraint of Trade' in 1889, Canada's first competition legislation and regarded by some as a 'political sham'.[5]

One hundred years later, a not so different yet seemingly opposite scenario was being played out in the Canadian airline industry. Deregulation of the domestic airline industry in 1988 was shortly followed by the elimination of Air Canada's government ownership. A previously protected business that enjoyed regulated market dominance was now subject to normal competitive forces. The inefficiencies created by 51 years of government ownership and regulated fares left Air Canada exposed to more efficient new entrants and only slightly more efficient existing competitors. This policy decision was made with an understanding that deregulation was expected to bring more competition, better service and lower fares, while at the same time removing the government from enterprise. However, to ensure Air Canada faced the competition with a strong advantage, it was privatized with a virtually debt free balance sheet and the proceeds of an equity issue.

While the National Policy led to inefficiency and attempts to concentrate industries years earlier, in this case, the inefficiency was already present in the airline industry. Consolidation ensued with vigour. Air Canada had an estimated 54% of the domestic market prior to deregulation, PWA Corporation ('Canadian Airlines') controlled an estimated 30%, and Ward Air had 9%.[6] Eleven years after deregulation, Air Canada purchased Canadian Airlines to form one dominant airline that controlled, at the time, an estimated 90% of the revenues in the Canadian market.[7] At the same time, there was also a recognition that stronger competition policy was required to encourage competition and prevent the newly combined carrier from abusing its government approved dominant position. With this in mind, the Commissioner of Competition (the enforcer of the Competition Act) had Air Canada commit to certain undertakings in an attempt to foster new entrants. Meanwhile, the legislators of the day were developing a bill to increase the powers of the Commissioner and design specific competition regulations for the airline industry.

---

[5] Michael Bliss (1973, p. 182) refers to the passing of the first competition legislation as a 'political sham'.

[6] September 15, 1987, Globe and Mail, p. B4.

[7] Konrad von Finkenstein, Q.C., in a letter regarding the 'Proposed Acquisition by Air Canada and 853350 Alberta Ltd. (the "Offeror") of Canadian Airlines Corporation' dated December 21, 1999 and addressed to Lawson Hunter, Q.C., Air Canada's legal counsel.

This look at government policy indicates that anti-competitive behaviour and changes in government policy appear to be inextricably linked. Where government regulation allows inefficiency to develop in industries, the casual evidence indicates that companies will resort to anti-competitive behaviour in order to concentrate the industry. The lesson for regulators is that an imbalance in the market created through government regulation will be fertile ground for predatory conduct when the mechanism that creates the imbalance is either put in place or removed.

Industry Structure

If consolidation of an industry leads to predation, and the ability to eliminate competitors encourages it, the Canadian airline in the last five years proves to be an ideal case study.

In mid 1999, two major network carriers dominated the airline industry in Canada: Air Canada and Canadian. Air Canada, as indicated above, was previously owned by the government, while Canadian could trace its roots back to the 1930s and was at the time the result of several airline mergers throughout its more than 60 year history. Air Canada and Canadian both provided domestic, transborder and international service, offering economy to first class products, flying from the most remote regions of Canada to international destinations sharing between them roughly 90% of the domestic market. During this period, a small domestic low-cost carrier was establishing itself in western Canada under the banner 'WestJet'. Several charter operators were also active, namely Canada 3000, Royal, Skyservice and Air Transat, who carried Canadians primarily to sun destinations and Europe with limited operations domestically.

After years of intense competition with Air Canada focused on market share, Canadian found itself at the breaking point. From 1988 (the year of deregulation) up until the second quarter of 1999, Canadian has lost $1.73 billion on revenues of $33.4 billion, a –5.2% net margin.[8] The slightly larger of the two, Air Canada, had not fared much better as it fought its way to a $22 million loss on revenues of $49.5 billion (–0.04% net margin). In this blow-for-blow battle of industry heavyweights, Air Canada had the upper hand, as Canadian could not stop the bleeding. With no significant assets left to sell, negative equity, and a debt burden that could not be sustained, Canadian's situation caught the government's attention. Certain sections of the Competition Act were suspended such that discussions could take place between Air Canada and Canada as the government was seeking a smooth transition for the near-insolvent Canadian Airlines.

On August 24th, 1999, Onex Corporation made an unsolicited offer to Air Canada shareholders with the intention of combining Canadian and Air Canada. Speculation suggested that Onex could see the benefits of rationalizing capacity in the domestic market. Air Canada, however, fought the Onex bid strenuously, forcing Onex to withdraw the offer, after which Air Canada made their own offer to the shareholders of Canadian, which was subsequently accepted.

[8] Net income after tax.

On December 23rd, 1999, Air Canada announced that all conditions had been met to proceed with the acquisition of Canadian Airlines, Canada's second largest airline. With this acquisition, Air Canada now controlled in excess of 90% of the Canadian domestic market, measured by revenue, and was dominant in US transborder and international operations. As the 7th largest airline in North America with more than $10 billion in revenue and 365 aircraft,[9] it dwarfed its nearest domestic competitor WestJet, which ended 1999 with 16 aircraft and $204 million in revenue.

Seizing on the opportunity created by a consolidated airline industry with very little domestic competition in the east, WestJet announced on January 20th, 2000 that it would be expanding to eastern destinations as early as March 9th, 2000, while establishing a hub in Hamilton. Shortly thereafter, in a February 9th press release, Air Canada stated:

> By reducing unprofitable flying in Canada, aircraft have been redirected on to profitable transborder flying, where yields have historically been higher. The reduction in domestic capacity of approximately 15% will be more than offset by increases in transborder and international capacity of 14% and 12% respectively.

This was consistent with expectations in the industry, and seemed to confirm the belief that the irrational domestic competition between Air Canada and Canadian focused on market share rather than profit had contributed significantly to the cumulative $1.75 billion in losses between the two since deregulation. Despite Air Canada's announced intentions, capacity (available seat miles) was only reduced by 6% in the June-September period of 2000, when compared to 1999. In addition, Canada 3000, Royal Airlines, and WestJet were increasing domestic capacity, albeit at a lower cost structure and were stimulating the market with lower fares. If there were any benefits to be gained by Air Canada reducing domestic capacity, they seemed reluctant to do so, unwilling to let the smaller domestic competitors replace high cost capacity with lower cost (and likely profitable) capacity.

In early 2000, while WestJet was negotiating with various airports in the east, including Halifax and Moncton, Air Canada declared in a press release that they would be *reducing capacity* on its Moncton to Toronto route (Toronto is in close proximity to Hamilton). On February 29th, 2000, WestJet announced it would begin flying from Hamilton to Moncton on April 19th, 2000 with fares ranging from $129-$339. Air Canada responded by matching the low fare classes, undercutting WestJet's 'walk-up' fare by $90 ($249 vs. $605 previously), and *adding 55% more capacity* than the announced reduction to the market. In addition, capacity was increased and prices were matched for all of the significant cities in the 'catchment' area of Moncton.

From April 2000 until September 2004, the Canadian airline industry has witnessed the expansion and fall of Canjet Airlines, Royal Airlines, Canada 3000 and Roots Air. Royal, Canjet, and Canada 3000 were all focusing on low fare domestic service in eastern Canada (although both Royal and Canada 3000 were

---

9 Air Canada presentation notes, CIBC World Markets First Annual Aerospace and Transportation Conference, Toronto, Ontario, September 21, 2000.

previously focused on charter opportunities) after the merger of Air Canada and Canadian. The intense domestic competition in eastern Canada resulted in Canada 3000 making the decision to consolidate the three airlines in the hopes that an integrated strategy would increase profitability. Canada 3000 purchased both Royal and Canjet in the spring of 2001, and the combined entity failed in November 2001 with a mixed fleet of roughly 40 aircraft. Roots Air came and went in 39[10] days in the spring of 2001. Roots Air ceased to operate the day after Air Canada announced an offer for an equity stake in the company that operated Roots' flights, an offer that never materialized in a purchase. Air Transat and Skyservice have survived the turmoil by remaining focused on their charter and tour package niche and continue to hold steady, while several other charter airlines/tour operators have started up with the view to replace the charter capacity lost by the failure of Canada 3000.

While all of this turmoil and seemingly difficult business environment might dissuade normal investors, this seems to have steeled the determination of those interested in owning airlines. As history has indicated, there is never a shortage of capital for start-up airlines. In June of 2002, Canjet re-started with basically the same 'low-cost' strategy, and the previous patriarch of Royal has started up a low fare airline under the trade name 'Jetsgo'. As at September 30, 2004, Canjet had six 737-200 aircraft and had plans to acquire four 737-500s, while Jetsgo had 14 MD-83 aircraft and had plans to introduce 18 F-100s into their fleet.

Not one to be left out of the low fare (but not necessarily low-cost) market, Air Canada started up not one, but two, airline-within-airline concepts. Responding aggressively to the significant drop-off in demand post September 11th, Tango was launched on October 10th, 2001 with mainly transcontinental routes operating point to point from Toronto, coincidentally overlapping many of Canada 3000's domestic routes. As mentioned previously, Canada 3000 failed in November 2001, reportedly one day before the Competition Bureau was about to place a cease and desist order on Air Canada's Tango operation. While the gloves may have been off, the weakness of Canada 3000 at the time necessitated little more than a gentle shove rather than a strong uppercut. Tango reached a peak of 20% of Air Canada's domestic capacity in the summer of 2002 with an estimated 21 aircraft in its schedule. Tango was eliminated and re-branded as a 'fare class' on Air Canada mainline in August 2003.

Spurred on by the success of Tango, Air Canada started its second low-fare airline, Zip in September 2002 based alongside WestJet in Calgary, Alberta. Zip operated with one class of service, slightly lower wages through union concessions, yet still interlined with the Air Canada network, provided snacks for free, and provided frequent flyer points. Zip was shut down in September 2004.

Notwithstanding the success of Zip and Tango in the low-fare market in early 2003, Air Canada apparently could not transition enough of its operations to the low-fare format quickly enough,[11] as the airline succumbed to bankruptcy with a filing under the Companies' Creditors Arrangement Act on April 1, 2003.

---

[10] The current Canadian record for the shortest life of an airline. The previous record holder, VistaJet, lasted 109 days and held the record since September 12, 1997. VistaJet themselves barely beat out Astoria, which flew for 117 days in the summer of 1995.

[11] Tongue firmly in cheek.

Even with the constant growth of WestJet and other new entrants more recently, Air Canada still dominates the domestic market with an estimated 60%[12] market share and has a commanding share of the transborder and international markets as of August 2004. Interestingly, Air Canada's share of the domestic market is still higher that it was just prior to deregulation in 1987.

In summary, the last five years of airline industry evolution in Canada demonstrates a consistent drive for consolidation with many failures and start-ups. Despite the assumption that market dominance should lead to higher returns through pricing power, there is no evidence to suggest that strategy has been successful in Canada, as the dominant, merged Air Canada has yet to post an annual profit. With respect to failures, whether they are the result of poor business models, external forces, or competitive behaviour, or more likely a combination of factors, the message is clear: it doesn't take much for new and smaller airlines to fail. Therefore, anti-competitive behaviour can quickly be rewarded in the short term in the airline industry. Mitigating this incentive should be the fact that there are very few barriers to entry in the airline industry, and no shortage of capital. Therefore, the elimination of a small competitor is not likely to result in any more than very short term benefits in the Canadian experience.

Airline Strategies

There is interplay between airline strategy and predation that is critical to understanding the causes of predation. This section will be limited to a comparison of low-cost airline strategy and network and full service airline strategy. The cornerstones of the different strategies is perhaps best demonstrated by the following lexicon (courtesy of David Gillen), under which we will name network, full service airlines Revenue Maximizing Carriers ('RMCs') and low-cost airlines Profit Maximizing Carriers ('PMCs'). It is important to note that these observations on revenue maximizing behaviour are based on the empirical financial results of many network, full service airlines, in addition to the testimony given by Air Canada witnesses during the Competition Tribunal hearings. RMCs are also coincidentally carriers that have been around since pre-deregulation and therefore developed in a regulated system where cost increases were simply passed on to the consumer. While not all network, full service airlines that were spawned pre-deregulation are inherently revenue maximizers, the description applies to a large number of these airlines.

The RMCs are just that, revenue maximizers with little regard for cost. The long term strategy is to extract as much revenue from a given geographic market as possible. The short term RMC strategy is designed around getting the most amount of revenue out of a given set of resources, and, given the financial history of the industry, apparently the long term as well. The RMC strategy starts with the assumption that the means of production, the aircraft, are fixed in the short term, as are the costs associated with operating that fixed number of aircraft. A schedule is

[12] Percentage of total available seat miles, using press releases from WestJet, Air Canada and Jetsgo, and an estimate for Canjet.

then designed that maximizes the opportunities for getting as much possible demand out of the given number of city pair options that the airline flies. Once the schedule is 'set', opportunities in the short term are justified by rationalizing that the incremental cost of flying a given aircraft for an extra cycle in a day is minimal – just the 'cash costs' associated with the flight (the staff are there anyway, the airport infrastructure is paid for, the spare parts are available, etc.). The logic for adding a flight works conversely to the logic of removing a flight, as it assumes that removing one flight from the system reduces very little costs, as the staff, aircraft, spare parts, etc., are being paid for regardless. Presumably, many flights are justified with both sides of this rationale and the airline suffers from chronic overcapacity through the false justification of current flights and incremental additions to the network.

The minimal incremental cost is then weighed against the revenue benefits of the flight. Here again, the measurement of revenue is biased towards the justification of incremental capacity. In addition to the revenue received for non-stop travel on the given city pair, the concept of 'beyond contribution' (also referred to as 'beyond revenue') is used as a guide to assess the flight's additional benefit to the entire network. Beyond contribution places a high value on the passengers that paid for a flight beyond the first point, by presuming that if the first leg of the flight was not flown, they would not have captured that passenger on the second leg of the flight. This same presumption is made for the second leg of the flight, that is, they wouldn't have received the person on the first leg if they didn't have the second leg.

Beyond contribution is calculated by taking the total fare paid for the two legs of the journey, assessing the costs associated with the passenger of roughly 20-30% of the revenue, and including the remainder in the assessment of the flight leg. As an example, for a passenger flying from A to B to C who paid $1,000, the beyond contribution for the A to B leg is $700-800, and is also $700-800 for the B to C leg. Thereby, the RMC would show a total beyond contribution on a passenger that paid only $1,000 to be between $1,400-$1,600. Flights that may be unprofitable on a pro-rated fare basis are then shown to 'contribute' revenue to the network and can be justified on this basis.

When these two concepts are combined, individual flights, even routes, are easily justifiable in the interests of 'network profitability'. In fact, Air Canada's Executive Vice-President, Commercial, stated recently at the Competition Tribunal hearings that 'the entire Jazz[13] operation would be considered unprofitable if it wasn't for beyond revenue'. In other words, he meant that the Jazz operation was not profitable based on a pro-rated amount of fares. Furthermore, he added that some '42% of the flights in their network would not pass the [Competition] Bureau's test'. That test, as defined by the Bureau, defines avoidable cost at a level below fully allocated costs and defines fares as only the pro-rated portion of the fare. The statement by Air Canada's EVP indicates that 42% of their routes are unprofitable not only on a fully allocated basis, but on a measure that doesn't even cover the avoidable cost as defined by the Bureau. While in the short-term, this

[13] Air Canada's regional airline.

strategy may minimize losses, the RMC's have not proven that they have the ability to adopt a longer term strategy of profitability.

The revenue based strategy is so well entrenched in the mind-set that in addition to justifying individual flights and entire regional feeder operations, RMCs have also been known to start 'fighting brands', well known in the airline industry as 'airlines within airlines'. Low fare airlines within high cost airlines that have been shut down by legacy airlines include: 'Continental Lite' by Continental, 'Shuttle' by United, 'Delta Express' by Delta, 'Metrojet' by US Airways, 'Go' by British Airways, and 'Zip' and 'Tango' by Air Canada. It is notable that United and US Airways are currently in Chapter 11 (bankruptcy reorganization), and many analysts expect Delta to file under Chapter 11 in the fall of 2004. Nonetheless, RMCs continue to revive the phenomenon is an attempt to maintain the grip on all revenue in their relevant market. The competitive arena has been described as the interaction between predator and prey, yet 'cannibal' is the more appropriate description for this behaviour. Losing less is not a successful long term strategy. Shareholders would be better served by eliminating the capacity and the infrastructure associated with flying unprofitable routes.

Given the focus on revenue maximization, the design of the RMC's business in both product and process becomes predictable. To soak up all demand in a geographic market, different aircraft types are required at different times of the day for different cities, and the hub and spoke system maximizes the opportunities to serve as many markets as possible. To segment the market by consumer preference and receive the maximum amount of revenue out of each potential customer, first class and business class seating is required as well as economy class, in addition to meals, lounges and entertainment. Further segmentation by consumer behaviour through Saturday night stay requirements, cancellation fees, and other ticketing restrictions allows more price discrimination and therefore the ability to charge even higher fares. Finally, a frequent flyer plan increases the switching costs and ensures a captive market of frequent travellers.

The RMC concept directly challenges the notion that all shareholder owned businesses are inherently in business to maximize profit. However, certainly in Canada, the resident RMC, Air Canada, has not proven any desire to maximize return for shareholders. As evidenced in the section in this paper on the structure of the industry, Air Canada, despite having market dominance with the unilateral ability to control capacity and therefore pricing in the market, has not posted an annual profit since the merger with Canadian. One could argue that the notion of revenue maximization was inherent in the business of all pre-deregulation airlines, therefore, the revenue maximization strategy was misconstrued to be the way to maximize profit. The change in thought and culture to challenge this assumption takes years, and the evidence is only just starting to show in Canada, that after bankruptcy, Air Canada may in fact be focused on profit.

Profit maximizing carriers, on the other hand, start with a review of the market that attempts to determine how many individual city pairs can be operated profitably with their chosen aircraft type, and then plans the aircraft requirements for the year based on that assessment and their ability to sustain the growth. Furthermore, fully allocated costs and only the pro-rated portion of revenues are used to justify flights, even if the airline operates with a network structure. Based on this requirement, each city pair has to meet certain profitability targets and if they are not met, individual flights are assessed on the same basis to determine what changes can be made to the capacity on that city pair. There is no complicated assessment of the 'contribution' to the network. Rather, very stringent requirements on capacity additions and keeping flights in the system are adopted, with profitability on each city pair flown being the overriding goal.

The new genre of PMCs that has been appearing regularly since deregulation are focused on certain principles of low-cost, with the understanding that lower fares stimulate more traffic. The basic competitive proposition is that with a lower cost structure, lower fares can be offered while still maintaining profitability, and the resulting stimulation in the market will allow room in the market for both the incumbent and the new entrant.

Given the focus on a lower cost structure and profit maximization, the evolution of product and process by the PMC's is predictable as well. The aircraft flown defines market opportunities, not the other way around, and most PMCs fly one aircraft type in order to minimize costs. A point-to-point route structure is favoured, since it not only lowers the cost of doing business, it maximizes the profitability by focusing efforts when the non-stop demand can sustain profitable service. Behavioural pricing segmentation is dramatically simplified, product differentiation focuses only on the generation of profits as the end goal, and simplification is emphasized over complication.

The dichotomy of airline strategies offers an interesting perspective on predation. Competition law was not designed to alter the normal business decisions of companies with no predatory intent. Given the rationale used by RMCs in the normal course of business, they do not perceive forms of selling below cost as a reasonable test of predatory behaviour, since this is something they do on a regular basis. Absent any competition, if RMCs are willing to sell seats below cost, it becomes a difficult proposition to find them guilty of predatory conduct since they are willing to do it to themselves. Despite the fact that Air Canada is a shareholder owned corporation, they have not proven that the company is in fact run for the shareholders.

The Competition Act

First, a quote from a perceptive academic and reformer, Frank Underhill (1889-1971):

> When we look around at the monopolies and near-monopolies in Canada which dominate so much of our economy, the Combines Act[14] resembles nothing so much as the bleating of a pathetic vegetarian lamb in the midst of a carnivorous jungle.[15]

It appears little has changed since Underhill uttered those words in 1944. The following will summarize Canada's 'vegetarian lamb' that is used to deter the carnivores in Canada's airline industry.

---

[14] Predecessor to the current Competition Act.

[15] Frank Underhill, 1944, pp. 114-115.

The legal framework governing predatory behaviour in Canada revolves around the Competition Act (the 'Act'), passed into law in 1986 as a replacement for the Combines Investigation Act. The Act is enforced through the Competition Tribunal Act, which outlines the role of the Commissioner of Competition and the legislative process through which the Act is enforced. There are several relevant sections in the Act, as well as certain regulations specific to the airline industry that will be explained below.

The following text is an excerpt from the beginning of the Act:

> The purpose of this Act is to maintain and encourage competition in Canada in order to promote the efficiency and adaptability of the Canadian economy, in order to expand opportunities for Canadian participation in world markets while at the same time recognizing the role of foreign competition in Canada, in order to ensure that small and medium-sized enterprises have an equitable opportunity to participate in the Canadian economy and in order to provide consumers with competitive prices and product choices.[16]

There are two main areas that address predatory behaviour. It is generally understood that 'predatory pricing' is dealt with in section 50, while 'abuse of dominance' is dealt with in sections 78 and 79, although there is overlap between the two areas. There were some changes made to the Act and the Competition Tribunal Act as a result of the merger between Air Canada and Canadian which will also be detailed below.

Section 50.(1)(c) of the Act defines the following as illegal, and liable to imprisonment for a term not exceeding two years:

> Every one engaged in a business who engages in a policy of selling products at prices unreasonably low, having the effect or tendency of substantially lessening competition or eliminating a competitor, or designed to have that effect.[17]

Section 78 of the Act describes various 'anti-competitive acts' that are to be considered when evaluating Section 79 of the Act, which details 'abuse of dominance'. Section 78(1)(j), introduced by Bill C-26 in May 2000 includes a reference to regulations specifically drafted for 'a person operating a domestic service', that details specific 'anti-competitive acts'. The most pertinent points of the Regulations Respecting Anti-Competitive Acts of Persons Operating a Domestic Service, enacted in August 2000 as a result of the merger of Air Canada and Canadian (the 'Regulations') are:

a) operating capacity on a route or routes at fares that do not cover the avoidable cost of providing the service;
b) increasing capacity on a route or routes at fares that do not cover the avoidable cost of providing the service;
c) using a low-cost second-brand carrier in a manner that is described in paragraph (a) or (b).

[16] Competition Act (Canada), Chapter C-34, Part I, Purpose and Interpretation, 1.1.
[17] Competition Act (Canada), Chapter C-34, Part VI, Offences in Relation to Competition, 50.(1)(c).

It is notable that having competition legislation specific to an industry is quite unique in Canada and in other jurisdictions. Furthermore, the Regulations delve into a level of detail that is not normally found in competition law.

Section 79 of the Act details 'abuse of dominant position' as:

a) one or more persons substantially or completely control, throughout Canada or any area thereof, a class or species of business,
b) that person or those persons have engaged in or are engaging in a practice of anti-competitive acts, and
c) the practice has had, is having or is likely to have the effect of preventing or lessening competition substantially in a market.[18]

The penalty for finding a person or persons guilty of Section 79 is an 'order prohibiting those persons from engaging in that practice', enforced by the Competition Tribunal. It is notable that the Regulations are regarded as some of the most prescriptive in terms of competition legislation, and are unique to Canada.

As a result of Canadian legislative and regulatory changes in the last half of 2000, the Commissioner of Competition under the Act was empowered to issue temporary orders in circumstances where the Commissioner is of the opinion that the conduct of a dominant carrier could constitute anti-competitive acts and that, in the absence of a temporary order, a competitor is likely to be eliminated, suffer a significant loss of market share, suffer a significant loss of revenue or suffer other harm that cannot be adequately remedied by the Competition Tribunal.[19] The Commissioner exercised this power on October 12, 2000, prohibiting Air Canada from offering a discount pricing structure on specific routes that were then also served by a start-up low fare competitor, Canjet Airlines. The initial 20-day order was extended twice, eventually expiring on December 30, 2000. Canjet was sold to Canada 3000 in March 2001.

On February 8, 2001, the Competition Bureau released for consultation, its Enforcement Guidelines on the Abuse of Dominance in the Airline Industry (the 'Guidelines'). The Guidelines set out the approach that the Competition Bureau will take in investigating and enforcing the *Competition Act* (Canada) and regulations relating to competition in the airline industry. In order to establish grounds for abuse of dominant position in the airline industry, the anti-competitive acts must be taken by a dominant airline in a relevant market and must 'prevent or lessen competition substantially' in that market. The Guidelines indicate that in establishing the lessening of competition, the Competition Bureau will be looking at whether the practice of the anti-competitive acts has one or more of the following effects: (i) raising rivals' costs or reducing rivals' revenues; (ii) foreclosing existing or potential rivals from essential services or facilities; or (iii) eliminating or disciplining competitors.

The most controversial addition to legislation was Bill C-23 in May 2002. Additional powers were granted to the Commissioner of Competition in the form of an expanded ability to prohibit alleged anti-competitive acts through a 'cease

[18] Competition Act (Canada), Chapter C-34, Part VIII, Matters Reviewable by Tribunal, 79(1).
[19] WestJet 2001 Annual Information Form, p. 3.

and desist order', without first consulting the Competition Tribunal. The caveat was that the Commissioner had to believe either that the absence of a temporary order would cause injury to competition that could not adequately be remedied by the Competition Tribunal or that a person would likely be eliminated as a competitor, suffer a significant loss of market share or revenue, or suffer other harm that could not be adequately remedied by the Tribunal. In addition, the Competition Tribunal was permitted to apply an administrative monetary penalty of $15 million against an airline found to have violated Section 79. Air Canada successfully appealed the cease and desist power of the Commissioner with the Quebec Court of Appeal in January 2003. Nonetheless, the cease and desist powers are still available, but must be demonstrably proven to the Chair of the Tribunal under the guidelines of the legislation, as was the case with the Canjet order.

On March 5, 2001, the Commissioner filed an application against Air Canada with the Competition Tribunal which alleges abuse of dominant position by Air Canada pursuant to Section 79 of the Act on several routes in eastern Canada, mainly applicable to routes that Canjet previously flew. WestJet applied for and received intervenor status on the basis that, among other things, the anti-competitive acts complaint involved Air Canada's pricing and capacity strategies affecting WestJet's Hamilton-Moncton route.

The hearing has been divided into two phases, recognizing the two-part test established by the legislation that first involves proving the predator sold 'capacity' at 'fares' below their 'avoidable cost', then demonstrating a 'substantial lessening of competition'. The hearing of the first phase of the Application commenced on August 27, 2001, but was adjourned on September 11, 2001 due to industry turmoil resulting from the terrorist attacks in the United States. On October 26, 2001, the Competition Tribunal ordered that the hearing recommence on April 15, 2002. Due to the withdrawal of a member of the panel, a motion was made by the Commissioner on February 19, 2002 to reconstitute the panel and re-commence the proceedings as soon as possible. The motion was heard and the judge ruled that the hearing re-commence in the fall of 2002. Phase 1 of the hearing re-commenced on November 25, 2002 and was completed on March 5, 2003.

Phase 1 focused on several words in the Regulations in order to properly apply the regulations as drafted. The meaning of 'avoidable cost', 'fares', and 'capacity' are all words with significantly different meanings between the Commissioner and Air Canada. These four words have created an inordinate amount of cost for all parties involved, as lawyers, economists, accountants and lay-experts present arguments over the appropriate definition. The Tribunal released their understanding of the Regulations on July 22, 2003, and found that Air Canada had in fact 'operated or increased capacity at fares that did not cover the avoidable costs of providing the service' on two routes in eastern Canada in 2000 and 2001. Despite the release of the decision in Phase 1, the Tribunal was not able to proceed with Phase 2 of the application as the action was stayed with the bankruptcy filing ('CCAA') of Air Canada on April 1, 2003. Air Canada emerged from CCAA on October 1, 2004, and now the Commissioner will still have to prove if there has been a 'substantial lessening of competition' in Phase 2 to prove abuse of dominance.

Over four and a half years after the alleged predatory conduct, competition legislation in Canada is proving that the law and process are unworkable. Phase 1 of the hearing is meant to define the boundaries of competition for the airlines in Canada, as they presumably would have an understanding of what is and isn't lawful in terms of prices and capacity. In reality, the precedent setting definitions would be argued forcefully by any defending party that *in this instance*, many of the parameters are different, costs act differently, and a whole new set of items are unavoidable. Once again, a painfully long process involving lawyers, economists, accountants and lay-experts would be presented in front of a panel with little experience on competition matters.

Not only is the law flawed in its design due to airline strategy, there is little use to most new entrants in the airline industry for a process that is measured in years. Most would have long since perished. The only way the Competition Act is ever going to be an effective deterrent to predatory conduct in the airline industry is if the Act moves to a 'per se' rule, whereby the incumbent airline is given clear rules about what it can and cannot do, and an automatic penalty mechanism is instituted if the rules are broken. Alternatively, a special tribunal that handles only competition law and related competition issues could be formed to handle these cases, with strict timelines given to both parties for submission of evidence and arguments, such that the decision would take no longer than a couple of months.

Not surprisingly, this example of poor law and poor process is not unique when it comes to competition legislation. Oddly enough, this case is rare in the fact that it is actually being tried under the Act. A study conducted over a five-year period beginning April 1, 1994 found that out of 382 complaints about predatory pricing with the Competition Bureau, there were no formal enforcement actions taken.[20]

Despite the prescriptive nature of Canada's Competition Act with respect to the airline industry, Frank Underhill's statement is as true today is it was almost 60 years ago – the legislation is a lamb in a jungle of carnivores.

Summary and Observations

The Canadian Experience Proves to be an Instructive Case in Airline Predation

It is ironic that the first competition legislation in Canada was put in place due to a market imbalance created by government policies. Similarly, deregulation in 1988 with an existing market imbalance ultimately led to anti competitive conduct. The lesson for regulators is that an imbalance in the market created through government regulation will be fertile ground for predatory conduct when the mechanism that creates the imbalance is either put in place or removed. In the search for the root cause of predatory behaviour, government policy is directly linked.

The evolution of the airline industry in Canada demonstrated a drive for consolidation after deregulation due to the inefficiency of the existing players and

[20] VanDuzer, Anthony J. and Paquet, Gilles, University of Ottawa, 'Anticompetitive Pricing Practices and the Competition Act Theory, Law and Practice'.

proved that there was an ability to eliminate smaller, weaker competitors with relative ease. These two ingredients would encourage predation in any industry. Thus, the industry structure has a direct causal effect on the appearance of predation. However, offsetting the ability to eliminate smaller competitors is the lack of barriers to entry in the airline industry, and the incessant capital put forward for new entrants. Perhaps bankruptcy and the continuous emergence of new entrants has had an impact on the behaviour of the dominant competitor in Canada, as Air Canada's share of the domestic market has declined from 90% in early 2000 to approximately 60% at the time of writing.

A review of airline strategy demonstrates that pre-deregulation airlines, with their focus on revenue maximization, will be prone to engage in predatory behaviour. Given the rationale used by RMCs in the normal course of business, they do not perceive forms of selling below cost as a reasonable test of predatory behaviour, since this is something they do on a regular basis. Absent any competition, if RMCs are willing to sell seats below cost, it becomes a difficult proposition to find them guilty of predatory conduct since they are willing to do it to themselves. In this case, strategy is definitely a root cause of predatory behaviour, especially in Canada.

Provisions in the Competition Act in Canada related to the airline industry are generally viewed as very proscriptive when compared to other jurisdictions. While the rules may in themselves prove to be an adequate test of predatory behaviour, the mechanism through which the rules are enforced is an extremely long process that would serve very few new entrant airlines, and for that matter, the welfare of the country. Obviously, if the Act is meant to prevent predatory conduct, it is failing, as any dominant competitor would know that the process takes far too long to be meaningful. A 'per se' rule or a process that has strict time limitations and an expert panel would much better serve the airline industry and the public.

This review of predation in the Canadian airline industry has demonstrated that the ingredients for predation were aligned in several ways, making the existence of anti-competitive conduct predictable, and the root causes of predation are not always immediately apparent.

Chapter 8

Most Low-Cost Airlines Fail(ed): Why Did Southwest Airlines Prosper?

Andreas Knorr and Andreas Arndt

Introduction

Southwest Airlines (SWA) was established in 1967 in Dallas, Texas, but unable to enter revenue service before June 14, 1971 due to a protracted legal battle with its incumbent competitors. Since then it has not only grown into the fourth largest airline in the USA, in terms of revenue passengers, but also became fourth largest worldwide by 2002.

While most other start-ups on the deregulated US market failed to survive even their first few years in the business, SWA's revolutionary formula (i.e. low-costs, low-fare, no-frills, high-frequency, short-haul, point-to-point-service instead of the traditional high-cost, high-fare, full-service, hub-and-spoke operation) turned the company into the most consistently profitable airline ever. It is hence not surprising that SWA became a role model for the growing number of low-cost carriers shaking up the industry around the globe. Despite their current spectacular growth rates, the final verdict on the long-term economic viability of this 'second wave' of low-cost carriers on the US domestic market is still out, given that all first-wavers except for SWA (e.g. People Express[1] and Air Florida[2]) folded quickly. Furthermore all the low-cost subsidiaries created by the large US carriers in reaction to SWA's and its various emulators have, in spite of their parents' financial muscle, either ceased operations or are being largely scaled down in the aftermath of 9/11.[3]

Hence, adopting a SWA-style business model *per se* is no guarantee for long-term success, as shown by the mixed fortunes of various other low-cost carriers. Although, while it is true that passengers in the US market have enjoyed significant benefits since deregulation; and according to a study by the US Department of Transportation, essentially because of a development it aptly named the 'low cost airline service revolution',[4] significant barriers to entry do persist in some key markets.[5] These barriers continue to effectively shield incumbents from low-fare competitors including, in some cases, SWA. Moreover, even SWA, which

---

[1] For a detailed analysis see *Gudmundsson* (1998).
[2] For a detailed analysis see *Gudmundsson* (1998).
[3] See *Woodyard* (2001).
[4] See *United States Department of Transportation* (1996).
[5] See *United States General Accounting Office* (1996).

currently accounts for around 90% (!) of all low-fare airline services in the USA,[6] also only narrowly escaped bankruptcy in its start-up period after competitors used price and non-price predation strategies against it. The lessons from SWA's rise – most of which challenge conventional wisdom about what corporate strategy would be appropriate in a deregulated business environment – therefore offer invaluable insights into the nature of competition in the airline industry.

SWA: A Company Profile

The Track Record: Key Corporate Data

The data in Table 8.1 reflects SWA's spectacular corporate growth from a small 4-aircraft airline in 1971 into one of the largest carriers on the US domestic market. Even the unprecedented crisis in the airline industry since 9/11, has been so far better endured by SWA than by any of its US competitors (e.g. both United and US Airways had had to resort to Chapter 11 creditor protection in order to stay in business (US Airways even twice within a year) and it appears very likely that Delta will follow suit).

An even more impressive picture emerges when some more detailed company-specific information is added:

- Since 1973 SWA has posted annual profits for 31 consecutive years and, in the third quarter of 2004, for a total of 113 consecutive quarters.[7]
- In 2000, it achieved a net margin of 11.1%, by far the best figure of any major US airline.[8] Even in the extremely difficult economic climate of 2001 and 2002, it managed to realize margins of 9.1% and 4.4% respectively.[9]
- SWA's current net debt/total capital ratio of 33% (excluding operating leases for aircraft and ground facilities) is the lowest in the US industry (American: 59%; Delta: 59%; United: 69%; Continental: 88%; USAirways: 92%; Northwest: 96% and AirTran, being the second largest low cost operator in the US market: 98%).[10]
- Standard and Poor upgraded SWA's credit rating for senior unsecured fixed-rate debt from A^{-} to A in 2000, which represents by far the best rating for any US airline (SWA is currently being rated Baa1 by Moody's, and A by Fitch).[11] After 9/11, it even increased its lead in this field compared to all other major US carriers.[12]

---

[6] See *Southwest Airlines* (2001: 4).
[7] See *Southwest Airlines* (2001: 4).
[8] See *Southwest Airlines* (2001: 1 and 7).
[9] Own calculations based upon *Southwest Airlines* (2002a and 2003).
[10] See *Cameron* (2001: 17).
[11] See *Southwest Airlines* (2001: 17 and 24; 2002b: F7).
[12] *Dallas Business Journal (Online Edition)* (2002); *Houston Business Journal (Online Edition)* (2002).

Table 8.1 Basic SWA Facts and Figures

	1971	2000	2001	2002
Size of fleet	4	344	355	375
Number of employees (full-time equivalent)	195	29,274	31,580	33,705
Number of revenue passengers	108,554	63,678,261	64,446,773	63,045,988
Number of cities[13] and (states) served	3 (1)	58* (29)	59* (30)	59* (30)
Total operating revenues	$ 2.13 mio	$ 5.65 bn	$ 5.55 bn	$ 5.52 bn
Net income (loss)	($ 3.8 mio)	$ 625.2 mio	$ 511.1 mio	$ 240.9 mio
Market share (as percentage of Domestic revenue passengers carried)	0 %	10.4 %	11.3 %	n/a yet
Market share (as percentage of domestic revenue passenger miles)	0 %	8.3 %	9.2 %	n/a yet

Sources: *Freiberg and Freiberg* (1996, p. 326); *SWA* (2001, 2002a, 2003a, 2003b); own calculations based on *Air Transport Association* (2002a and 2003).

- At 7.41 cents/ASM[14] in 2002 (slightly down from 7.48 cents/ASM in 2001) SWA boasts the lowest operating costs of all major US carriers, outperforming the competition by a margin of 20 to 40%.[15]
- At US$ 11.08 B its market valuation is almost five times (!) as high as its largest US competitors' valuations – American, Delta, United, Northwest, and Continental – combined (US$ 2.12 B).[16]
- Never in its corporate history; not even in the crises years of: 1979 (second oil price shock), 1982/83 (US recession), 1991/92 (Gulf War) and 2001-02 (9/11) which forced large industry-wide lay-offs, has SWA had to dismiss any staff. By comparison, in 2001, the major US airlines had to lay off a total of 80,300 employees from the industry's total of 567,800 as a consequence of 9/11 (the latter figure includes SWA's roundabout 31,000 staff as of 2001!).[17] While

---

[13] Houston is counted as two cities as SWA serves both Houston Lobby Field and Houston George Bush Intercontinental Airport.

[14] See *Southwest Airlines* (2003).

[15] See *Doganis* (2001: 131). If adjusted for SWA's significantly much lower average trip length compared to the competition's, its cost advantage per ASM is even greater.

[16] See Yahoo! Finance (2003). Before 9/11, it was only equivalent to the aggregate market capitalization of the other big six (then also including pre-Chapter 11 USAirways). See *Done* (2001: VII).

[17] See *Freiberg and Freiberg* (1996, p. 7); *Air Transport Association* (2002b: 6).

the competition has not yet announced any new hiring, SWA has added roughly 4,000 employees to its workforce in 2002 alone.[18]

- Aside from paying at least market wages, SWA was the first airline to introduce as early as 1974 a profit-sharing plan for its employees.[19]
- Setting another industry record, around 81% of SWA's staff is unionized.[20]
- According to US Department of Transportation statistics (the Air Travel Consumer Report) SWA has had the lowest number of complaints per 100,000 passengers carried in the last ten consecutive years.[21] Moreover, setting another unprecedented industry record, between 1992 and 1996 it was named the best US carrier in all three 'quality of service' categories by the surveying agency.[22]
- Finally, having had only one hull-loss and no fatality on record so far, SWA boasts (alongside America West, another post-deregulation start-up) the best safety record of all US airlines and one of the best safety statistics worldwide.[23] A may come at the cost of flight safety.

SWA's Business Model

It appears that SWA has managed to identify a sustainable competitive advantage, which its network-based rivals have found impossible to emulate. As mentioned above, analysts and academic researchers have long argued that SWA's success is largely due to the carrier having identified a market niche neglected by the traditional network carriers – such as Delta, United, American, US Airways, Continental and Northwest – and more importantly the carrier proceeded with this winning formula ever since. SWA formula is believed to be made up of the following pillars: low-cost, low-fare, no-frills, high-frequency, short-haul, point-to-point service between secondary or uncongested airports and the airline primarily aims at price-sensitive (i.e. leisure) travellers. By contrast the traditional network carriers' product is seen as high-cost, high-fare, full-service, medium to long-haul, medium-to high-frequency (mostly) connecting service via hubs, catering for the special needs of price-insensitive business travellers.

However this view has always been an incomplete and hence somewhat misleading description of both SWA's and the traditional carriers respective markets. In fact SWA has continually aimed strongly at the business travel segment, in particular by being the only low-cost carrier offering high-frequency

[18] See *Flug Revue* (2002: 25), *Lewis* (2002: 7).
[19] See *Southwest Airlines* (2003a). Employees currently own around 10% of SWA's stock.
[20] Ibid.
[21] See *Southwest Airlines* (1997: 16).
[22] The three categories then were: mishandled baggage, on-time performance, and overall customer satisfaction as measured by the number of complaints per 100,000 passengers carried. Recently, a fourth category – oversales (i.e. overbookings) – has been added.
[23] See *Richter and Wolf* (2002: 86f).

service,[24] and the majority of network carriers' passengers are in fact price-sensitive leisure travellers). Indeed the gap between SWA's and the network-based carriers' business models has narrowed over time.

For example:

- Originally conceived as the unrestricted low fares archetype with a simple peak-off-peak fare structure, SWA had to react in 1986 to the network carriers' increasingly sophisticated yield management techniques by also introducing basic yield management.[25] It offers ever since discounted restricted fares (significantly undercutting its own low unrestricted fares) with similar advance purchase and minimum stay requirements as the traditional carriers. Given SWA's substantial cost advantage, no major carrier has been able to match these extremely low fares for any extended period of time, while SWA consequently managed to improve and stabilize its profits. However, SWA still retains the lowest fare dispersion of all major US carriers.[26]
- In 1987, SWA created a frequent-traveller bonus scheme of its own (Rapid Rewards). As the programme is based on the number of trips taken instead of accumulated mileage, it is widely perceived to be the least restrictive and most generous loyalty programme of all US airlines.
- After taking delivery of its first Boeing 737-700 aircraft, the only 737-plane with transcontinental range, SWA entered selected long-haul markets, such as Baltimore–Los Angeles (at 2,329 miles the longest SWA route so far), offering low-frequency service (usually 2 or 3 round trips per day).[27]
- Between 20 and 30% of SWA's customers are 'through-passengers' on one – or multi-stop services or are on-line connecting (flow) traffic at its primary airports.[28]
- In turn, in order to compete head on with SWA four of the network carriers, set up their own low-cost divisions (these are United's Shuttle by United, USAir's MetroJet, Continental's Continental Lite, and Delta's now defunct Delta Express). However, this 'airline within the airline'-model failed deplorably for various reasons and most low-cost divisions were either discontinued or largely scaled down.[29]

---

[24] The same is true for most European low-cost carriers like easyJet which reports a business traveller share of 60%. See *The Economist* (2001: 69); *Rubython* (2001: 25). For recent empirical analyses on the ever increasing attractiveness of low-cost carriers for business travellers on the European market see *Mason* (2000 and 2001).

[25] See *Petzinger* (1995: 329f).

[26] The highest fares on SWA are about twice the median fare for all passengers. For the domestic services of all other US carriers, it is at least three times the median. See *Transportation Research Board* (1999: 1-10).

[27] See *Southwest Airlines* (1998: 18; 2001: F22f; 2003a).

[28] See *Southwest Airlines* (2001).

[29] To escape from their current crises, some of these carriers – in particular Delta with an outfit called Song and United's TED – are trying again to transfer substantial parts of their operations to low-cost subsidiaries.

- As part of a wider cost reduction plan to counter SWA's cost advantage the network carriers also eliminated most in-flight food services in their economy class cabins (traditionally, none of the big US carriers has ever offered free alcoholic beverages in economy class on domestic services).
- Most of the specific amenities offered by network carriers (e.g. lounge access or preferred check-in facilities) are generally only available to the small number of full-fare first- or business-class ticket-holders and to those passengers who have reached elite status in their frequent traveller programmes.

Due to the above reasons it will therefore be argued in the following that SWA's business model needs to be more broadly defined. The definition should not only include the standard features of the low-cost, low-fare model, which was put to perfection by SWA, but also its specific approach to labour relations, financial and strategic management.

The Particular Sources of SWA's Lasting Competitiveness

Evidently SWA has taken advantage of low input costs wherever possible and reasonable and most noteworthy in this respect is the standardization of its fleet around the Boeing 737 aircraft leading to large maintenance, training, staffing and scheduling-related cost savings. SWA has been among the launch customers for three of the four 737-variants (B737-300, B737-500, and B737-700), which it is currently operating, hence profiting not only from significant manufacturer's discounts, but also from the very favourable economics associated with operating the youngest fleet of all major US airlines. Moreover, as stated above, SWA has also managed to reduce infrastructure access costs by focusing on underutilized airports.

Apart from these permanent cost-saving efforts on the input side, SWA has throughout its history primarily relied on the more sustainable strategy of becoming the industry's cost leader by maximizing its overall productivity through clever product and process design:

- Partly these productivity-induced savings too are due to the company's one-class, point-to-point service to/from secondary airports,[30] leading to the industry's fastest turnaround times (including a mean ground taxi time of only 3 minutes 44 seconds[31]), which translates (at an average of 7.2 flight or almost 12 flight hours per aircraft per day[32]) into the highest industry-wide aircraft utilization rate. These operation strategies maximize fleet, infrastructure and staff productivity.[33]

---

[30] Whenever SWA serves busier airports, including other carriers' hubs, it avoids scheduling its flight during peak periods, opting for the periods between connecting banks instead. See *Transport Research Board* (1999: 1-15).

[31] See *Southwest Airlines* (2003a).

[32] See *Southwest Airlines* (2003a).

[33] Although SWA's pilot earn market wages, they cost around 40% less to employ per block hour than the major US airlines' average. See *Barkin, Hertzell and Young* (1995: 89).

- SWA has the lowest distribution costs of all the major US airlines, which can be largely attributed to its pioneering use of technology to automate sales and ticketing. More than 70% of all seats are sold directly by SWA (industry average: 20 to 25%) through two different distribution channels: phone reservation centres and, increasingly, the Internet. Having been the first US airline to adopt online booking, SWA is currently earning 49% of its ticket revenues from Internet sales,[34] while its competitors only reach figures between 5 and 10%.[35] SWA introduced ticketless travel system-wide on January 31st, 1995 long before its competition and in 2000 more than 80% of SWA's customers used this service.[36] Moreover in 1980, SWA was the first US airline to install automated ticket vending machines at its main airports.[37]

These obvious factors aside, SWA's low-cost business model is unique in that it is built on at least five more mutually reinforcing cornerstones, which its various competitors have either failed to notice and/or imitate: a high quality of service, excellent labour relations, unconventional advertising and public relations, conservative corporate finance, and organic corporate growth. In the following each factor will be considered in turn.

High Quality of Service Due to their point-to-point operation, low-cost carriers stand a good chance of offering better quality of service from the passengers' perspective than their traditional network competitors in some crucial areas, such as on-time performance and mishandled baggage (a small percentage of flow traffic inevitably decreases the risk of missing and misdirected baggage). In contrast to its imitators SWA has always been careful to add another quality dimension to its low-fare product: the same high-frequency service as offered by its network competitors in order to minimize overall travel time for all customers and to fully meet the specific needs of the short-haul business traveller.

Excellent Labour Relations In an industry notoriously plagued by labour strife, SWA has been unique in that it has the highest degree of unionization but has never been subjected to industrial action. SWA's management has been much more successful than its competition in aligning the interests of all employee groups with the company's interests, thus creating and preserving a strong corporate culture (the underdog 'Southwest Spirit'). This success can largely be attributed to the following: its no-furlough policy and non-hierarchical, highly decentralized organization, the payment of standard wages and establishment of the industry's first profit-sharing plan in 1974, careful recruitment and comprehensive training, and, with one exception (see below), the stringent pursuit of a strict strategy of internal growth. With the unions' consent, management has been able to implement many important productivity-enhancing and highly flexible operational processes throughout the entire organization. Visible signs for the excellent labour-management relationship firstly include an unprecedented 10-year labour contract

---

[34] See *Southwest Airlines* (2003a).
[35] See *Pappas* (2001: 51).
[36] See *Southwest Airlines* (1996: 15; 2001: 15).
[37] See *Freiberg and Freiberg* (1996: 139).

between SWA and the Southwest Airlines' Pilot Union signed in 1995 and secondly that SWA was repeatedly voted one of the most admired US companies as well as one of the best US companies to work for.[38]

Unconventional Advertising and Public Relations SWA has always made good use of advertising in general and the media in particular to cultivate its underdog image and to promote its brand and product. Moreover, in an effort to counteract decreasing service quality in the US airline industry, the United States Department of Transportation began, in the late 1980s, to track the major carriers' performance in three categories (on-time-performance, mishandled baggage, and customer complaints per 100,000 passengers boarded). These statistics have been published ever since on a monthly and yearly basis and SWA was named 'best carrier of the year' in all three categories for the consecutive five years from 1992 to 1996.[39]

Conservative Corporate Finance In stark contrast to the other major US airlines, SWA has always pursued a very conservative approach to corporate finance. It mainly relies on internally generated funds for its investment needs instead of taking on too much external debt. This approach did not only lead SWA to obtain and maintain the industry's best credit rating, thereby reducing its costs of finance, but moreover this self-imposed restriction also shored up the company's conservative growth strategy. This strategy shields it effectively from the potentially disastrous consequences of excessive debt-financed expansion that turned out to be a key factor in many of its less prudent competitors' decline or even demise.

Organic Corporate Growth From its inception, SWA has always regarded ground transportation to be its main competitor and hence tried to win over customer through low-fare, high-frequency and short-haul services. The clear focus on this specific market niche also explains why the airline has always opted to deploy newly acquired aircrafts to build up frequencies on existing city-pairs before expanding into new markets. As a result, SWA has on average added two new destinations to its network every year since 1971 – offering high-frequencies from the beginning. In 2002, more than 140 cities had requested SWA to serve them.[40] Moreover, SWA has never expanded by withdrawing resources from its Texas home market (or other established markets) to employ them in other regions. A strategy used to preserve a strong and very profitable home base and keeping customers loyal. Basically remaining faithful to its short-haul niche, SWA also expanded rapidly into specific geographical regions – once it had decided to serve them. Throughout its history it very cautiously added one region after another to its network, in roughly the following sequence: Texas' neighbouring states, Midwest, Southwest, California, Pacific Northwest, Florida, and, from the mid-1990s, the Northeast.[41]

[38] For a survey of all pertinent recognitions SWA has received in its corporate history see *Southwest Airlines* (2003a).
[39] See *Southwest Airlines* (1997: 16).
[40] See *Southwest Airlines* (2003a).
[41] See *Southwest Airlines* (2000: 14f).

Finally, aside from the 1993 takeover of Morris Air, a small low-cost carrier based in Salt Lake City, operating a homogenous fleet of 21 B737 aircraft at the time on a complementary network in the US northwest, SWA has only grown internally. It was thus able to avoid the manifold pitfalls inevitably associated with the subsequent need to harmonize vastly different corporate cultures, seniority schemes, pay scales etc. As the fate of many of its less prudent competitors (e.g. USAirways, Northwest and the defunct Texas Air) shows, SWA's conservative approach to corporate growth may have shielded it from permanently tense labour relations and 'imperial and managerial overstretch', and clearly helps to sustain its very healthy financial position and competitiveness to date.

To summarize, SWA's specific business model safeguards the carrier far better than its more vulnerable competitors from the recurrent crises of the airline industry and lets it emerge ever stronger after each downturn. In fact, SWA expanded very strongly both during and immediately after the recession years 1980-82 and 1993-94, adding a total of 10 new destinations in each of these periods.[42]

Moreover, some exogenous factors have fuelled SWA's and the entire low-cost segment's spectacular growth worldwide, while at the same time exposing a large structural weakness of the traditional network carriers' business model: their dangerous over-dependence on passengers willing and able to pay an ever increasing premium for their service.[43] In recent years this once very profitable market segment has, however, dramatically shrunk in the US (and is elsewhere showing clear signs of saturation, too). This is in particular the result of three interdependent developments:

- widespread corporate downsizing which, in the light of tighter travel budgets, makes business travellers' demand for airline service increasingly price-sensitive;
- the substantial price hikes for full-fare tickets, which network carriers regularly imposed upon their most loyal customers – only to use the additional proceeds to substantially expand, and finance through internal cross-subsidization, their own low-fare offerings to counter the onslaught of low-cost carriers in many of their core markets;[44] and
- the phenomenal boom of fractional ownership (e.g. a company's right to use a corporate aircraft jointly owned by a number of enterprises for a pre-determined number of flight hours a year).[45]

---

[42] See *Southwest Airlines* (2000: 14f).

[43] *Martin*, for example, argues that many traditional airlines still cling to the obsolete and dangerous notion of air travel being an élite form of travel instead of a mass market. See *Martin* (2002: 13).

[44] See *Petzinger* (1995: 460).

[45] For a survey of fractional ownership growth trends and a detailed discussion of its advantage and disadvantages see *aerokurier-online* (1999), *Andersen* (2001).

The 'Southwest'-Effect: Empirical Evidence

By the end of the 1980s SWA was still widely being perceived as a regional phenomenon and a mere non-threatening niche player. This perception has dramatically changed since. SWA is now being credited with being the single most important competitive force in the US airline industry. Its increasing market penetration is widely being regarded as the principal source of the significant savings and service improvements, which consumers have benefited of in the post-deregulation era. In turn qualitative and quantitative analysis of the 'Southwest Effect' has recently become the object of intense economic research.

SWA's Impact on Fares and Passenger Volumes

The reviewed literature offers several distinct definitions of the 'Southwest Effect' and alternative methodologies to quantify it. However there is widespread consensus that SWA has been the single most important stimulus of competition in the post-deregulation US airline industry. Furthermore its impact has been growing substantially since the early 1990s, in parallel with its ongoing network expansion to cover an ever-larger portion of the nation's prime aviation markets.

Initially the term 'Southwest Effect' was used to describe the positive change in the number of enplanements plus the induced decrease of the average fare at the airports and/or on the specific markets SWA had entered.[46] In a 1996 study, the United States Department of Transportation identified three distinct facets of the 'Southwest Effect' without, however, trying to quantify them or assess their importance relative to each other:

- The *direct competitive effect* – lower fares, higher passenger volumes – on the specific city-pair market, SWA has chosen to enter.
- The *'halo'-effect* with SWA's entry lowering fares at nearby airports.
- The *role model-effect* with SWA's successful business formula serving as a blueprint for newcomers, whose entry in other markets, in turn, has intensified competition there.[47]

Vowles (2000 and 2001) found by using regression analysis in two recent studies, a statistically significant effect of low-cost carriers (including SWA) on fare levels at most neighbouring airports[48] and that SWA's presence at one airport in multi-airport regions lowered fares in the affected area as a whole.[49] Vowles (2001) paper includes nine case studies revealing interesting fare and enplanement trends. According to his calculations using the data contained in his study, after

---

[46] See *Windle and Dresner* (1995); *Southwest Airlines* (2000).
[47] See *United States Department of Transportation* (1996: 5 and 11). In 1993 the U.S. DOT had already commissioned a report on SWA's procompetitive impact in the regions it had served at that time. See *Bennett and Craun* (1993).
[48] See *Vowles* (2000).
[49] See *Vowles* (2001).

one year of operation, SWA's entry had on average lowered fares on 13 city-pairs to or from the served multi-airport regions by 47% (outliers: –11%; –82%).

The most comprehensive and ambitious attempt to define and quantify the 'Southwest Effect' to date is Morrison's 2001 study.[50] Through disaggregation he distinguishes three different types of SWA's competitive impact:

- *Actual competition*: SWA is a direct competitor on the route. As a result, both SWA's and the incumbents' passengers profit from lower average fares.
- *Adjacent competition*: SWA operates on an adjacent route, which is being considered by a significant proportion of consumers as a good substitute for the incumbent's offering.
- *Potential competition*: this includes incumbents' fare reductions to deter SWA – already present at the airport in question or at a nearby facility – from entering new routes which would put the two carriers into actual or adjacent competition.

Using a simulation model, he estimates that, in order to enter new markets SWA undercuts incumbents by an average of 46.2%,[51] which translate into overall savings to US airline passengers of US$ 3.4 B per year (figure for 1998). Further, he estimates the savings for consumers through the effects of other carriers lowering their fares in reaction to SWA's actual, adjacent, and potential competition to another US$ 9.5 B yearly. Hence, he estimates that a grand total of US$12.9 B or around 20% of the entire industry's 1998 annual revenues and slightly more than half of all fare savings have accrued to passengers after and because of airline deregulation.[52] He also believes that SWA, as an actual and adjacent competitor, is now able to directly influence fares on 44.8% of the domestic US market in terms of passenger miles flown. In addition to that, he ascribes to SWA a similar influence, as a potential competitor, on another 49.4% of the market – in sum an amazing 94% of the total.[53]

Morrison's most interesting finding however is that some key routes between major US airports,[54] among them a number of network carriers' hubs, are beyond SWA's reach. As will be argued below, this is largely due to the persistence of severe barriers to entry prevalent in some of these key markets. Although SWA has meanwhile – the study is based on 1998 data – entered four network carriers' hubs (Buffalo, Hartford, Raleigh–Durham and recently in May 2004 on a massive scale Philadelphia, which is one of the two principal USAirways hubs) and started to provide some adjacent competition to New York LaGuardia (a slot-constrained field) and Newark (a prime Continental hub) through its new service to Islip near New York City,[55] other network carriers' hubs such as: Atlanta (the principal Delta

[50] See *Morrison* (2001). It is partly based upon research presented before the US Senate in 1998 already. See *Morrison* (1998).
[51] See ibid., p. 250.
[52] See ibid., p. 239.
[53] See ibid., p. 243.
[54] See ibid., p. 243 (footnote 6).
[55] In addition, rapidly expanding JetBlue has offered low-fare service from its base at New York's Kennedy Airport from 2000.

hub, with Delta facing some low-fare competition through Air Tran, however), Charlotte (a USAirways hub), Memphis and Minneapolis/St. Paul (both major Northwest hubs), Richmond, Pittsburgh and Pittsburgh (the second principal USAirways hubs) remain largely out of SWA's reach.

SWA's Impact on Air Service Quality

As shown in the preceding section, all studies on the 'Southwest Effect' exclusively focus on SWA's impact on fare levels and passenger volumes. In order to provide a more complete picture it is therefore suggested to add another component to the 'Southwest Effect', which is service quality as measured in the United States Department of Transportation's Air Travel Consumer Reports.[56] Since these statistics have been available since 1987, it can be shown that SWA has, as mentioned above, consistently performed extremely well in all three categories: mishandled baggage, on-time performance, and overall customer satisfaction (number of complaints per 100,000 passengers), especially in the period from 1992-1996.

While SWA still remains among the top performers in overall customer satisfaction, it has recently been outdone by some network carriers in the two other categories. More importantly however, is the fact that these statistics witness a steady narrowing of the gap between individual carriers' service standards – at generally higher levels of service quality. Furthermore, while SWA's product quality has remained constantly superior over time, its unique combination of low-cost, low-fare *and* high-quality service made public by the Department of Transportation statistics has forced the more expensive network carriers to significantly improve their performance.

Southwest's Strategies to Overcome Entry Barriers

Barriers to entry in an industry are generally defined as cost advantages enjoyed by incumbents over newcomers as the former are already present in the market.[57] With respect to the post-deregulation airline industry, it is widely held that effective barriers to entry may only exist due to infrastructure bottlenecks and as a result of strategic behaviour by established carriers (in practice the distinction between both types of entry barriers may be much less clear-cut, since some infrastructure bottlenecks only exist due to strategic hoarding by incumbents).[58] In the following SWA's strategies targeting these specific entry barriers will be discussed and more importantly, where they failed, and why. Further legal impediments and their impact on SWA too will be highlighted, as SWA was

---

[56] The most recent issues of the Air Travel Consumer Report are available for download at the United States Department of Transportation's homepage (http://www.dot.gov/airconsumer).

[57] For a general discussion on the economics of entry barriers see the seminal work by *Bain* (1956).

[58] For an overview see *Transportation Research Board* (1991: 134ff); *Knorr* (1998: 448f); *Kummer and Schnell* (2001).

established in the pre-deregulation period when some types of legal barriers to entry existed even in such fairly liberal states as Texas (SWA's home state) and which represented a significant impediment even to intrastate airline competition.

Barriers to Entry in the Airline Industry

Legal Barriers Potentially, there are four different types of legal entry barriers in the airline industry: ownership rules – which do not play a role on the domestic US market,[59] operating licences, route-specific traffic rights, and perimeter rules at airports. Moreover, in order to set up an airline, all countries require a formal operating licence granted by their civil aviation authorities. As this requirement, however, did not affect SWA's entry, it will not be further discussed in the present paper.

The requirement to obtain route-specific traffic rights was the single most important impediment to domestic airline competition in the US market before deregulation.[60] Route-specific traffic rights were issued by the Civil Aeronautics Board (CAB) for all intrastate services and by its state counterparts (e.g. the Texas Aeronautics Commission (TAC) or the California Public Utilities Commission (CPUC)). Newcomers though were required to produce in conjunction with their application for route authority some convincing evidence that their proposed service was necessary to meet pent-up demand, which had been so far neglected by the incumbent carrier(s). The incumbent however was allowed to take legal action for strategic reasons against the decision by the authorities to grant any newcomer this 'certificate of necessity and convenience'. As for interstate services, the CAB refused to certify[61] any new trunk carriers – i.e. a new competitor in the category of the major airlines such as American, Delta, Northwest, Continental, United (late) Eastern etc. – aside from the 16 airlines which had obtained their certification as early as 1938, the year the CAB was established. In the 1950s, after the major airlines had begun to either withdraw from a significant number of short-distance city-pairs or to substantially reduce frequencies, the Board reacted by creating a new airline category to fill this gap: 'local carriers', under which such airlines as Alaska Airlines, Allegheny Airlines, Ozark Air Lines or Piedmont Airlines have been classified. In addition, 'Commuter Airlines' and 'Supplemental carriers' were certified for non-scheduled air taxi and package tour-services respectively. With the exception of only three states – Texas, California, and Florida – state authorities were as restrictive as the CAB in granting intrastate routes.

Even after deregulation 'perimeter rules' restrict entry to some important US airports – Washington National, New York LaGuardia, and Dallas Love Field – either by forbidding revenue services to destinations beyond a certain geographical

[59] See for example *Knorr and Arndt* (2002).

[60] For details see *Biedermann* (1982) and *Bailey, Graham and Kaplan* (1985).

[61] NB: Legally speaking, only the CAB was allowed to certificate an airline. In other words, all intrastate airlines permitted to operate *within* their home states boundaries by their respective state authorities, including SWA, did not have nor require the status of a (CAB-) certificated carrier. As we will discuss below, this distinction turned out to be of vital importance for SWA's successful implementation of its business model in the start-up phase.

distance[62] or, as in the case of Dallas Love Field, by only allowing commercial flights to/from airfields in specific US states.[63]

Infrastructure Bottlenecks The absence of legal barriers to entry is only one prerequisite for free, undistorted competition. At an increasing number of major airports infrastructure bottlenecks, which are often due to a command-and-control approach to infrastructure allocation and/or strategic behaviour of incumbents, with the legal power to control access, impede new competitors from entering the market. Basically, two sources of potentially anti competitive infrastructure-related entry barriers can be distinguished:

- The lack of slots (at popular times) due to insufficient runway capacity or strategic hoarding of available slots by incumbents,[64] or
- the lack of access to vital ground facilities such as gates, largely due to long-term 'exclusive-use leases' between airport operators and incumbent airlines.[65]

Strategic Barriers to Entry Aside from the strategic use of infrastructure bottlenecks or the legal system, research has identified a number of other strategic barriers to entry in the airline industry.[66] In particular these are:

- the incumbents' strategic use of the most important distribution channels, i.e. computer reservation systems (CRS)[67] and travel agencies;
- passenger loyalty schemes (e.g. frequent flyer programmes); and
- (the threat of) price and non-price predatory action targeting newcomers.

Other Barriers to Entry Other barriers to entry include for one that newcomers have to overcome their critical lack of brand and product awareness through advertising in order to make an impact. Furthermore the flying public may be reluctant to use low-cost airlines for fear of perceived lower safety standards as compared to the major carriers. For example, after the fatal May 1996 crash near Miami of a DC-9 operated by the low-cost airline ValuJet (which, in turn, went bankrupt as a result) many no-frills carriers suffered cancellations, a significant decrease in new bookings and a substantial decline of their share prices. By contrast, established full-service carriers with well-known brands such as

---

62 The thresholds for incoming and outgoing flights are 1,500 miles at LaGuardia and 1,250 miles at Washington National. See *U.S. General Accounting Office* (1996: 13). Only recently have flights to/from Denver been exempted from the perimeter rule in force at LaGuardia. See *Transportation Research Board* (1999: 3-5).

63 See below at subchapter 3.2.1.

64 For a detailed discussion of US slot allocation procedures see *Langner* (1995).

65 For details see *FAA/OST Task Force* (1999); *U.S. General Accounting Office* (1996: 13).

66 See for many *Levine* (1987); *Weinhold* (1995); *United States General Accounting Office* (1996); *United States Department of Transportation* (1999); *Meyer and Menzies* (2000).

67 In the US, all CRS are operated by major airlines: SABRE by American, APOLLO by United, WORLDSPAN by Delta and SYSTEM ONE by Continental, to mention just the most important ones.

USAirways, which recorded between 1991 and 1995 a series of five deadly crashes, did not see their business being seriously affected by these accidents.

How Did SWA Overcome These Entry Barriers – and Where Did It Fail?

Legal Entry Barriers In its start-up phase SWA faced a series of legal hurdles in three areas, route authority, airport selection, and perimeter rules, which were more often than not erected by incumbent carriers trying to prevent the new competitor from 'getting off the ground'.[68]

Route Authority After filing the documents to incorporate the airline on March 15, 1967 and obtaining the operating licence soon after, SWA applied to the TAC on November 27, 1967 for traffic rights between three cities in Texas – Dallas, Houston, and San Antonio. The rights were granted on February 20, 1968. However on February 21, 1968, local incumbents 'Braniff', 'Trans Texas' (later to become Texas International) and 'Continental' obtained a temporary restraining order from the courts on the grounds that the markets SWA wanted to serve were already saturated. After lengthy legal proceedings, ending in late 1970 when the U.S. Supreme Court refused to hear the case, after the Texas Supreme Court had ruled in SWA's favour, SWA emerged as the winner. 'Braniff' and 'Trans Texas' however filed another complaint, this time with the CAB, the body which only regulates interstate airline services, to prevent SWA from finally 'taking off'. They creatively argued that SWA was not an intrastate carrier as some of SWA's passengers *might* inter-connect with other (CAB-certificated) carriers to travel on to destinations outside Texas, hence leaving SWA outside the TAC's legislative powers. On June 16, 1971, two days before SWA's first scheduled flight, the two carriers had succeeded in obtaining another restraining order from a lower Texas court, arguing that SWA's had deviated from the original operating plan approved by the TAC. However on June 17, 1971, the Texas Supreme Court ordered that restraining order to be dissolved.[69]

The main outcome of the legal controversies for SWA was that it only narrowly escaped bankruptcy. Not only did these legal delays deprive SWA of commercial revenues for almost four years. More importantly, the incurred legal expenses had largely depleted SWA's financial assets (US$ 543,000 at the date of its incorporation). By January 1971, with only US$ 142 (!) left, SWA had accumulated liabilities of around US$ 80,000.[70] This is not surprising as SWA, due to the uncertain outcome of the trials, had been unable to attract any new investors before the U.S. Supreme Court's final decision.

Airport Selection and Perimeter Rules In June 1972, the cities of Dallas and Fort Worth as well as the Regional Airport Board filled a suit against SWA, as the company had refused to move its operations from Dallas Love Field to the new

---

68 See *Freiberg and Freiberg* (1996: 15ff) and *Calder* (2002: 22ff) for a detailed history of SWA's start-up phase.

69 Furthermore, both the CAB and the Court of Appeals in the District of Columbia rejected the plea. See *Petzinger* (1995: 29).

70 See *Freiberg and Freiberg* (1996: 19).

Dallas-Fort Worth Regional Airport (DFW). The construction of the new airfield had been largely financed by municipal bonds containing the provision that all 'certified' carriers serving Love Field would instead use the new facilities after its completion. SWA, which in late November 1971 had moved its Houston operation from the remote Houston Intercontinental airport to Hobby Field in downtown Houston – and as a result was experiencing a spectacular boost in load factors which made this route SWA's first big success, successfully argued before the courts that the provision in question exclusively referred to (CAB-)certified (interstate) carriers, but not to intrastate operations under TAC jurisdiction like theirs.[71]

After deregulation took effect in 1978, SWA immediately applied for interstate route authority to New Orleans, but again they were opposed by former DFW supporters, amongst them Fort Worth congressman Jim Wright. Though incompatible with essence of the Airline Deregulation Act 1978, he convinced a majority of the US Congress to support his 'Wright Amendment' of 1979. This piece of legislation; which was slightly modified by the Shelby Amendments still in effect today, although it was recently held to be illegal by the US Department of Transportation,[72] bans any airline from offering any nonstop or through-plane interstate revenue service out of Dallas Love Field except to airports in the four neighbouring states of Texas: Arkansas, Louisiana, New Mexico and Oklahoma. The Shelby Amendment later added the states of Mississippi and Alabama. While being only a negligible impediment to SWA's expansion and competitiveness during its first three decades of operation, the act has so far prevented the airline from expanding into the long-haul market[73] from its Dallas base, thereby putting it at a clear disadvantage vis-à-vis its DFW-based competitors in its home market.

Infrastructure Bottlenecks As quick turnaround times, a high average daily fleet utilization and high frequency services on its routes make up the three key ingredients of SWA's outstanding overall performance, the carrier had to largely concentrate its operations on secondary and uncongested airports. Given the abundance of underutilized airfields in all major US population centres, scarcity of slots have so far never been an important entry barrier for the airline or an impediment to its expansion into any geographical area in general.

However with respect to ground infrastructure bottlenecks a different picture emerges, i.e. gate-constraint airports, some of which SWA had tried to enter unsuccessfully. Northwest's Minneapolis hub is a case in point. At this airport, all 70 gates have been leased to incumbents under 'exclusive-use' agreements, 54 of them to Northwest. While since 1999, 32 of the 70 gate leaseshave been subject to

---

[71] See *Petzinger* (1995: 31).

[72] See *FAA/OST* (1999: 24). In late 1998 the US Department of Transportation issued an order allowing airlines to serve any US city out of Dallas Love Field provided that the aircraft's capacity does not exceed 56 seats. A complaint against this ruling filed by American Airlines is still pending. For details see *American Bar Association* (2000: 7).

[73] Aside from the above mentioned long-haul services from the East Coast to Phoenix, SWA has recently introduced, and significantly expanded, transcontinental services from Chicago Midway and other East coast airfields, most of them to California. See *Southwest Airlines* (2002c).

gradual conversion from exclusive use to preferential use (by their current exclusive lessee), the first 22 leases of the remaining 38 gates will only expire in 2015.[74] All carriers wishing to serve this airport – newcomers as well as incumbents wanting to increase their market presence – will therefore have to enter into a sublease agreement with Northwest or any other incumbent. In order to obtain the needed gates SWA would have had to begin serving Minneapolis. Northwest further insisted in its proposed sublease agreement that SWA had to exclusively use Northwest's staff for all ground-handling services.[75] However SWA realized that given Northwest's staff lower productivity and the obvious conflicting loyalties inevitably arising from this arrangement this legal condition would have put SWA at a clear competitive disadvantage vis-à-vis Northwest. Hence SWA decided not to enter the Minneapolis market and as a matter of fact it still has not entered most other gate-constrained airports.

Strategic Entry Barriers Above three possible strategic barriers incumbents might use to deter newcomers from entering the industry and/or specific markets were identified: control of the main distribution channels, loyalty schemes for passengers and (the threat of) predatory action.

Distribution Channels Notwithstanding the increasing number of Internet bookings, airlines have traditionally relied on the intermediation of travel agencies to distribute the bulk of their tickets (around 80 to 90%[76]) Travel agencies, in turn, usually subscribe to one of the airline-operated CRS for flight and fare information and for automated bookings. Apart from the standard commissions, most airlines also offer 'override commissions' as a special incentive to those travel agencies which generate a high (pre-set) volume of annual revenues for their carrier. Hence ticket distribution costs, including overrides, via travel agencies typically make up between 17 to 25% of the ticket price.[77]

While the commercial importance of unbiased access to CRS and travel agencies should not be underestimated, SWA's experience shows that these distribution channels are not as relevant to the survival of an airlines as previously thought. In May 1994, after having set up their own low-cost subsidiaries (Shuttle by United and Continental Lite) in direct competition with SWA, both United and Continental used their respective CRS's ('APOLLO' and 'SYSTEM ONE') to disable the highly important automated ticketing function for all SWA flights.[78] Moreover, Delta downgraded SWA on its WORLDSPAN CRS after the SWA had entered Delta's Salt Lake City hub following its acquisition of Morris Air. Although already selling a significant portion of its tickets at that time (around 45%) directly, it is obvious that SWA had to react to this ouster with a comprehensive strategy to diversify its distribution channels in order to remain in business:[79]

[74] See *U.S. General Accounting Office* (1999: 17).
[75] See *U.S. General Accounting Office* (1996: 13).
[76] See *Grant* (1996: 179); *Bouvard and Somosi* (1997: 173).
[77] See *Bouvard and Somosi* (1997: 173); *Pappas* (2001: 50).
[78] See *Southwest Airlines* (1995: 13); *Freiberg and Freiberg* (1996: 136).
[79] See *Southwest Airlines* (1995: 6, 13 and 19).

- SWA created a simplified CRS of its own (SWAT – Southwest Airlines Air Travel) as an interim solution. This new system provides direct access and ticketing services to fifty of its highest-volume travel agencies.
- Guaranteed overnight ticket delivery was introduced for the around 300 next largest travel agencies in business with SWA.
- The 'Ticket-by-Mail-Service' for direct customers who had booked through SWA's phone reservation centres was sped up to a guaranteed three-day delivery.
- A new phone reservation centre was opened at Little Rock.
- On January 31, 1995 system-wide 'ticketless' travel was introduced on SWA flights; in addition, SWA was the first US airline to offer (ticketless) on-line booking on its homepage.
- As of May 1, 1995, SWA negotiated a no-frills access for travel agencies and direct customers ('BASIC SABRE') with American in addition to American's SABRE CRS. (American had long pursued a strategy of coexistence with SWA in their joint Dallas home market, focusing on flow traffic, i.e. on connecting medium- to long-haul and premium passengers, and leaving the short-haul market largely to SWA.)[80]

Through this plethora of defensive actions SWA not only significantly reduced its dependence on its principal competitors' CRS, but more importantly, it also managed to substantially lower its own distribution costs via these innovations. The lowered distribution costs represent a competitive advantage, which SWA's competition has so far not been able to fully harness, which in turn further strengthens SWA's cost leadership position.[81]

By contrast, SWA could not overcome successfully all barriers to entry imposed on its operations. For example, in one of the few cases where SWA had to withdraw its services from a market, it cited Northwest's override policy as the critical factor in its decision to withdraw from the Detroit–Indianapolis route in the mid-1990s. As a direct consequence of Northwest's override policy local travel agencies' booking patterns had substantially shifted in Northwest's favour.[82] Although this is the only known incident of that kind, it cannot be ruled out (given SWA's and other low-cost carriers' conspicuous absence from two of Northwest's three main hubs – Minneapolis and Memphis – and its negligible presence at Northwest's third hub in Detroit) that the combination of an aggressive override policy and Northwest's dominance of gates at its hub airports has proven to be an insurmountable barrier even for SWA.[83]

---

[80] See *Bartkin, Hertzell and Young* (1995: 86). In 1993, when SWA had decided to enter California, American immediately scaled down its local short-haul operation to avoid a direct confrontation with SWA. See *Petzinger* (1996: 458).

[81] In general, direct selling of tickets by the airline, instead of using intermediaries, cuts distribution costs by half. Direct selling via the internet reduces these costs by at least another 50%. See *Pappas* (2001: 50).

[82] See *U.S. General Accounting Office* (1996: 16), *U.S. General Accounting Office* (1999: 19).

[83] This might also in part explain *Morrison*'s findings that some major US airports still lack low-fare services. See *Morrison* (2001: 243).

Loyalty Schemes Frequent flyer programmes are generally considered to be the most powerful marketing tool used to increase passenger loyalty. Frequent travellers are not only rewarded with 'free flights' – although in fact those rewards are not free but just a different kind of discount – but also with other amenities such as lounge access, waiting-list priority, preferred check-in etc. Usually, full-fare tickets earn their holders a multiple of the mileage available in return for flying on discount fares (with some discount fares being not eligible at all for mileage accrual). Hence all those passengers who do not have to pay for their travel themselves, but who are allowed to redeem their frequent flyer awards for private purposes – i.e. most passengers travelling on a corporate budget, will have no incentive to use, ceteris paribus, the least expensive airline. Instead they will choose the fare/airline that will earn them the maximum mileage possible for any given itinerary. Therefore it is unlikely that they will travel with a low-cost, low-fare carrier even though this would lead to significant savings for their companies. In order to solve this principal-agent problem however, corporations increasingly require their employees to relinquish all frequent flyer benefits earned on business trips or alternatively to use the lowest available fare. Further, privately used award tickets resulting from company paid flights, are now widely considered as a specific form of non-wage income, which is subject to income tax.

Another point against frequent flyer programmes is that they are a two-edged sword even for large airlines. Although these programmes are proven to increase passenger loyalty, they are a very expensive marketing tool. Substantial administration costs aside, airlines incur large (opportunity) costs as they have to reserve on average around 10% of their load capacity for passengers wishing to redeem their mileage for 'free' flights. In addition to the aforementioned developments, the loyalty schemes' attractiveness and hence their anticompetitive potential has further declined due to:

- the introduction of an ever increasing number of restrictions, such as stricter blackout dates during periods of peak demand (holidays, special events etc.), imposed on travellers wanting to redeem their points for award tickets; and
- the decision by most US carriers to eliminate the 'use-by' restriction of accumulated mileage.[84] In the past unused mileage credit would typically expire after an average of three years. However the decision to honour mileage for good in turn relieves passengers from the need to earn as many miles as fast as possible in order to obtain the desired award.

Although SWA did create its own frequent flyer programme in 1987, this programme not only operates vastly differently from industry standards in some crucial aspects, but is also much simpler than other programmes.[85] Most importantly, award tickets do not require the accrual of a specific pre-determined amount of mileage but instead the programme member has to have flown a minimum number of trips per year. Moreover, aside from defining only a very small number of blackout dates, the company does not set aside capacity for award

[84] Some airlines require their passengers to accumulate mileage at least once in three years lest they lose their credit after this period.
[85] See *Southwest Airlines* (2002d).

travellers. Instead it allocates all available seats on a first come, first served basis, i.e. it does not discriminate between award travellers and revenue passengers in the booking process. In sum, these features allow SWA to offer its most loyal customers all the benefits of a fully-fledged frequent flyer scheme – with the exception of international award travel given that SWA is a purely domestic airline, in a much more cost-effective manner than its competitors.

(Threat of) Predatory Action

Predation is traditionally defined as an incumbent's attempts to:

- either drive more efficient actual competitors out of the market, or to coerce them into accepting the incumbent's price leadership,
- or to deter potential competitors from entering the market, either by temporarily charging consumers prices below (variable) costs ('predatory pricing') or by artificially raising rivals' costs.

The key objective of this behaviour is to either defend or conquer a dominant position – and the monopoly profits associated with it – in the market in question.

In other words, if the predator prevails, this would not be due to its superior efficiency but of its ability to abuse its market power – with ultimate detrimental consequences for consumers who will have to bear all the negative effects of the ensuing monopoly. While some schools of thought, most notably the Chicago School of Antitrust Analysis, dismiss predation as an irrational strategy for any company to pursue, more recent research, based on game theory and the theory of industrial organization, has clearly demonstrated that predatory pricing may indeed be a viable option for incumbents.[86]

This would be the case where:

- the market in question is characterized by substantial barriers to entry, and
- recoupment of the losses incurred due to temporary predatory pricing is possible.

Recoupment could either take place by charging monopoly prices once the newcomer has exited the market or by overcharging consumers in other markets also served by the incumbent, but where no effective competition takes place ('multi-market recoupment').

Not only has the allegedly increasing number of incidents of predatory behaviour in the US airline industry become the object of scholarly research recently,[87] but it has also prompted the United States Department of Transportation to draft – based upon a 1998 white paper (revised in 1999)[88] – an '*enforcement policy regarding unfair exclusionary conduct in the air transportation industry*'.[89]

---

[86] See, for example, *Ordover and Saloner* (1989), *Milgrom and Robert* (1990), *Roth* (1995) and *Lott* (1999) for a detailed discussion of the economics of predation.

[87] See *Morrison and Winston* (2000: 7ff); *Oster and Strong* (2001).

[88] See *United States Department of Transportation* (1998 and 1999).

[89] See *United States Department of Transportation* (2001).

This policy was devised in order to protect start-up airlines against unfair competitive practices by incumbents. Furthermore the Department of Justice filed an antitrust complaint against AA, which however was unsuccessful as the Department failed to convincingly prove its accusations.[90]

Although predatory behaviour is notoriously difficult to prove in practice – due to the lack of comprehensive cost and fare data, the intricacies of the diverse cost and recoupment tests and the large difficulties involved in proving predatory intent – SWA's history offers some interesting lessons, both with regard to possible types of predation in the airline industry and how to counter them.

While only narrowly surviving price and non-price predation attempts by incumbents in its start-up phase, SWA, unlike many other smaller low-cost carriers, has not been subject to such practices for long. It appears that the incumbents' recent restraint towards SWA is largely due to their experiences gained from attacking SWA. SWA's specific business model, its significantly lower cost base and its current sheer size, are all features protecting the airline from predatory action against them. These features make the failure of predationary actions not only highly likely, but also prove extremely costly for the incumbent.

SWA's reputation as being rather immune to predation rests upon the following pillars:[91]

- Historical precedent: In February 1971, incumbent 'Braniff' attacked SWA by cutting its one-way fare between Dallas and Houston – then SWA's only profitable market – by 50% to $ 13. Aside from playing the David-vs-Goliath card in its advertising and in the media, SWA reacted by offering its passenger a choice of two fares: $ 13 with the standard no-frills service, or $ 26 *including* a free bottle of liquor (that did not have to be consumed on board). Initially, 76% of all passengers, most of them businesspeople travelling on expenses, choose the latter option, temporarily rendering SWA the biggest distributor of spirits in Texas.[92]
- Having evolved into the most consistently profitable US airline as well as the industry's undisputed cost leader, SWA has accumulated the resources to sustain a price war for an even longer period than most established network carriers. This fact is convincingly demonstrated by SWA's success in fending off the incumbents' low-cost subsidiaries.
- Whereas many low-fare start-ups only offer a small number of daily round trips, SWA usually enters a new market with high-frequency services. Consequently, it would be prohibitively costly for incumbents to apply a 'bracketing' strategy against SWA. ('Bracketing' means that incumbents schedule additional flights closely around the new entrants' departure times and match or even undercut its fares on just these flights while maintaining their normal fares on all other flights to the new entrants' destination(s).)

---

90 All pertinent legal documents are available at Americans's (http://www.aadoj.com) and the United States Department of Justice's homepages (http://www.usdoj.gov/atr/cases/indx199.htm).

91 See *U.S. Department of Transportation* (2001: 29); *Oster and Strong* (2001).

92 See *Southwest Airlines* (1997: 10); *Freiberg and Freiberg* (1996: 33).

- Typically incumbents only offer connecting or through-plane services, hence competing with an inferior product against SWA's nonstop service.
- Serving 59 airports in most parts of the USA, SWA now competes, directly or indirectly, in many geographical markets with the network carriers. This multi-market approach allowa SWA to potentially retaliate over a broad range of city-pairs, which creates strong incentives for network carriers not to engage in predation against SWA in the first place.

It is further noteworthy that SWA was able to survive the incumbents' predatory attacks in its fledgling years without the support of regulators or antitrust authorities. This was not least due to its marketing savvy but most of all by avoiding competing head on with potential predators on large parts of its network – a strategy bluntly ignored by most other low-cost carriers which failed as a result.

Conclusions

SWA's is not the only low-cost business model in the airline industry. For example, AirTran is competing against Delta in Atlanta by emulating the latter's hub-and-spoke model and by offering two classes of service and JetBlue is operating from New York's high-cost (though uncongested) JFK airport. However what disinguishes SWA from other low-cost airlines is that it is by far the most consistently successful one. Moreover, SWA flourished although (or rather because?) it defied '*every success maxim of the post-deregulation world*':[93] such as the alleged need to erect strategic barriers to entry by establishing a hub-and-spoke operation, by using sophisticated yield management techniques, by relying on CRS and travel agents as the principal distribution channels and by creating passenger loyalty via frequent flyer schemes. Although SWA smartly adopted some very basic versions of the above tools where needed, it never abandoned its original business formula, but refined it further by adding said features to its product.

In other words, the SWA story not only shows that the effectiveness of strategic barriers to entry has long been overestimated by economists, airline professional, and policymakers alike, but on the contrary, it clearly proves that infrastructure bottlenecks – which, in turn, are overwhelmingly caused or at least amplified by ill-designed allocation rules and access regulations – must be seen as the only true barrier to entry in the airline industry. Not only do these rules and regulations shield inefficient incumbents directly from leaner competitors, but more crucially experience showed that established carriers can only erect meaningful strategic entry barriers upon this base.

---

[93] *Petzinger* (1995: 319).

References

Aerokurier-online (1999), FACTS – Bestand Fractional Ownership Flugzeuge. (November 24, 1999.) (www.aerokurier.rotor.com/akDaten/Aefracte.htm.)

Air Transport Association (2002a), 2002 State of the U.S. Airline Industry: A Report On Recent Trends For U.S. Air Carriers. Washington DC.

Air Transport Association (2002b), *Annual Report 2001*, Washington DC.

Air Transport Association (2003), Annual Traffic and Capacity. U.S. Schedules Airlines – Schedules Services. Washington DC. (http://airlines.org/public/industry/display 1.asp?nid=1032).

American Bar Association. Section of Public Utility, Communications and Transportation Law (2000), *Report of the Aviation Committee: 2000 Spring Council Meeting*, Washington DC.

Andersen (2001), Business aviation in today's economy: A guide to the analysis of business aircraft uses, benefits and effects on shareholder value. White Paper Series No 9. (http://www.andersen.com/website.nsf/contents/IndustriesProductsBusiness TravelForum!OpenDocument.)

Bailey, E., Graham, D. and Kaplan, D. (1985), *Deregulating the Airlines*, Cambridge (Mass.).

Bain, J. (1956), *Barriers to New Competition*, Cambridge (Mass.).

Barkin, T., Hertzell, O.S. and Young, S. (1995), Facing low-cost competitors, Lessons from US airlines, *The McKinsey Quarterly*, Vol. 1 (4): 87-99.

Bennett, R.D. and Craun, J.M. (1993), The Airline Deregulation Evolution Continues, The Southwest Effect, Washington DC.

Biedermann, P. (1982), The U.S. Airline Industry – End of an Era, New York.

Bouvard, F. and Somosi, A. (1997), Europe's airlines choose between two ticket distribution strategies, *The McKinsey Quarterly*, Vol. 3 (1): 173-177.

Calder, S. (2002), No Frills, The Truth Behind the Low-Cost Revolution in the Skies, London.

Cameron, D. (2001), Air turbulence could mean a bumpy ride for investors, *Financial Times (European Edition)* (September 19, 2001): 17.

Dallas Business Journal (Online Edition) (2002), S&P downgrades AMR's rating (June 28, 2002) (http://dallas.bizjournals.com/dalls/stories/2002/06/24/daily55.html).

Doganis, R. (2001), *The airline business in the 21st century*, London and New York.

Done, K. (2001), Tried, Tested and ahead, *Financial Times (European Edition)*, Supplement World's Most Respected Companies (December 17, 2001): VII.

FAA/OST Task Force (1999), Airport Business Practices and Their Impact On Airline Competition, Washington DC.

Flug Revue (2002), News Luftfahrtindustrie, No 4. (April 2002): 25.

Freiberg, K. and Freiberg, J. (1996), NUTS! Southwest Airlines' Crazy Recipe for Business and Personal Success, Austin.

Grant, B. (1996), Trends in US airline ticket distribution, *The McKinsey Quarterly*, Vol. 2 (4): 179-184.

Gudmundsson, S. (1998), New-entrant airlines' life-cycle analysis: growth, decline and collapse, *Journal of Air Transport Management* 4: 217-228.

Houston Business Journal (Online Edition) (2002), S&P cuts credit rating on Continental (July 1, 2002) (http://houston.bizjournals.com/houston/stories/2002/07/01/daily2.html).

Knorr, A. (1998), Zwanzig Jahre Deregulierung im US-Luftverkehr – eine Zwischenbilanz, *ORDO* 49: 419-464.

Knorr, A. and Arndt, A. (2002), Entwicklungstendenzen im Weltluftverkehr und deren Auswirkungen auf den Verkehrsmarkt, *in: DVWG (Hrsg.): Europäischer Luftverkehr – wem nützen die strategischen Allianzen?*, Schriftenreihe B der Deutschen Verkehrswissenschaftlichen Gesellschaft e.V. Band 246, Bergisch Gladbach: 7-20.

Kummer, S. and Schnell, M. (2001), *Strategien und Markteintrittsbarrieren in europäischen Luftverkehrsmärkten: Theorie und neue empirische Befunde*, Schriftenreihe D der Deutschen Verkehrswissenschaftlichen Gesellschaft e.V. Heft 70, Bergisch Gladbach.

Langner, S. (1995), Contractual aspects of trading slots in the United States, *Journal of Air Transport Management* 2: 151-161.

Lawton, T. (2000), Flying Lessons: Learning From Ryanair's Cost Reduction Culture, *Journal of Air Transportation World Wide* 5 (1): 89-106.

Levine, M. (1987), Airline Competition in Deregulated Markets: Theory, Firm Strategy and Public Policy, *Yale Journal on Regulation* 4: 393-493.

Lewis, P. (2002), Southwest to add 737s, *Flight International* (January 1-7): 7.

Lott, J. (1999), Are Predatory Commitments Credible?: Who Should the Courts Believe?, Chicago.

Martin, P. (2002), Travelling with excess baggage, *Financial Times (European Edition)* (January 8, 2002): 13.

Mason, K. (2000), The propensity of business travellers to use low cost airlines, *Journal of Transport Geography* 8: 107-119.

Mason, K. (2001), Marketing low-cost airline services to business travellers, *Journal of Air Transport Management* 7: 103-109.

Meyer, J. and Menzies, T. (2000), The Continuing Vigil. Maintaining Competition in Deregulated Airlines Markets, *Journal of Transport Economics and Policy* 34, Part 1: 1-20.

Milgrom, P. and Roberts, J. (1990), New Theories of Predatory Pricing, in: Bonanno, G. and Brandolino, D. (eds): *Industrial Structure in the New Industrial Economics*, Oxford: 115-137.

Morrison, S. (2001), Actual, adjacent, and Potential Competition, Estimating the Full Effect of Southwest Airlines, *Journal of Transport Economics and Policy* 35: 239-256.

Morrison, S. (1998), New Entrants, Dominated Hubs, and Predatory Behavior, Statement before the Subcommittee on Aviation, Committee on Commerce, Science, and Transportation, U.S. Senate, Washington DC. (April 23, 1998).

Morrison, S. and Winston, C. (2000), The Remaining Role for Government Policy in the Deregulated Airline Industry, in: Peltzman, S. and Winston, C. (eds): *Deregulation of Network Industries: What Next?*, Washington DC: 1-40.

Ordover, J. and Saloner, G. (1989), Predation, Monopolization and Antitrust, in: Schmalensee, R. and Willig, R. (eds): *Handbook of Industrial Organization*, Amsterdam: 537-596.

Oster, C. and Strong, J. (2001), *Predatory Practices in the U.S. Airline Industry* (January 15, 2001). (http://ostpxweb.dot.gov/aviation/domestic-competition/.)

Pappas, G. (2001), The impact of the internet on the airline industry, in: Euromoney Yearbooks (ed.): *Airfinance Annual 2000/2001*, London: 49-52.

Petzinger, T. (1995), *Hard Landing*, New York.

Richter, J. and Wolf, C. (2002b), Unfall-Statistik. Wie sicher ist meine Fluggesellschaft?, *AERO international*, No 3. (März 2002): 86-87.

Roth, T. (1995), Predation, Cost-Based Tests and Predatory Intent, *Journal of Legal Economics* 5: 35-47.

Rubython, T., The difference is winning, *EuroBusiness* 3, No. 7 (December 2001): 24-26.

Southwest Airlines (1995), *Annual Report 1994*, Dallas (http://www.southwest.com/about_swa/annual_reports.html).

Southwest Airlines (1996), *Annual Report 1995*, Dallas (http://www.southwest.com/about_swa/annual_reports.html).

Southwest Airlines (1997), *Annual Report 1996*, Dallas (http://www.southwest.com/about_swa/annual_reports.html).

Southwest Airlines (1998), *Annual Report 1997*, Dallas (http://www.southwest.com/about_swa/annual_reports.html).

Southwest Airlines (2000), *Annual Report 1999*, Dallas (http://www.southwest.com/about_swa/annual_reports.html).
Southwest Airlines (2001), *Annual Report 2000*, Dallas (http://www.southwest.com/about_swa/annual_reports.html).
Southwest Airlines (2002a), *Southwest Airlines Reports Fourth Quarter Earnings and 29th Consecutive Year of Profitability*, News Release, Dallas (January 17, 2002) (http://www.southwest.com/about_swa/press/prindex.html).
Southwest Airlines (2002b), *Annual Report 2001*, Dallas (http://www.southwest.com/about_swa/annual_reports.html).
Southwest Airlines (2002c), *Southwest Announces Further Growth Plans; Airline Plans More New Flights for June*, News Release, Dallas (March 13, 2002) (http://www.southwest.com/about_swa/press/prindex.html).
Southwest Airlines (2002d), *About Rapid Rewards*, Dallas (http://www.southwest.com/rapid_rewards/about_rr.html).
Southwest Airlines (2003a), *Southwest Airlines Fact Sheet (Updated February 18, 2003)*, Dallas (http://www.southwest.com/about_swa/press/factsheet.html).
Southwest Airlines (2003b), *Southwest Airlines Reports Fourth Quarter Earnings and 30th Consecutive Year of Profitability*, News Release, Dallas (January 22nd, 2003) (http://www.southwest.com/about_swa/press/prindex.html).
The Economist (2001), Low-cost airlines: No frills, plenty of promise (November 3, 2001): 69.
Transport Research Board, National Research Council (1991) *Winds of Change: Domestic Air Transport Since Deregulation*, Washington DC.
Transport Research Board, National Research Council (1999) *Entry and Competition in the U.S. Airline Industry: Issues and Opportunities*, Washington DC.
U.S. Department of Transportation (Several years), *Air Travel Consumer Report*, Washington DC. (http://www.dot.gov/airconsumer).
U.S. Department of Transportation (1998), Competition in the U.S. Domestic Airline Industry: The Need for a Policy to Prevent Unfair Practices, Revised, Washington DC. (August 1998).
U.S. Department of Transportation (1999), *Competition in the U.S. Domestic Airline Industry: The Need for a Policy to Prevent Unfair Practices, Revised*, Washington DC. (May 1999) (http://ostpxweb.dot.gov/aviation/domestic-competition/).
U.S. Department of Transportation Enforcement Policy Regarding Unfair Exclusionary Conduct in the Air Transportation Industry, Findings and Conclusions on the Economic, Policy, and Legal Issues (Docket OST-98-3713), Washington DC (January 2001) (http://ostpxweb.dot.gov/aviation/domestic-competition/).
U.S. Department of Transportation, Office of the Assistant Secretary for Aviation and International Affairs (1996), *The Low Cost Airline Service Revolution*, Washington DC.
U.S. General Accounting Office (1996), Airline Deregulation. Barriers to Entry Continue to Limit Competition in Several Key Domestic Markets (GAO/RCED-97-4), Washington DC. (October 18, 1996).
U.S. General Accounting Office (1999), Airline Deregulation. Changes in Airfares, Service Quality, and Barriers to Entry (GAO/RCED-99-92), Washington DC (March 4, 1999).
Vowles, T. (2001), The Southwest Effect in multi-airport regions, *Journal of Air Transport Management* 7: 251-258.
Vowles, T. (2000), The effect of low fare air carriers on airfares in the US, *Journal of Transport Geography* 8: 121-128.
Yahoo! Finance (2003), AMR CORP (March 20, 2003) (http://finance.yahoo.com/q?s=amr&d=1y).
Yahoo! Finance (2003), CONTL Airlines B (March 20, 2003) (http://finance.yahoo.com/q?s=cal&d=1y).
Yahoo! Finance (2003), DELTA AIR LINES (March 20, 2003) (http://finance.yahoo.com/q?s=dal&d=1y).

Yahoo! Finance (2003), NORTHWEST AIR (March 20, 2003) (http://finance.yahoo.com/q?s=nwac&d=1y).

Yahoo! Finance (2003), SW AIRLINES (March 20, 2003) (http://finance.yahoo.com/q?d=c&c=&k=c1&t=1y&s=luv&a=v&p=s&1=on&z=m&q=1).

Yahoo! Finance (2003), UAL CORP (March 20, 2003) (http://finance.yahoo.com/q?s=ual&d=1y).

Yahoo! Finance (2003), US AIRWAYS GROUP (March 20, 2003) (http://finance.yahoo.com/q?s=UAWGQ.OB&d=t).

Weinhold, M. (1995), Computerreservierungssysteme im Luftverkehr, Baden-Baden.

Windle, R. and Dresner, M. (1995), The Short and Long Run Effects of Entry on U.S. Domestic Air Routes, *Transportation Journal* 35 (2): 14-25.

Woodyard, C. (2001), United drops shuttle flights, low-fare model, *USAToday (Online Edition)* (December 4, 2001) (www.usatoday.com/money/biztravel/2001-0-04-united-shuttle.htm).

Part D:
Europe

Chapter 9

The Lufthansa-Germania Case at a Glance

Kai Hüschelrath

1. Introduction

Cases of alleged predatory pricing in the airline industry are not rare and often follow a similar structure: A new and usually small airline enters a route thus threatening the incumbent operating in that city-pair. Consequently – and according to standard oligopoly theories – it is rational for the incumbent to adjust the pre-entry (monopoly) price downwards 'significantly'.[1] The question one needs to ask here is if the kind and extent of the incumbent's adjustment(s) can be explained by 'competitive behaviour in the presence of market entry' or if it is motivated by 'anti-competitive' purposes aiming at maintaining market power.

Much less common than cases of alleged predatory pricing are successful antitrust cases of predatory pricing – successful from the point of view of the alleged prey. Morrison and Winston (2000) for example report that, despite several attempts, no alleged predator in the US airline industry has ever been prosecuted so far. The enforcement records in Europe and other parts of the world look very similarly.

Against this background, it is not the observed market conduct as such that makes the Lufthansa-Germania case special and hence worth describing, but the way the antitrust authority reacted to it: The incumbent – Deutsche Lufthansa – was sanctioned for its attempt to rule out its competitor – Germania Fluggesellschaft – from the Frankfurt–Berlin–Frankfurt market. The following Section is a short description of the case at hand. Section 3 presents the decision and clarifies the major arguments of the German Federal Cartel Office (FCO). Section 4 concludes with some interesting economic questions, which were raised by the case decision.

2. The Facts

On the 12th of November 2001, Germania Fluggesellschaft ('GER') began its scheduled flights between Frankfurt/Main and Berlin/Tegel. At the time GER –

[1] This is normally not true for – often observable – increases in capacities and/or flight frequencies in response to the market entry of a rival (see Morrison (2004), p. 91).

which is privately owned by the German millionaire Hinrich Bischoff – was only operating in the German charter market. The price for a flexible one-way economy ticket without restrictions was €99 (including passenger fees and value added tax) and a round-trip was €198. GER offered four flights in each direction per working day.

The previous monopolist on this route – Deutsche Lufthansa ('DLH') – offered fourteen daily flights (in both directions) at a price of €485 for a flexible economy round-trip ticket. After DLH became aware of the imminent market entry by GER, it reacted – before market entry actually took place. On the 9th of November 2001, DHL introduced a new one-way tariff of €100 for economy class, which translated into a round-trip ticket cost of €200. This 58% price reduction was limited to a relatively small number of seats per flight (see Bundeskartellamt (2002), pp. 2-3).

On the 12th of November 2001, GER responded to the move by DLH with a tariff reduction from €99 to €55. According to an official of GER, the airline was forced to undertake this step as otherwise potential customers would have had no incentive to fly with GER given that DLH offers several major auxiliary services such as free on board services, airport lounges, a frequent flyer programme and the important convenience of more than triple the number of daily flight connections.

At the beginning of 2002 both airlines changed their prices again. By mid February 2002 DLH increased its €100 tariff slightly to €105.11 for a flight from Berlin to Frankfurt and €105.31 for a flight from Frankfurt to Berlin. Therefore, a round-trip was available for €210.42. Furthermore, DLH introduced several restrictions for this reduced tariff, e.g. a fee for reclassification of €22 and a duty to book outbound and inbound-flights separately. However it is important to note that the new tariff was not subject to the typical restrictions, such as a minimum stay, a reclassification ban or a Sunday rule (see Bundeskartellamt (2002), p. 3), which traditionally separate business travellers from leisure travellers.

GER raised its fares on 1st January 2002 from €55 to €99 aiming to reach the break-even point which was – according to them – impossible by maintaining the low fare. Following the tariff increase, the demand for GER tickets dropped about 39%, whereas the demand for DLH flights had not experienced much movement between December 2001 and January 2002.

3. The Antitrust Case

The market conduct described above led to the DLH-GER antitrust case, which is outlined in the following. First, the final decision of the German Federal Cartel Office (FCO) is reiterated, followed by an overview of the general way of (economically relevant) reasoning by the antitrust authority.

The Decision

On the 18th of February 2002, the FCO decided, based on Section 19 of the German '*Act Against Restraints of Competition*' ('Abuse of a dominant position'), that DLH is prohibited from offering a price (including passenger fees) for a one-way ticket per passenger on the Frankfurt–Berlin route, which is not at least €35 above GER's price, as long as DLH does not have to charge more than €134 as a result.[2]

This rule however is inoperative for tickets with 'hard restrictions' like a reclassification ban and/or a minimum stay and/or a Sunday rule. The prohibition remains valid for two years, as it was reasoned that within this time frame GER should have gained adequate brand name recognition and have established a sufficiently large clientele base to ensure survival (see Bundeskartellamt (2002), p. 1). Figure 9.1 outlines the price changes in the Frankfurt–Berlin–Frankfurt market prior to the decision of the FCO as well as the general impacts of this decision on future prices.

The Economic Reasoning of the FCO

The decision of the Bundeskartellamt is separated into two (economically relevant) parts. First, some introductory remarks by the FCO – which strengthen the assumption that DHL abused its dominant market position – are given. The second part contains the economic reasoning in detail.

Introductory Propositions Before beginning the analysis of the abuse of a dominant position, it is important to highlight the following three propositions given by the FCO:

- Proposition 1: With the introduction of the new tariff, DLH did not merely undertake an adjustment to GER's price but factually undercut it.

In its explanatory statement, the FCO stated that the reduced fare of DLH contains several auxiliary services including: on board service, better seats in the airplane, a frequent flyer programme, more daily flight connections, access to airport lounges, better links with travel agencies and hence access to computer reservation systems, an existing extensive airline network, the advantage of a strong brand name recognition particularly among business travellers and travel commission override programmes between DLH and the majority of the 100 biggest firms in Germany (see Bundeskartellamt (2002), pp. 11-12).

---

[2] Section 19 of the German 'Act Against Restraints of Competition' says in Paragraph 1 that 'The abusive exploitation of a dominant position by one or several undertakings shall be prohibited. [...] (4) An abuse exists in particular if a dominant undertaking, as a supplier or purchaser of certain kinds of goods or commercial services, 1. impairs the ability to compete of other undertakings in a manner affecting competition in the market and without any objective justification; 2. demands payment or other business terms which differ from those which would very likely arise if effective competition existed; in this context, particularly the conduct of undertakings in comparable markets where effective competition prevails shall be taken into account; [...]'.

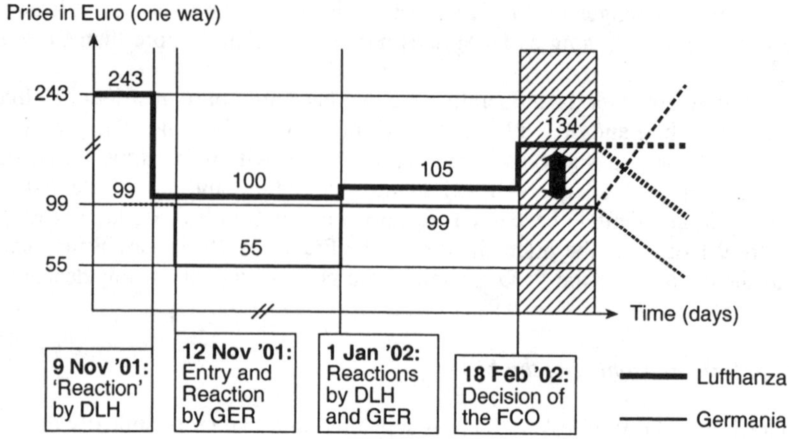

Figure 9.1 Price changes in the Frankfurt–Berlin–Frankfurt market (November-February 2002) and the decision of the FCO

In its analysis, the FCO tried to calculate the pecuniary benefit of some of these auxiliary services. The motivation for this calculation was largely driven by the belief that an essential part of predation analysis is to determine if 'enough' people are solely driven by monetary considerations when choosing a flight and hence would readily switch from DLH to GER due to GER's lower fares (see Bundeskartellamt (2002), pp. 6-13).

To give an example, the monetary benefit of the frequent flyer programme was calculated as follows: Every flight with the new fare is rewarded with at least 500 frequent flyer miles. A normal flight within Germany is rewarded with 1000 miles. Once 20,000 miles have been reached a free bonus flight (worth on average: €488) is offered, hence a free flight is due after 40 flights. Therefore, the pecuniary benefit of the frequent flyer programme for one flight can be calculated to be €12 (see Bundeskartellamt (2002), pp. 7-8).

The FCO also assessed the monetary benefit of the higher daily flight frequencies of DLH. A time advantage of about one hour for customers of DLH was assumed, which translates into an estimated benefit of between €25 and €60. Consequently the FCO concluded that DLH's fare price must be at least between €25 and €60 higher compared to GER's price in order to compensate for the higher frequencies of DLH (see Bundeskartellamt (2002), pp. 10-11).

- Proposition 2: Due to the low DLH tariff, GER is stripped of its only competitive advantage with respect to DLH.

The FCO argued that GER has clear disadvantages (compared to DLH) in the auxiliary service sections including flight frequencies, seat comfort, frequent flyer programme, access to lounges and on board service. In the service aspects of safety, punctuality, efficient check-in and free choosing of seats, the FCO evaluated that neither DLH nor GER has a clear advantage. Therefore, GER only has competitive potential in the aspects of price and cabin personnel. Furthermore, the FCO argued that if DLH matches these two competitive aspects, it would be impossible for GER to attract enough passengers in order to operate profitably (see Bundeskartellamt (2002), pp. 9-10).

- Proposition 3: The new low fare tariff of DLH does not cover the average costs per passenger.

Following the (unpublished) calculations of the FCO, DLH does not cover the average costs per passenger with the low fare tariff of €105. This is the result of an analysis of route profitability data provided by DLH (see Bundeskartellamt (2002), pp. 12-13).

The economic reasoning in detail Following the guidelines of Section 19 of the German 'Act against Restraints of Competition', the investigation had to firstly define the relevant market followed by the determination that the company (or companies) of interest had a dominant position in this market. Then an abusive exploitation of this dominant position had to be established.

- relevant market and dominant position

The FCO provided only a short analysis of the relevant market. It interpreted business flights between Frankfurt and Berlin as a relevant market on its own, which consequently had to be separated from the less time-sensitive leisure travellers. The FCO particularly mentioned that possible substitution alternatives such as car or rail travel are no real alternatives for business travellers due to the significantly longer travel time. By excluding rail and car travel alternatives from the overall travel demand between Frankfurt and Berlin, GER had (at the time of the decision) a market share of about 10% in the Frankfurt–Berlin Route. Therefore, the FCO concluded that DLH was the dominant firm in this market. The general resources of DLH compared to GER's were used as another argument to reinforce this point of view (see Bundeskartellamt (2002), p. 14).

- determination of the abuse of a dominant position

The determination whether there had been an abuse of market power usually takes place in three steps. After establishing the allegations, the FCO has to show that the dominant firm reduced its price sufficiently to reach its presumed (anti-competitive) goal. Finally, the FCO needs to prove the actual existence of such an anti-competitive strategy.

  - Alleged violation of Section 19

  DLH is abusing its dominant position in the Frankfurt-Berlin-Frankfurt market when offering airfares (with similar restrictions) to the public which are not at least €35 above GER's price, as long as DLH does not have to charge more than €134.[3]

o Suitability of the price decrease to reach the assumed aim

'Predatory prices'[4] damage GER's competitive potential and hence could have significant effects on future competition in the market for scheduled flights between Frankfurt and Berlin. DLH had a monopoly position before GER's market entry and would in all likelihood regain it after the exit of GER. It can be further argued that the defence attempts undertaken by the newcomer would cause considerable obstructions to competition in this and presumably also in other markets, as future entry attempts could potentially be deterred by the DLH's aggressive behaviour.

'Predatory prices' can under certain conditions (e.g. a high degree of market power, deep pockets and high entry barriers) be an effective way to reach the assumed aim of regaining the monopoly position in the Frankfurt–Berlin–Frankfurt market (see Bundeskartellamt (2002), p. 17).

o Actual existence of the presumed predation strategy

The FCO first assessed whether the new DLH tariff affects the competitive potential of GER. The reaction of DLH seriously endangered the GER's prospects of establishing a new profitable service on this route. The use of a tariff of €105.21 by DLH constituted – in conjunction with its auxiliary services – a factual underselling of at least €35 and the acceptance of operative losses. The revenue per passenger is significantly smaller than the average costs per passenger.

It is argued that such a strategy is only rational where the key objective is to force the newcomer (GER) out of the market. After the newcomer's exit the losses can be recouped by increasing prices up to the previous (monopoly) level. Moreover, the actual behaviour of DLH might set a deterrent signal to other potential entrants in this and other routes. Therefore, the FCO interpreted the introduction of the low tariff as an attempt to force GER out of the market (see Bundeskartellamt (2002), pp. 17-21). The FCO's judgement is further strengthened by the following arguments:

➢ DLH introduced the new tariff selectively only on the one route, where the new entrant began to offer its service. The tariff was aimed at the same group of passengers (business travellers) and DLH introduced the same tariff restrictions as GER.

---

[3] In other words, if GER continues to charge €99, DLH must charge at least €134 for a one-way economy ticket in order to avoid abusing its dominant position in that market. If GER lowers its price, DLH is allowed to meet the price drop proportionally. If GER raises its price, DLH is allowed to keep a price of €134.

[4] The FCO defines predatory behaviour as follows: 'Predatory behaviour is aggressive market conduct [...] by dominant firms with the aim of ruling out competitors from the market, to discipline them or to deter them from new market entry' (Bundeskartellamt (2002), p. 15, translated by the author).

- DLH's price reduction – almost up to the level of GER's – is a factual clear underselling, as DLH's tariff includes several auxiliary services (see above). DLH's low tariff foreclosed GER's prospect to reach a sufficiently high switching rate of price-sensitive business travellers from DLH to GER.
- Experiences (by other competitors) with market entries in other DLH monopoly markets (e.g. after entry of 'Go-Fly' in the London/Stansted-Munich market) have shown that predatory behaviour is generally a suitable strategy in the German aviation market to rule out competitors.
- Price comparisons with the only other competitive route within Germany (between Berlin and Munich), where DLH competes with 'Deutsche BA', showed that DLH's Frankfurt-Berlin tariff was significantly below the price of a ticket in the Berlin-Munich market (€441).
- Furthermore, DLH is financially strong and hence able to survive a sustained period of low (predatory) prices. Consequently it can be argued that the 'predatory investment' was a plausible option for DHL.

Due to these reasons, the FCO acted on the assumption that the pricing policy of DLH in the Frankfurt–Berlin–Frankfurt market had as its sole aim to force GER out of the market.

- 'no objective justification'

The FCO asserted that no objective justification for the underselling strategy existed. DLH's strategy aims at recovering and consolidating its monopoly position in the market. Even the particular importance of the DLH Hub in Frankfurt (and the corresponding economies of scope and scale) cannot justify DLH's pricing strategy.

- Fixing the sanctions

  - Fixing the minimum price distance

  The fixing of the minimum price distance of €35 occurred according to the best judgement of the FCO. It was based on the calculations of the monetary benefit of some of the auxiliary services mentioned above.[5] The FCO expected that after the introduction of the minimum price distance rule, the predatory effect would disappear and that normal competitive pressures would dominate the market. The FCO further notes that the price rule as such does not deliver unlimited protection for GER, as GER stated that it can reach break-even with a price of €99, hence the protective aspect of the price rule vanishes with increasing prices (see Bundeskartellamt (2002), pp. 21-22).

---

[5] The FCO assumed three Euro for drinks and newspaper, €12 for the value of frequent flyer miles and €25 for the estimated value of the flight frequency. The total of €40 was reduced by five Euro, presumably as some kind of ad hoc adjustment.

- Fixing of the validity of the minimum price distance

The sanction was set for two years. The FCO justified this fixing by stating that the need for a minimum price distance disappears after the new competitor has established itself in the market (e.g. reputation, optimization of competition parameters) and hence the potential for predatory behaviour by the incumbent is naturally reduced (see Bundeskartellamt (2002), pp. 22-23).

4. Outlook

The DLH-GER case raises at least two distinct economic questions. The first concerns the general question of whether the behaviour of DLH was really predatory. As the FCO did not deal with a complete theory-based framework for predation (e.g. like the most recent one by Bolton et al. (2000/2001)), there is space for criticism and different opinions concerning the case. Unfortunately, due to the confidentiality of all cost and demand data, a rigorous and comprehensive analysis seems impossible. Furthermore, the case renews the question of 'if' and 'how' price-matching should be adjusted for service quality in predatory pricing cases (see Morrison (2004) for a discussion).

The second interesting question raised by the case concerns how to sanction predators after their predatory conduct is proven. This question seems to be relatively new, as most (recent) predation cases in the EU as well as in the US got stalled in the detection phase and hence never reached the 'sanction phase'. Consequently the DLH-GER decision treads onto 'virgin territory', given that a predator is convicted and a (behavioural) sanction is imposed. With respect to the FCO's 'minimum-price-distance rule' – as a resolution to the case at hand – one needs to ask whether such a rule is a) helpful in fighting predators and b) the best possible answer to the problem of sanctioning predation (see Hüschelrath (2004) and especially Edlin (2002) for an ex-ante predation rule).

Independent of the answer to these questions, the DLH-GER decision certainly initiated a new discussion about the optimal antitrust answer(s) to predation strategies. From a practical point of view, the FCO has shown its willingness to intervene against predators. This might help the FCO to create (or maintain respectively) a tough reputation concerning the enforcement of competition rules generally and in particular concerning Section 19 violations.[6]

Acknowledgement

I'm indebted to Peter Forsyth for many valuable remarks.

---

[6] It is obvious that the overall success of the whole FCO intervention depends on future competition in the market. In May 2004 – after the expiration of the FCO intervention – GER was still operating in the Frankfurt–Berlin–Frankfurt market. The one-way tariff was €88 and the customers were able to choose between four flight connections per day (Morning/evening from/to Frankfurt). The cheapest online price for a round-trip DLH ticket in the economy class was €156 (off-peak) and €285 (peak).

References

Bolton, P., J. Brodley and M. Riordan (2000), Predatory Pricing: Strategic Theory and Legal Policy, *Georgetown Law Review*, 88, 2239-2330.

Bolton, P., J. Brodley and M. Riordan (2001), Predatory Pricing: Response to Critique and Further Elaboration, *Georgetown Law Review*, 89, 2495-2529.

Bundeskartellamt (2002), *Beschluss in dem Verwaltungsverfahren gegen Deutsche Lufthansa AG*, Köln, B 9-144/01, Bonn.

Edlin, A. (2002), Stopping Above-Cost Predatory Pricing, *Yale Law Journal*, 111, 941-991.

Hüschelrath, K. (2004), *How to Prey on Predators?*, Unpublished Working Paper, Vallendar.

Morrison, S.A. and C. Winston (2000), The Remaining Role for Government Policy in the Deregulated Airline Industry, in: Peltzman, Sam and Clifford Winston, *Deregulation of Network Industries: What's Next?*, Brookings Institution Press, Washington D.C.

Morrison, W. (2004), Dimensions of Predatory Pricing in Air Travel Markets, *Journal of Air Transport Management*, 10, 87-95.

Chapter 10

Predatory Pricing in the Airline Industry as a Challenge to Competition Law Enforcement

An Assessment of the Current Legal Practice in the U.S. and Germany

Christian Ewald

1. Introduction

Two recent cases in the U.S. and Germany have once again raised interest in the economic and legal assessment of predatory pricing, putting this issue back on the current agenda of the competition policy debate. In both cases the pricing strategies of dominant airlines were the test bed of competition law.

In 2001 the District Court for the District of Kansas in the U.S rejected a motion for summary judgement[1] filed by the Department of Justice (DoJ) against American Airlines (AA).[2] The U.S. Court of Appeals confirmed this legal assessment in July 2003.[3] The lawsuit initiated by the DoJ almost coincided with the proposal of a regulatory policy approach against predatory practices in the airline industry by the Department of Transport (DoT).[4]

By contrast, in Germany, the Oberlandesgericht Düsseldorf (Higher Regional Court of Düsseldorf) provided a preliminary positive opinion on a decision of the Bundeskartellamt (Federal Cartel Office; FCO), concerning predatory pricing in the airline sector. The FCO's decision prohibited the German flag carrier Deutsche Lufthansa AG (DLH) from using a low price strategy on particular fare tickets without any significant restrictions (e.g. minimum stay etc.), which were targeted exclusively on a route also served by its competitor, the Low Cost Carrier (LCC) Germania.[5] DLH was forbidden to charge airfares on the Frankfurt–Berlin route below a specific 'minimum price distance' compared to the airfares of Germania. Furthermore, the FCO ordered the immediate enforcement of this decision. DLH

---

1 District Court for the District of Kansas (2001).

2 Department of Justice (1999).

3 U.S. Court of Appeals (2003).

4 Department of Transportation (2001a).

5 Bundeskartellamt (2002); English version not available; an English summary of the proceeding and the final decision is presented by Hüschelrath (2003), 5 ff.

lodged an appeal at the Court of Düsseldorf seeking for the suspension of the effects of the decision. The Court, however, rejected this application in March 2002.[6]

The different outcomes of competition law enforcement in the U.S. and Germany raise interesting economic as well as legal questions. One might ask for example if there are any criteria provided by economic theory to give competition authorities guidance in identifying predatory pricing strategies? Against this setting, what may explain the diverging conclusions of the courts in the U.S. and Germany? Were factual differences decisive or are there different standards of proof or 'models' of competition law enforcement?

In order to answer these questions it is necessary to first define the prevailing conflicts in the theory of predatory pricing. Subsequently, Joskow and Klevorick's two-tier approach will be used as a framework for a comparative assessment of the current legal practice in the U.S. and Germany.

2. Lines of Conflict in Predatory Pricing Theory

Defining predatory pricing is one of the few aspects of predatory pricing theory that seem to be undisputed:

> In most general terms predatory pricing is defined in economic terms as a price reduction that is profitable only because of the added market power the predator gains from eliminating, disciplining or otherwise inhibiting the competitive conduct of a rival or potential rival. Stated more precisely, a predatory price is a price that is profitable only because of its exclusionary or other anticompetitive effects.[7]

What is disputed however is whether predatory pricing can be 'rational' (i.e. profitable) and whether competition authorities should expend much attention on it. Furthermore, many different theoretical approaches are concerned with the appropriate standard of proof.

Rationality Debate: Chicago-School vs. Game-Theoretic Approaches

Predatory pricing theory is – like many other theoretical debates in the field of competition law and policy – characterized by taking issue with positions of the Chicago School. The skeptical view of Chicago scholars' (e.g. McGee, Koller, Bork or Easterbrook) about government intervention in the competitive process and the adherence to the freedom of price setting manifests itself in the opinion that 'attempts of predation have been rare and that successful attempts will be found to be still rarer'.[8] In their view it is quite questionable that predation can be profitable at all, because, among other reasons, superior (i.e. less costly) monopolization strategies are available, e.g. merging or colluding with

---

[6] Oberlandesgericht Düsseldorf (2002); English version not available.
[7] Bolton et al. (1999), 3.
[8] McGee (1980), 292.

competitors. This argument may be easily refuted by referring to effective competition law enforcement in the fields of merger control and prosecution of cartels and tacit collusion.

Another core argument of the Chicago School seems however to be more convincing, at least at first sight.

From an economic point of view predatory pricing may be seen as analogous to an investment. Under the classical assumption of long-run profit-maximization predatory behaviour is only rational if future profits due to subsequent market dominance exceed the costs of predation. This prerequisite, however, may be rarely satisfied. Due to higher and even increasing market shares, the predator's costs (foregone profits) are significantly higher than the costs carried by the prey. Since mobilizing additional capital is not costless, this observation cannot be refuted by pointing to the bigger financial resources ('deep pocket' or 'long purse'[9]) of the incumbent. Furthermore, future profits of market dominance (recoupment) are far from guaranteed. Higher prices make further attempts of market entry more attractive. This incentive might be offset by high barriers to entry, which are, however, synonymous with high barriers to exit, so that the prey might also be willing to endure predation strategies for a longer period of time.

As *Hemphill* points out, this line of argument has the shortcoming of suggesting that predatory pricing can only be rational if the incumbent is able to reap profits from supra-competitive or monopolistic pricing once the competitor has exited the market. This assumption however neglects efficiency gaps between the predator and the prey. If the newcomer is more efficient than the incumbent, the incumbent risks continuous erosion of its market position. Taking this into account 'the incumbent might be willing to engage in costly predation, even if total profit is negative, since the alternative to predation is not a zero baseline, but something much worse'.[10] Furthermore, the positions of the Chicago School came under fire from game-theorists.[11] Most of the game-theoretic models of predation have been developed in response to *Selten*'s chain-store paradox in which predation was shown to be always irrational.[12] Contributions by Milgrom and Roberts[13] and by Kreps and Wilson[14] indicate that the irrationality-hypothesis loses strength as soon as the assumption of complete and symmetric information of all 'players' is relaxed. If a newcomer has imperfect knowledge of the willingness of the incumbent to use predation as a strategic option, predatory pricing might be a suitable instrument to deter market entrance ('reputation effects'). Even the 'deep-pocket' or 'long purse' argument regains plausibility, since the costs of withstanding predation may be significantly higher for the prey due to imperfect knowledge of investors or banks about the real reason for its rising losses ('financial market predation').[15]

---

9 Telser (1966).

10 Hemphill (2001), 1594.

11 Kate, Niels (2002), 6 f.; Phlips (1995), 185 f.

12 Selten (1978).

13 Milgrom, Roberts (1982).

14 Kreps, Wilson (1982).

15 Bolton et al. (1999), 10; Blair, Harrison (1999), 497.

Appropriate Standards of Proof: Per-Se Rules vs. Two-Tier Approach

The rationality debate between Chicago and post-Chicago was initiated and driven by the discussion about the appropriate standard of proof to be used by competition authorities. In this context, the cost based approach developed by Arreda and Turner[16] is often used as a reference point. Their approach is based on the assumption that deviations from short-run profit-maximization are a necessary condition for predatory pricing. Since in a world of perfect competition the short-run profit maximum is defined by the familiar 'marginal-cost-equals-price' rule, each price undercutting marginal cost (MC) is assumed to be predatory. Arreda and Turner however take account of the difficulties inherent in measuring MC and hence propose to use average variable costs (AVC) as the deciding indicator. Due to the difficulties of defining fixed and variable costs, Posner[17] puts forward to take long-term variable costs into account, at least as a supplementary indicator. Furthermore, applied to multi-product firms the Arreda-Turner Test risks giving wrong signals. Therefore, Baumol states that 'average avoidable cost is crucial in testing predation'.[18] Avoidable costs would have to be determined by simulating the cost levels, which would arise if the product in question was not offered.

All standards of proof presented so far exclusively assess price reductions of the incumbent as one of the core characteristics of predatory pricing strategies. Other advocates of a per-se rule, however, stress the importance of capacity adjustments. Following Williamson's approach[19] a dominant firm should not be allowed to respond to market entrance with an expansion of its capacity. Starkie, on the other hand, advocates compelling the incumbent 'to operate at least at the same capacity as it did immediately prior to the withdrawal of an entrant'.[20] Obviously, these 'output-restriction-rules' would involve major state intervention, a criticism that also applies to Baumol's approach, which focuses on the price increase carried out by the incumbent after the newcomer has been forced out of the market.[21] According to Baumol, the incumbent should be allowed to reduce prices and adjust capacity in response to market entrances. Once a competitor has exited the relevant market, however, the incumbent should be prohibited from increasing prices for a reasonably long period of time.

Unlike the '*output- or price-increase-restriction rules*', cost-based approaches are widely accepted by courts in several countries. With regard to the U.S. experience Niels and Kate identify three different stages in the development of American case law.[22] Until the end of the 60s the courts focused on predatory intent, however from the beginning of the 70s until the mid-80s the aspect of selling below costs dominated the legal assessment of predatory pricing strategies. Following the decisions of the Supreme Court in the cases '*Matsushita*' (1986) and

[16] Arreda, Turner (1975).
[17] Posner (1976).
[18] Baumol (1996), 49.
[19] Williamson (1977).
[20] Starkie (1999), 284.
[21] Baumol (1979).
[22] Niels, Kate (2000), 880.

'*Brooke Group*' (1993) the cost-based approach has been supplemented by the recoupment test (an assessment of the likelihood that the incumbent might be able to recoup its losses after the prey has exited the market). Like the courts in the U.S. in the 70s and 80s, the European Court of Justice (ECJ) also affirmed in its '*AKZO*'-judgement,[23] that if AVC has been undercut there is evidence of an abusive predatory pricing strategy. Furthermore, even a price exceeding AVC but undercutting average total cost (ATC) may be abusive if there is evidence of predatory intent. Contrary to the case law developed in the U.S., the ECJ expressly stated in 'Tetra Pak II' that it would not be appropriate to require additional proof that the potential offender has a realistic chance to recoup its losses.[24] In the '*Deutsche Post*' case the European Commission, on the other hand, specifically referred to Baumol's '*per-se rule*'. 'Deutsche Post' infringed Article 82 EC when it decreased its prices for business-to-consumer parcel delivery services below incremental costs, i.e. costs that only arise due to a specific activity (product line), which would be avoided if the activity ceased.[25]

Following Joskow and Klevorick, the shortcomings of each of the '*per-se rules*' presented above can be summarized as follows:

Due to the disregard of structural market differences, '*per-se rules*' are not sufficient to cope with the 'Scylla-and-Charybdis'-challenge of separating fierce but healthy price competition from predatory pricing strategies. Competition authorities – just as the ancient Greek hero Odysseus on his journey home – have to find a safe passage avoiding the 'monsters' of negative welfare effects due to the possible suppression of fierce but healthy price competition on the one hand ('Scylla') and the potential misinterpretation of predatory pricing strategies as healthy price competition on the other ('Charybdis'). Therefore, an approach is needed 'that would minimize the sum of the expected costs of error and the costs of implementation that would result if the policy were applied to the market'.[26] In other words, while Chicago scholars give the advice to 'Odysseus' (the competition authorities) to refrain from navigating through the narrow strait between 'Scylla' and 'Charybdis' (i.e. assessing and intervening against predatory pricing) because there might be the risk of drowning (i.e. doing more harm than good), the conclusion of Joskow and Klevorick is to provide 'Odysseus' with some sort of compass to find his way through at minimum risk.

The decision-theoretic framework of Joskow and Klevorick is based on (1) the analysis of the costs (i.e. welfare losses) of misinterpreting fierce but healthy price competition as predation ('false positive') and failing to identify a predatory pricing strategy as an abusive conduct ('false negative') and (2) the evaluation of the core factors which determine error cost levels. Their conclusion can be summarized as follows:

---

[23] European Court of Justice (1991).
[24] European Court of Justice (1996).
[25] European Commission (2001); for an assessment see: Nicolaides (2001); Bergman (2001).
[26] Joskow, Klevorick (1979), 218.

> Since structural factors that raise the costs of false positive errors lowers the costs of false negatives and vice versa, an approach to predatory pricing should employ standards that are carefully conditional on structural factors. Thus, we suggest a two-tier approach that is sensitive to differences in market structure and allows us to screen out market situations where a reasonable threat of predatory pricing exists from those where it does not.[27]

Joskow and Klevorick classify the relevant structural characteristics into three categories: (1) short term monopoly power, (2) conditions of entry and (3) several dynamic effects of competition.

The thesis that the costs of false positive errors (false negative errors) decrease (increase) with the degree of market dominance constitutes the core justification for those competition law regimes in which only dominant firms are subject to abuse control by competition authorities. With respect to the second category it has to be mentioned that Joskow and Klevorick intentionally do not use the term 'entry barriers' since 'the traditional examples of "entry barriers" are both too restrictive and too easily misinterpreted. In particular, they do not capture the dynamics of entry process in different markets.'[28] They do not just refer to widely accepted entry barriers like economies of scale and scope or high marketing expenses, but also highlight some dynamic and strategic aspects of market entry and entry deterrence (e.g. reputation effects), without making reference to the game-theoretic approaches, which were developed later. In the third category of structural characteristics Joskow and Klevorick consider that negative welfare effects of reduced dynamic or technological competition may even exceed those of rising prices, in particular if process and product innovations are predominantly developed by newcomers.

Due to the basic idea of the two-tier approach, it is no surprise that – once the first threshold of state intervention has been overcome – Joskow and Klevorick assess the pricing strategy of the scrutinized incumbent in a rather strict manner. Hence where AVC has been undercut, predatory pricing should be assumed *per se*. If the prices are higher than AVC but lower than ATC the incumbent should be required to prove the suitability of its prices to reach the objective of short-run profit-maximization. Even prices exceeding ATC should be assessed according to Baumol's '*per se rule*' which imposes very restrictive conditions on price increases in the transition period after a competitor has exited the market.

3. The Airline Industry on the Test Bed of the Two-Tier Approach: A Comparative Assessment of Legal Practice in the U.S. and Germany

The two-tier approach seems to be a suitable framework for the comparative assessment of the current legal practice in the U.S. and Germany. The following paragraph analyses the relevant differences relating to the legal assessment of the

---

[27] Joskow, Klevorick (1979), 223.
[28] Joskow, Klevorick (1979), 226/227.

market structures and of the competitive conditions in the U.S. and in Germany. The analysis is carried out on the basis of an identification of those specific structural characteristics of the airline industry, which have an influence on the probability of success of predatory pricing strategies. Whether the pricing strategies of the incumbents are judged differently in the two countries will be analysed in the following section.

Assessment of Structural Characteristics of the Airline Industry

Deregulation, liberalization and market structures The airline industry is one of those economic sectors, which were opened up for competition after long periods marked by monopolistic or oligopolistic market structures and state owned enterprises. Although the regulatory frameworks in the U.S. and Europe differ greatly,[29] the market structures on both sides of the Atlantic are quite similar. In the U.S. the effects of market deregulation at the end of the 70s were seen as overall positive. Nevertheless, the rather optimistic expectations have not been fulfilled.[30] Immediately after deregulation the number of market entries increased significantly, but only a few newcomers survived. The established network airlines, which dominated the market before deregulation, successfully defended their market position. As the DoT points out, one of the main reasons for this was due to the restructuring of the route systems carried out by the major airlines:

> The operation of the deregulated industry has been quite different from the forecast. [...] Industry analysts had not expected every major airline to adopt a hub-and-spoke-route system. Airlines generally operated linear route systems during the regulatory era, as did the intrastate airlines [...].
> [...]
> Because the hubbing airline obtains [...] competitive advantages, its hub markets tend to have little competition. [...] Each network airline focuses on routes that have one of its hubs as an endpoint, and it rarely serves a competitor's non-stop hub markets.[31]

In Europe a market structure characterized by national or regional hub-and-spoke-'fortresses' was primarily the result of a long lasting protectionist legal framework in favour of state owned national flag carriers. The different steps of the EU liberalization process,[32] aimed at dismantling national market barriers, have not led to a significantly higher intensity of competition in interstate airline services:

[29] Grundmann (1999), 27 ff./142 ff.
[30] Levine (1987); US General Accounting Office (1999); Oster, Strong (2001).
[31] Department of Transportation (2001a), 19 and 26; the economic background of this tendency will be discussed in depth later.
[32] Grundmann (1999), 142 ff.

> The European Commission's own study concluded that competition had little effect on routes run as a monopoly or duopoly, which represents 94% of intra-Community routes. [...] Following European liberalization, established flag carriers move to consolidate their position by share purchase of and alliances and franchising with smaller airlines. Together with the strengthening of their hubs through flight coordination, these changes might be seen as defensive (i.e. to deter entry) in the intra-EU context.[33]

Consequently, in the U.S. as well as in many European countries competitive drives in the airline industry mainly come from Low Cost Carriers (LCC), whose business concept is based on reduced service features ('no frills'), very low airfares and a route network centring around smaller hub airports. While in the U.S. this '*Low Cost Airline Service Revolution*'[34] had already taken place in the 90s, in most European countries it is in its infancy. Before 2002 there were no German newspaper headlines such as 'LCCs conquer the skies' or 'A lot of LCC on German skies'. Against this setting, it is no surprise that the German FCO's current assessment of the competitive situation in the German airline industry is quite similar to that of the DoT in the mid 90s:

> While there is reason to be encouraged by the progress low cost carriers have made in penetrating markets at some concentrated network hubs, and the fact that this new service makes economic sense, there remains cause for concern. [...] Allegations of predatory behavior have increased as a result of the recent emergence and growth of a number of low cost, low fare new entrant airlines. Given the incentives and the reality of very high prices for local passengers at certain network hubs, we have to be concerned about possible predatory behavior or unfair competition practices.[35]

The main reasons for these market developments and the justification for growing concerns about predatory practices can be identified in both the supply side or cost related peculiarities and the demand side characteristics of airline passenger services.

Capacity restraints, cost structure and hub-and-spoke 'fortresses' First, despite the deregulation and liberalization, many effective economic and administrative entry barriers still remain. Capacity constraints on essential infrastructures and in particular the limited availability of take off and landing rights (slots), are very relevant entry barriers. The negative economic effects are intensified by national regulatory regimes of slot allocation, which grant special favour to established network airlines.[36] In addition other airport services, such as check-in or ground handling services, turn out to be bottlenecks for successful market entry.

Furthermore, the cost structures of the provision of airline passenger services differ largely from those of other industrial sectors.[37] Given a certain flight

---

[33] Morrell (1998), 48/49.
[34] Department of Transportation (1996).
[35] Department of Transportation (1996), 31.
[36] The relevant legal framework is described by Grundmann (1999), 48 ff./162 ff.
[37] Ehmer (1998), 58 ff.; Wiezoreck (1998), 13 ff.

frequency and seat capacity, the ATC per passenger decreases, with MC at a very low level. Therefore, a high utilization of total seat capacity (load factor) is a crucial element for profitability. However the option to manage a low or decreasing load factor using a reduction of flight frequency is rather limited, since ATC increases when flight frequency and size of aircraft decrease (and vice versa). These 'economies of density'[38] are reinforced by the fact that a reduced flight frequency also lessens the 'quality' of airline passenger services, particularly in the eyes of time-sensitive business travellers. Overall the provision of airline passenger services is characterized by decreasing ATC per passenger, economies of density and the requirement to maintain a minimum flight frequency. Due to these structural features the likelihood that predatory pricing strategies will succeed increases when the costs of predation are being constrained. When a newcomer enters the market ('city pair') at minimum flight frequency, the incumbent may only need to reduce prices for a relatively small portion of total seat capacity.[39] Since the newcomer's option to respond to this strategy via an adjustment of capacity is limited, the newcomer will be confronted with a decreasing load factor, rising ATC per passenger and/or decreasing revenues per passenger. Both the DoJ[40] and the FCO[41] identified these effects resulting from a specific cost structure as core elements and main consequences of predatory pricing strategies of incumbent airlines. The DoJ even referred to internal documents of American Airline (AA), which described and assessed this strategic approach as very promising and profitable.

In addition, the cost structure of airline passenger services has to be seen as one of the main reasons for the evolution of hub-and-spoke route systems by dominant network airlines. Contrary to linear route systems with non-stop flights connecting each point of origin and destination, a hub-and-spoke system is characterized by flights on different spoke routes via the hub. This enables the hubbing airline to attract both local traffic (passengers flying from/to the hub airport as their final destination) and flow traffic (passengers transfering to other flights bound for their destination) and to generate a higher flight frequency in the spoke markets. The implied cost advantages (decreasing ATC per seat and passenger) might be a sufficient incentive to engage in predatory practices. Furthermore, the probability of successful loss recoupment is reasonably high because airfares in spoke markets tend to be significantly more expensive without competition from LCC ('hub premium'[42]). In a recent study the DoT concluded for the US market that 'in dominated hubs as a whole [...] passengers pay on average 41% more than do their counterparts flying in hub markets with low-fare competition'.[43] In its complaint the DoJ particularly mentioned the hub-and-spoke route system of AA as one of the main reasons for its dominant position on the

[38] Ehmer (1998), 63; Crocioni (2000), 28.
[39] The possibility to discriminate between categories of passengers ('yield management') will be discussed in depth in Section 3.2.
[40] Department of Justice (1999), No. 32 ff.
[41] Bundeskartellamt (2002), 5 f.
[42] Department of Transportation (2001a), 27; Oster, Strong (2001), 32.
[43] Department of Transportation (2001b), 1.

routes to and from Dallas/Ft. Worth (DFW).[44] Furthermore, the DoJ notes that the higher probability of successful recoupment is proven by the output and price adjustments of AA after the market exit of competitors:

> In September 1996 Vanguard announced that it would expand its DFW operations [...] Within days of Vanguard's announcements American planned [...] schedule changes [...]. In addition to carrying out the plan American also matched Vanguard's fares [...]. After Vanguard announced that it would be exiting the three DWF spoke routes American promptly increased its fares on those routes. By June 1997 American's [...] average fare had increased by more than 50%.[45]

The FCO also referred to price and capacity adjustments by DLH before and after the market exit of competitors to highlight the prospects of successful recoupment.[46]

Elasticity of demand and marketing instruments With regard to demand side characteristics Joskow and Klevorick stressed the importance of assessing the elasticity of demand of the dominant firm's products.[47] The less elastic the demand for the product the greater the opportunities to extract monopoly profits after the market exit of the prey. Since it seems to be undisputed that the elasticity of demand for passenger airline services is very low for some groups of travellers (in particular business travellers),[48] this prerequisite may be satisfied for airline passenger services. Due to the possibility of engaging in price discrimination in the airline passenger market, the higher demand elasticity of other market segments (leisure travellers) does not lead to a lower overall elasticity of airline passenger services, an aspect which will be discussed later.

In addition dominant hub-airlines have developed marketing instruments to capture business travellers and reduce elasticity of demand even further. The most important instruments in this context are the familiar Frequent Flyer Programmes (FFP), i.e. rebates on ticket purchases merely in place to increase customer loyalty. Furthermore specific marketing instruments targeting travel agents are used to reduce elasticity of demand indirectly. Some airlines pay extra commissions to travel agents (Travel Agents Commission Overrides, TACO's) to encourage them to book passengers on the offerings of a specific airline. In this context TACO's based on the share of the total tickets sold are most effective, since they discourage entry of new competitors. Also preferential access to Computer Reservation Systems (CRS) used by travel agents may have the effect of weakening the allocative function of airfare levels.[49]

The DoJ, as well as the FCO, referred specifically to these demand side characteristics of the airline industry in order to prove that predatory pricing strategies are very likely to be a profitable undertaking in this sector. The DoJ

---

[44] Department of Justice (1999), No. 17.
[45] Department of Justice (1999), No. 36-39.
[46] Bundeskartellamt (2002), 20.
[47] Joskow, Klevorick (1979), 226/227.
[48] Bundeskartellamt (2002), 20 (making reference to some empirical studies).
[49] Grundmann (1999), 86; US-General Accounting Office (1997), 19.

mainly stressed the relevance of FFP and TACO's as entry barriers and factors of market dominance,[50] while the FCO emphasized the FFP as a quality feature of airline passenger services. These elements, together with other service features such as flight frequency and catering, make it unnecessary for a hubbing airline like DLH to match the airfares of LCCs to remain competitive. Consequently, by matching the airfares of its competitor Germania, DLH is effectively undercutting the LCC.[51] This consideration seems to be shared by the DoT:

> Various factors – the unmatchable advantages offered by the dominant airline's more frequent flights and its more attractive override commission and frequent flyer programmes, for example – make it difficult for airlines to compete in another airline's hub markets. An airline without an offsetting advantage of its own, such as the low costs of the low-fare airlines, will find it difficult to operate profitably in such a market.[52]

Not only do the DoT and FCO both regard demand side characteristics of the airline industry as important structural factors leading to higher probability of a successful loss recoupment, but they also share the opinion that these characteristics reduce the costs of predation since the incumbent airline does not have to match the airfares of LCCs in order to drive them out of the market.

Assessment of the case law in the U.S. and Germany According to the two-tier approach the airline industry is largely characterized by the probability of successful (i.e. profitable) predatory pricing strategies. Therefore the negative welfare effects of failing to identify predation are more relevant than the error costs of misinterpreting fierce price competition as predatory pricing. Against this setting the assessment of the pricing strategy of AA by the District Court for the District of Kansas is surprising, in particular the opinion that 'there is no dangerous probability of recoupment'.[53] Nevertheless, this assessment of the market structures in the airline industry seems to be in line with the standards of proof developed by the Supreme Court in the '*Matsushita*' and '*Brooke Group*' cases.[54] Referring to these decisions the Court stated:

> Even assuming American engaged in below-cost pricing on relevant routes, the government must prove a dangerous likelihood that American would subsequently recoup its losses by supra-competitive pricing. [...] In Brooke Group the Court noted the 'general implausibility' of predatory pricing and the underlying notion of recoupment. Accordingly the Court set forth a standard of proof which is expressly 'not easy to establish'. In setting this high standard, the Supreme Court cleared the way for summary rejection of most predatory pricing claims. That is also the appropriate result in this case.[55]

---

[50] Department of Justice (1999), No. 17.
[51] Bundeskartellamt (2002), 5/7.
[52] Department of Transportation (2001a), 38.
[53] District Court for the District of Kansas (2001), 140.
[54] Section 2.2.
[55] District Court for the District of Kansas (2001), 121.

Even internal documents presented by the DoJ did not convince the Court since 'evidence alleged to recoup is insufficient where objective evidence fails to indicate the dangerous probability of recoupment'.[56] The most important point of the analysis, however, is the implicit assertion of the Court that successful recoupment requires '*supra-competitive pricing*'. This standard can not only be refuted by theoretical arguments[57] but also implies the necessity to prove abusively high prices to affirm a predatory pricing strategy.

In addition, the District Court does not take into account most of the other structural features of the airline industry presented above. First, the Court did not accept the fact that predation might be profitable due to the reputation effects in other relevant markets. In line with the Supreme Court it restricts its assessment to the probability of recoupment on the specific relevant market where the alleged predatory pricing takes place.[58] Furthermore, the District Court interprets the '*Meeting Competition Defence*' in a rather strict way. The assessment of the price levels is limited to 'published airfares' and does not consider the quality of the services provided.[59] Moreover, the Court not only neglects the network airlines' strategic room to manoeuvre given their hub-and-spoke route systems, but also does not further consider structural features like the low elasticity of demand, the common marketing instruments and, as a consequence, the option to drive a LCC out of the market without undercutting its published airfares. In its appeal of the District Court's decision the DoJ has reinforced its arguments concerning the dangerous likelihood of recoupment and the irrelevance of the meeting competition defence in the specific case.[60] Just as the FCO in the DLH case the DoJ stated 'American cannot avail itself of any meeting competition defense, because it undercut its rival's prices by offering greater value. Its frequent flyer programme offers significant value to passengers'.[61] However as the Court of Appeals primarily focused on which may be the relevant cost-price-test applied, the DoJ's arguments were not discussed in any depth in the final decision.

In contrast, the regulatory approach of the DoT against predatory practices in the airline industry is expressly based on Joskow and Klevorick's two-tier approach.[62] Such different approaches in decision making have important implications, which become clear when considering for example the divergent interpretation of the '*Meeting Competition Defence*'. On the one hand, the DoT concludes that 'the Department should be reluctant to pursue cases where the incumbent has only matched low airfares offered by a new entrant'.[63] On the other hand, the DoT states that 'given the other service features offered by network airlines [...] matching the airfares offered by a low-fare airline may amount to

---

[56] District Court for the District of Kansas (2001), 122.
[57] Section 2.1.; Hemphill (2001), 1594.
[58] District Court for the District of Kansas (2001), 131/132.
[59] District Court for the District of Kansas (2001), 114.
[60] Department of Justice (2002),
[61] Department of Justice (2002), 28.
[62] Department of Transportation (2001a), 16.
[63] Department of Transportation (2001a), 175.

undercutting the fare'.[64] In addition, the approach of the DoT seems to be closely linked to a shift of the burden of proof from government to the investigated airlines.

> If the Department is considering enforcement action, it should always give the airline being investigated the opportunity to show that its competitive response was legitimate, even if aggressive. [...] If the airline can show that it responded to entry in ways that were economically rational whether or not the entrant ended or reduced its services, the Department should not take action.[65]

The DoT's approach towards predatory practices seems to be a reaction to the case law developed by the courts and in particular to the development of an almost prohibitive standard of proof. Since the DoT has not yet had the opportunity to take action against alleged predation in a particular case, it is an open question whether this approach risks high error costs of state intervention.

Competition law enforcement in Germany is based on a very different standard of proof. In accordance with the case law of the Bundesgerichtshof (Federal Court of Justice, FCJ) the Court of Düsseldorf stated in the DLH case that 'the assessment whether Lufthansa abused its dominant position through predatory practices targeted against Germania [...] depends on circumstantial evidence as well as on an overall consideration of both subjective and objective criteria'.[66] Consequently, the competition authority in Germany does not have to give evidence regarding each detail of the case to prove a predatory pricing strategy. According to Section 19 Par. 4 No. 1 of the Act against Restraints of Competition (ARC), an 'impairment of the ability to compete of other undertakings' is prohibited as an exploitation of a dominant position only if it takes place 'without any objective justification'. Therefore, conforming with German competition law, the assessment of a specific pricing strategy crucially depends on whether an 'objective justification' can be assumed.

According to case law this assessment has to be based on,

> striking the balance between the interests of the undertaking affected (the prey) and those of the dominant firm by particularly emphasizing the objective of the ARC to safeguard market structures indispensable for effective competition and the chances of successful market entrance.[67]

Evidently an evaluation of the specific structural features of the relevant market (i.e. the first stage of the two-tier approach) must be one of the core elements of this assessment. The specific structural characteristics of the affected market are expressed via the relative weight given to the interests of predator and prey. In the DLH case the Court of Düsseldorf concluded that 'taking into account

---

[64] Department of Transportation (2001a), 175.

[65] Department of Transportation (2001a), 67.

[66] Oberlandesgericht Düsseldorf (2002), 601/602 (WuW/E DE-R 867, 869); English version not available; translation by the author.

[67] Oberlandesgericht Düsseldorf (2002), 602 (WuW/E DE-E 867, 870); English version not available; translation by the author.

the objective of the ARC to safeguard competition and uphold the probability of future market entrances, the interests of the newcomer Germania outweigh (…) the interests of the dominant firm Lufthansa'.[68] Contrary to the American District Court, the Court of Düsseldorf largely accepted the FCO's assessment of the specific structural characteristics of the airline industry, implying that successful predatory pricing strategies are highly probable.

As mentioned above, one of the main consequences of the different standard of proof used in Germany and in the U.S. relate to the different interpretation of the '*Meeting Competition Defence*'. Contrary to the judgement of the District Court, the Court of Düsseldorf accepted the FCO's view resulting from taking into account the different service features of the offerings of DLH, DLH's nominal matching of its competitor's airfares, DLH reacted in an objectively unjustified manner. Consequently, the Court of Düsseldorf mainly focused on the examination of the minimum price distance to be kept by DLH. Accepting the assessment of the FCO in general, the Court only slightly reduced the amount defined by the FCO.[69]

Assessment of Price Levels

Peculiarities of the airline industry and suitability of cost indicators Models of yield and revenue management developed by most dominant network airlines largely characterize price setting in the airline industry.[70] The aim of these pricing strategies is to offer seats at different prices and conditions and to manage the number of seats in each of these categories so as to obtain maximum revenues per flight, the ultimate goal being to skim off the maximum willingness to pay off each group of passengers. The different price categories of these resulting complex fare structures depend on such aspects as: how far in advance the customer books his flight or on other restrictions imposed on the client (for example weekend or minimum stay and restricting the ability to change the booking without penalty). Therefore, a price reduction may result from merely dropping a specific restriction, which used to characterize a specific price category. The number of available seats in each price level is flexible. On the one hand, this enables the airline to adjust seat capacities according to actual demand and consequently to effectively control and optimize the utilization of total seat capacity. On the other hand, the same mechanism represents the lever to influence the load factor of competitors with the aim of driving them out of the market. Therefore, in order to correctly assess the airfares of network airlines, it is always necessary to take ticket restrictions into account. In the DLH case, the new tariff introduced by DLH on the Frankfurt–Berlin route entailed the same restrictions as DLH's fully flexible business tariff but was significantly cheaper.[71] Consequently, by introducing the new tariff DLH

[68] Oberlandesgericht Düsseldorf (2002), 605 (WuW/E DE-R 867, 873); English version not available; translation by the author.
[69] Oberlandesgericht Düsseldorf (2002), 603 ff. (WuW/E DE-R 867, 871).
[70] Wiezorek (1998), 21; Oster, Strong (2001), 22 ff.; Department of Transportation (2001a), 23 ff.
[71] Bundeskartellamt (2002), 3/4.

'cannibalized' its revenues from less price sensitive business traveller tickets without getting a better utilization of seat capacity.

The appropriate definition of the prices to be compared is, however, not the only difficult issue in the application of cost-price tests to airline passenger services. These tests are also closely linked to limitations concerning the suitability of the different cost indicators discussed. MC, i.e. the additional cost necessary to produce one additional unit of output, may send wrong signals or even contradict the purpose of the ARC to protect competition by the means of abuse control. Since MC per seat is close to zero, the incumbent might even make free tickets available without violating the MC-rule. Furthermore, to the extent that the incumbent adds capacity in response to market entry, the question arises whether MC per seat or MC per flight should be the relevant indicator. The application of the AVC – or avoidable-cost test is complicated by the fact that seat capacities can quite easily be reallocated to another route of the network. While, for the route system as a whole, capital costs of aircraft deployed appear fixed, on a specific route they have to be seen as variable costs. For the same reason it is not easy to define these costs, which can be avoided once an airline stops to provide passenger services on one route of a complex and interrelated route system.

Current legal practice in the U.S. and Germany The pricing strategy of AA was characterized by matching the airfares of competitors and by expanding flight capacities. Against this setting the DoJ primarily focused on marginal revenues and MC of an added flight:

> The additional revenues American obtained as a result of adding capacity [...] were less than American's costs of adding the flights as demonstrated in various ways, including the following: a) On each route, American's revenues on one or more of its added flights were below that flight's variable costs; b) On each route, as a result of adding capacity, American's total revenues on the route fell below American's total cost of serving the route [...]; and c) On each route, American's operation were unprofitable under its own measure of route profitability after it added capacity.[72]

The District Court did not accept this cost related assessment of AA's pricing strategy. Taking account of different expert opinions and referring to the case law of the Supreme Court the District Court stated: 'American priced its products above its average variable costs, the only appropriate, credible measure of costs in the present case'.[73] Despite the fact that the case law of the Supreme Court refers to prices 'below an appropriate measure of costs'[74], the District Court qualified the assessment of the DoJ as 'inconsistent with existing law',[75] an assessment that was confirmed by the Court of Appeals.[76]

The regulatory approach of the DoT, on the other hand, abstains from using cost-price tests. The DoT concluded 'that a profit-maximization standard would

---

[72] Department of Justice (1999), No. 50.

[73] District Court for the District of Kansas (2001), 98.

[74] District Court for the District of Kansas (2001), 99.

[75] District Court for the District of Kansas (2001), 99.

[76] U.S. Court of Appeals (2003), 17.

give better guidance than a cost standard in determining when a hubbing airline's response to new competition appears to be an unfair method of competition'.[77] Given that the burden of proof is on the incumbent airline it will need to prove that a specific pricing strategy is compatible with the aim of profit maximization. At this point it becomes obvious that the almost prohibitive standards of proof developed by U.S courts triggered the development of a risky regulatory approach as it may lead to a hardly justifiable intensity of market intervention.

In its decision against DLH, the FCO stated that the new tariff introduced by DLH on the Frankfurt–Berlin route was below ATC per passenger.[78] But as mentioned above, the main focus of the decision centres around the fact that DLH did not meet competition by matching Germania's airfares. Furthermore, it has to be borne in mind that the interpretation of German competition law by the courts requires an overall assessment of the competitive response of an incumbent airline. The Court of Düsseldorf accepted the assessment of the FCO and stated:

> Based on a cost analysis and referring to average cost per passenger the previous calculations show that by applying its new tariff DLH runs a deficit. In assessing the tariff as an isolated measure of a dominant firm it has to be qualified as a loss-price strategy being characteristic for predatory pricing. This assessment can not be weakened by DLH by referring to the fact that due to the high quota of fixed costs the marginal costs of an additional passenger are very low.[79]

Obviously, the Court of Düsseldorf did not only accept the assessment of the FCO based on ATC per passenger, but also expressly rejected the MC-test as relevant benchmark for the airline industry. Once again it needs to be stressed though that German law and the courts' case law require an overall assessment of a pricing strategy to be qualified as predatory pricing.

4. Conclusion

Measured against the criteria of the two-tier approach, the airline industry seems to be an economic sector characterized by a higher probability of success of predatory pricing strategies than most other sectors. Therefore, following the conclusions of Joskow and Klevorick, the airline industry is a sector in which predatory pricing proceedings carried out by competition authorities are less likely to be more harmful than positive. The causal structural features of the market are: economic and regulatory capacity restraints, specific cost structures, hub-and-spoke route systems and low elasticity of demand tied to common marketing instruments such as FFPs or TACO.

---

[77] Department of Transportation (2001a), 73.

[78] Bundeskartellamt (2002), 12.

[79] Oberlandesgericht Düsseldorf (2002), 605 (WuW/E DE-R 867, 871); English version not available; translation by the author.

The current case law in the U.S. neither consider the structural peculiarities of the airline industry, nor the consequences of an assessment of potentially abusive predatory pricing strategies. The courts mainly refer to the '*Brooke Group*' and '*Matsushita*' judgements of the Supreme Court, which are explicitly based on the theoretical framework of the Chicago School. The resulting high standards of proof led to the development of a regulatory policy approach by the Department of Transportation against predatory practices. This approach however risks leading to an intensity of state intervention far greater than would be resulting from the enforcement of competition law which specifically considers the structural peculiarities of the relevant market. The current German enforcement practice and case law take better account of the intrisic structural features of airline passenger services., The more realistic standards of proof however are not the result of a reference to the two-tier approach, but are largely the consequence of a 'model' of competition law, which substantially differs from the Chicago-Model. Due to the explicit need to show that the incumbent has an 'objective justification' for its potentially abusive competitive conduct, courts will, in case of doubt, give priority to the protection of competitive market structures and the maintenance of future market entrance over the incumbent's freedom of price setting.

References

Articles Published in Journals or Anthologies

Areeda, Phillip, Turner, Donald (1975), Predatory Pricing and Related Practices under Section 2 of the Sherman Act, in: *Harvard Law Review*, 88 (1975), S. 697-733.

Baumol, William J. (1979), Quasi-Permanence of Price Reduction: A Policy for Preventing Predatory Pricing, in: *Yale Law Journal*, 89 (1979), S. 1-26.

Baumol, William J. (1996), Predation and the Logic of the Average Variable Cost Test, in: *The Journal of Law and Economics*, 39 (1996), S. 49-72.

Bergman, Mats (2001), A Prohibition against Losses? The Commission's Deutsche Post Decision, in: *European Competition Law Review*, 2001, S. 351-354.

Blair, Roger, Harrison, Jeffrey (1999), Airline Price Wars: Competition or Predation?, in: *The Antitrust Bulletin*, 1999, S. 489-518.

Bolton, Patrick, Bradley, Joseph F., Riordan, Michael H. (1999), Predatory Pricing: Strategic Theory and Legal Policy, Boston University School of Law Working Papers, Nr. 5-1999, Boston 1999; *Georgetown Law Journal*, August 2000, S. 2239 ff.

Crocioni, Pietro (2000), Defining Airline Markets: A Comparison of U.S. and EU Experiences, in: *The Antitrust Bulletin*, 2000, S. 1-54.

Easley, David, Masson, Robert, Reynolds, Robert (1985), Preying for Time, in: *Journal of Industrial Economy*, 33 (1985), S. 443 ff.

Ehmer, Hansjochen (1998), *Der Wettbewerb im Linienluftverkehr mit kleinen Verkehrsflugzeugen*, Göttingen 1998.

Grundmann, Silvia (1999), *Marktöffnung im Luftverkehr – Hoheitliche Eintrittsbarrieren in den USA und in der EG*, Baden-Baden 1999.

Hemphill, Scott (2001), The Role of Recoupment in Predatory Pricing Analysis, in: *Stanford Law Review*, 53 (2001), S. 1581-1612.

Hüschelrath, Kai (2003), *How to Prey on Predators? – The Lufthansa-Germania Case*, WHU Graduate School of Management, Institute for Industrial Organization, unpublished paper, March 2003.

Joskow, Paul, Klevorick, Alwin (1979), A Framework for Analysing Predatory Pricing Policy, in: *Yale Law Journal*, 89 (1979), S. 213-270.

Kate, Adriaan Ten, Niels, Gunnar (2002), On the Rationality of Predatory Pricing: The Debate between Chicago and Post-Chicago, in: *The Antitrust Bulletin*, Spring 2002, S. 1-24.

Klevorick, Alwin (1993), The Current State of the Law and Economics of Predatory Pricing, in: *American Economic Review*, 83 (1993), S. 162-167.

Kreps, D.H. and Wilson, R. (1982), Reputation and Imperfect Information, in: *Journal of Economic Theory*, 27 (1982), S. 253-279.

Levine, Michael E. (1987), Airline Competition in Deregulated Markets: Theory, Firm Strategy and Public Policy, in: *Yale Journal of Regulation*, 1987, S. 393-494.

McGee, John (1958), Predatory Price Cutting: The Standard Oil Case, in: *Journal of Law and Economics*, 1 (1958), S. 137-169.

McGee, John (1980), Predatory Pricing Revisited, in: *Journal of Law and Economics*, 23 (1980), S. 289-330.

Milgrom, Paul, Roberts, John (1982), Predation, Reputation and Entry Deterrence, in: *Journal of Economic Theory*, 27 (1982), S. 280-312.

Morrell, Peter (1998), Air Transport Liberalization in Europe: The Progress So Far, in: *Journal of Air Transport World Wide*, 3 (1998), S. 42-60.

Nicolaides, P. (2001), Effective Competition in Network Industries: An Assessment of the Commission Decision 2001/354 Imposing a Fine on Deutsche Post for Abusing its Dominant Position, in: *European Competition Law Review*, 2001, S. 390-393.

Niels, Gunnar, Kate, Adriaan Ten (2000), Predatory Pricing Standards: Is There a Growing International Consensus?, in: *The Antitrust Bulletin*, 2000, S. 787-809.

Oster, Clinton V., Strong, John S. (2001), *Predatory Practices in the U.S. Airline Industry, Washington 2001* (abrufbar auf der Internetseite des Sekretariats des DoT unter: http://ostpxweb.dot.gov/aviation/domestic-competition; Zugang über: www.dot.gov).

Phlips, Louis (1995), *Competition Policy: A Game-Theoretic Approach*, Cambridge 1995.

Posner, Richard (1976), Antitrust Law: An Economic Perspective, Chicago 1976; *Journal of Reprints for Antitrust Law and Economics*, 10 (1980), S. 83-97.

Selten, Reinhard (1978), The Cain-Store Paradox, in: *Theory and Decision*, 9 (1978), S. 127-159.

Starkie, David (1999), The US Department of Transportation's Statement on Predatory Conduct in the Airline Industry: An Alternative Approach, in: *European Competition Law Review*, 20 (1999), S. 281-286.

Telser, Lester G. (1966), Cutthroat Competition and the Long Purse, in: *Journal of Law and Economics*, 9 (1966), S. 259-277.

Wiezorek, Bernd (1998), Strategien europäischer Fluggesellschaften in einem liberalisierten Weltluftverkehr, *Diskussionsbeiträge zur Wirtschaftspolitik*, Nr. 87, Dortmund 1998.

Williamson, Oliver E. (1977), Predatory Pricing: A Strategic and Welfare Analysis, in: *Yale Law Journal*, 87 (1977), S. 284-339.

Zerbe, Richard, Mumford, Michael (1996), Does Predatory Pricing Exist? Economic Theory and the Courts after *Brooke Group*, in: *The Antitrust Bulletin*, 1996, S. 949-985.

Decisions and Papers of Governmental Institutions

Bundeskartellamt (2002), *Beschluss im Verwaltungsverfahren gegen Deutsche Lufthansa AG*, Köln, vom 18.2.2002, Geschäftszeichen: B9-144/01 (www.bundekartellamt.bund.de).

Department of Justice (1999), *Complaint, Civil Action No. 99-1180-JTM*. United States of America vs. AMR Corp., American Airlines Inc., American Eagle Corp., Washington 1999 (www.usdoj.gov/atr).

Department of Justice (2002), *Reply brief for appellant United States of America*; Case-No. 01-3202, Washington 2002 (www.usdoj.gov/atr).

Department of Transportation (1996), *The Low Cost Airline Service Revolution*, Washington 1996 (http://ostpxweb.dot.gov/aviation/domestic-competition; Zugang: www.dot.gov).

Department of Transportation (2001a), *Enforcement Policy Regarding Unfair Exclusionary Conduct in the Air Transport Industry – Findings and Conclusions on the Economic, Policy and Legal Issues*, Docket AST-98-3713, Washington 2001 (http://ostpxweb.dot.gov/aviation/domestic-competition; access: www.dot.gov).

Department of Transportation (2001b), *Dominated Hub Fares*, Washington 2001 (http://ostpxweb.dot.gov/aviation/domestic-competition; access: www.dot.gov).

European Commission (2001), Entscheidung der Kommission vom 20.3.2001, Fall COMP/35.141 – Deutsche Post AG, *Amtsblatt der EG*, L 125 vom 5.5.2001, S. 27 ff.

European Court of Justice (1991), Urteil der Gerichtshofes vom 3. Juli 1991, Rechtssache C-32/86 (AKZO/Kommission), *Sammlung 1991*, S. I-3359 ff.

European Court of Justice (1996), Urteil der Gerichtshofes vom 3. Juli 1991, Rechtssache C-333/94 (Tetra Pak II), *Sammlung 1996*, S. I-5951 ff.

Oberlandesgericht Düsseldorf (2002), Beschluss im Kartellverwaltungsverfahren Deutsche Lufthansa AG/Bundeskartellamt vom 27.3.2002 (Kart 7/02 (V)); *Wirtschaft und Wettbewerb*, 6/2002, S. 599-607 (WuW/E DE-R 867).

OECD (1989), *Predatory Pricing*, Paris 1989 (www.oecd.org).

U.S. Court of Appeals – 10th Circuit (2003), Order, *Case-No. 01-3202*, July 2003 (www.kscourts.org).

US-District Court for the District of Kansas (2001), Memorandum and Order, Civil Action No. 99-1180-JTM, *United States of America vs. AMR Corp.*, American Airlines Inc., Eagle Holding Corp., April 2001 (www.usdoj.gov/atr).

US-General Accounting Office (1990), *Airline Competition – Industry Operating and Marketing Practices Limit Market Entry*, Dokument GAO-RCED-90-147, Washington 1990 (www.gao.gov).

US-General Accounting Office (1997), *Airline Deregulation: Barriers to Entry Continue to Limit Competition in Several Dominated Hubs*, Dokument GAO-RCED-97-4, Washington 1997 (www.gao.gov).

US-General Accounting Office (1999), *Airline Deregulation – Changes in Airfares, Service Quality, and Barriers to Entry*, Dokument GAO-RCED-99-92, Washington 1999 (abrufbar unter: www.gao.gov).

Chapter 11

The European 'No-Frills'-Aviation Market: Current and Future Developments

Bernard Michael Gilroy, Elmar Lukas and Tobias Volpert

1. Introduction

In the course of the last thirty years the volume of global airline passengers has increased six fold. Currently some one and a half billion passengers travel by air annually. Historically, growth in the aviation industry has often been limited by restrictive governmental legislation. In recent times, however, the deregulation of the aviation industry has encouraged underlying competition among airlines.

In this innovative competitive environment new structures and strategies are emerging as reflected by the activities of the low cost carriers (LCCs) or low cost airlines, or simply 'No-Frills' airlines. In the USA during the eighties, 'Southwest Airlines' held a predominant position in these observed developments. Similarly, after the European aviation deregulation process began in 1991, the Irish airline 'Ryanair' excelled to be followed by other airlines such as easyJet, Go-fly, germanwings, or Buzz as they competed for market share and passengers successfully triggering a downright aviation boom. Berster and Wilken (2004), taking account of various information sources on seat capacity and load factors cautiously estimate the market penetration of LCCs in Germany at the beginning of the year 2004 to be some 20%. Indeed, they find strong indications that LCC type services will strongly increase in Germany and Europe and forecast a market share of 40 to 50% in the long run.

The focus of this chapter is thus to offer a concise survey on these dramatic developments exemplified by the European LCC market. In section two the historical path of the European aviation deregulation process is briefly discussed, which ultimately resulted in the formation of over twelve 'no-frill' airline carriers (mainly in Great Britain and Germany). Section three subsequently illustrates the activities of these airline carriers and highlights the major European LCCs: Ryanair, easyJet, Germanwings, Hapag-Lloyd-Express (HLX), Air Berlin and Germania. Section four brings to mind the fundamental economic foundations inherent in the current day activities of LCCs. The analysis characterizes and compares the situation of the LCCs and traditional scheduled carriers with regard to future chances and dangers involved from a competitive economic viewpoint, and pays specific attention to the possibility of predatory behaviour. Illustrative examples are presented where possible. Although many of the presented arguments and elements are applicable to other non-European markets, important distinctions remain which have to be taken into consideration.

2. Deregulation of European Airlines

The liberalization of American and European air transport took place independently and during different time periods. North-America initiated its deregulation process in 1978 by launching the 'Airline Deregulation Act'. In Europe, however, the deregulation process took another direction. The first step towards liberalization in Europe was when new air services agreements were negotiated between pairs of European countries in the early eighties. In particular, Great Britain led the way in renegotiating new bilateral agreements. The UK-Netherlands agreement, signed in 1984, in particular set the pattern for all subsequent European bilateral agreements thereafter. These agreements consist of elements that alleviate market entry, partially offset capacity and frequency restrictions as well as allow prices to be freely set by the airline companies.[1]

During the same period, the European Commission's Directorate General of Transportation adopted the July 1983 Council Directive on Inter-regional Air Services. Subsequently, airlines were allowed to freely develop air routes between European airports. However, the routes were restricted to those between regional airports and thus excluded air routes from regional cities to capitals or major hubs.[2]

However, neither of these actions significantly affected the liberalization of European air transport as a whole. It was not until the European Council of Ministers decided to launch three 'liberalization packages' that basic conditions evolved setting the pace for liberalization on a macro scale. The first liberalization package, the 'December 1987 package', relaxed the quota sharing of capacity by airlines of each state and allowed the multi-designation of frequent routes. In addition, airlines were allowed to establish fare zones in order to take advantage of pricing flexibilities.[3]

On November 1st 1990, the second liberalization package came into operation. It allowed for the designation of several carriers on particular routes and further relaxed requirements for fare approvals given by national governments. In addition, quota sharing of passengers was progressively abolished. Notwithstanding these actions, the second liberalization package accommodated only for marginal improvements.[4]

However, it was not until January 1st 1993, when the third package was launched, that the European air transportation market really achieved market liberalization. From then on airlines from the corresponding member states could operate with full traffic rights on any route within the European Union. Even on routes outside the airlines' home country the airline faced no capacity restrictions. The key elements of the third package are as follows:[5]

---

[1] See Doganis 2002, p. 55f.
[2] See Doganis 2001, p. 57.
[3] See Council Directive 83/416/EEC.
[4] See Niejahr 1999, p. 59.
[5] See Doganis 2001 p. 66f.

- Governments may impose restrictions only on environmental, infrastructure capacity, regional development or public service grounds.
- There are no price controls and capacity restrictions.
- Complete freedom exists with respect to the determination of fares and cargo tariffs.
- Companies from member states are granted the right to set up and operate an airline in any EU member state or to buy such an airline.
- Criteria that grant operating licences and air operators' certificates are harmonized.

However, due to the resistance of the stronger EU countries, the remaining cabotage limitations were not resolved from the beginning.[6] It took four more years until all restrictions were abolished.

Parallel to these developments the European Union designated competition rules in order to prevent monopolistic practice or anti-competitive behaviour. Three areas were primarily targeted, namely cartels and restrictive agreements, monopolies and mergers and state aid/subsidies to producers.[7] The following Table 11.1 summarizes the main distinctive features of the three liberalization packages.

Today, the European airline-market is fully deregulated. Theoretically speaking, every airline is allowed to offer services within the EU with no restrictions either on price and routes provided. However, reality is full of market entry barriers such as e.g. provisioning, diversified booking systems or frequent flyers programmes. Governmental restrictions are tolerated only with respect to scarcity of infrastructure, technical standards, or environmental standards. In suspicious cases regarding predatory pricing, however, the liberty of unrestricted fare price-setting may be amended.[8]

It is worthwhile noting that the European liberalization policy has only further accommodated existing inter-European bilateral agreements and international regulations. Air traffic from and to the EU is still very restricted, although 'open-sky' agreements with several countries, such as the United States and Canada, also exist.

As expected, liberalization increased competition intensity in the European airline market. Between 1993 and 1996, 80 new airline companies started operations, while 60 others abandoned their operations and services or merged with other companies.

Finally, by the implementation of full cabotage in 1997, the European air transport market counted some 155 airline companies. These companies covered 9.1% of the world-wide air traffic from and to Europe. In contrast, the North-American market hosted 55 airlines which shared 31% of the global air traffic market.[9] Beside this, another driving force giving rise to more airlines was the success of charter airlines, which specialized on targeting tourist destinations. With respect to the total amount of passengers, Thomas Cook, Britannia and LTU are

---

[6] See Lewis 1992, p. 42.

[7] See Doganis 2001 p. 67.

[8] See Sterzenbach 1999, p. 40.

[9] See AEA Jahrbuch 1998 and 1999, p. 20 and p. 4.

among the market leaders.[10] Apart from this, a number of independent airlines exist, serving the interregional traffic between secondary airports.[11] These airlines however, strongly depend on the travel business and on regional airline companies like e.g. the German company Eurowings, as well as the Austrian Tyrolean Airways.

Table 11.1 Constitutional features of the three European liberalization packages

Feature	1. Package (1.1.88)	2. Package (1.11.90)	3. Package (1.1.93)
Tariff Approval	Double approval	Double disapproval from 1.1.1993 on	Double disapproval (only both community members can block)
Fare Zones	Economy ---- Discount zone 95-65% Super discount zone 65-45%	105-95% 94-80% 73-30%	No limitations
Capacity	55:45 (until 30.09.89) 60:40 (until 31.10.90)	67.5:32.5 (until 31.10.91) 75.0:25.0 (until 31.10.92)	No capacity controls
Market Access	Relaxation of access for Community air carriers to intra-community air routes (except routes between primary airports), more regional services	Relaxation of access for Community air carriers to intra-community air routes. Limitations valid in case of capacity shortage.	Free entry, limited cabotage (until 31.03.1997)
5th Freedom	Up to 30% of supplied capacity	Up to 50% of supplied capacity	Approved
Designation	Multiple designation: 1988 > 250,000 1989 > 200,000 1990 > 180,000 Passengers per flight per route	Multiple designation: 1991 > 140,000 1992 > 100,000 Passengers per flight per route	Multiple designation

Source: Grundmann, 1998, p. 136.

[10] See Baker 2000, p. 78.
[11] See Wiezorek 1998, p. 114.

In addition, deregulation increased the privatization of companies. What followed was that the state-owned Flag Carriers were ousted by private companies. It is noteworthy that airlines like Air France (56%), Aer Lingus (95%) or Alitalia (53%) are still under governmental control, while airlines like Lufthansa, British Airways, or Iberia are not. The abolition of air tariff approval procedures due to liberalization drastically lowered prices. This put price competition in operation, establishing the image of air travel as a common mass transportation means instead of the historical image of being a luxury service item. Consequently, consumers have become more price sensitive than in the past.

Since the mid 90s these prerequisites gave rise to a new form of airline companies. Starting in Great Britain and Ireland, these 'no frills' airlines or Low Cost Carriers (LCC) started penetrating the European continent.

3. The European 'No-Frills' Market Players

Overview and Owner Structure of European Low-Cost Carriers

Ten years ago, it was not predictable that with the foundation of Ryanair and its low cost/no frills strategy, an irreversible process would take place that would reshape the European airline industry. The LCCs' path of success has its origin in the routes between Great Britain and Ireland. In early 2001 these companies began to expand their activities to continental Europe by establishing several bases.[12] Due to further deregulation efforts, the capacity supplied within this segment rose on average 48% annually between the years 1994 and 2001. The current market share of no-frills airlines with respect to inter-European air traffic is 7%.[13]

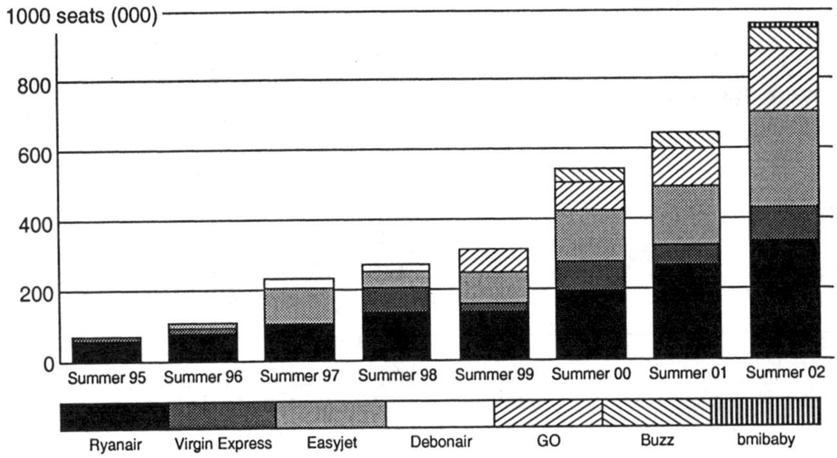

Figure 11.1 'No-Frills' Carriers' Weekly Seats, Source: AEA Yearbook, 2002, pp. I – 4, Note: Airline Summer Season is Apr. 1 – Oct.

[12] See Borghetto et al. 2002, p. 2.

[13] See Sparaco 2002a, p. 60.

Serving as an example for the growth of European LCCs, Figure 11.1 depicts the expansion of the British no-frills airlines with respect to their total annual capacity provided. Similarities are apparent between the European LCCs strategies; they all rest on the original basic business model of the US Southwest airlines implemented by Herb Kelleher and Rollin King in 1971.

Table 11.2 Overview of European no-frills airlines

Airline	Fleet size (type)	Destinations	HUBS	Owner
easyJet Go-fly	64 (B737)	35	London Gatwick/ Luton/Stansted, Liverpool, Amsterdam, Genf, Paris Orly/Charles de Gaulle, East Midlands, Bristol	Independent
Ryanair	52 (B737)	58	London Stansted, Glasgow Prestwick, Brussel Charleroi, Frankfurt Hahn, Milan Orio al Serio, Paris Beauvais, Shannon, Dublin, Stockholm Skavsta	Independent
Buzz	10 (8 BAE146 and 2 B737)	22	London Stansted	Ryanair, formerly KLM
Virgin Express	14 (B737)	15	Bruxelle	Virgin Group
Germanwings	10 (A319)	19	Cologne-Bonn	Eurowings/Lufthansa
Hapag-Lloyd Express	10 (B737)	17	Cologne-Bonn	TUI
My Travel Lite	2 (A320)	7	Birmingham, Belfast	My Travel
Bmi-baby	7 (B737)	26	East Midlands, Cardiff, Manchester	BMI British Midland
Basiq Air	26 (B737)	13	Amsterdam, Rotterdam	KLM
Air Berlin	32 (B737)	6 in City Shuttle as LCC	10 German airports	Independent
Deutsche BA	16 (B737)	10	Munich	British Airways
Germania	8 (B737) 19 (Fokker-F-100)	84	Berlin	20% TUI

According to Doganis (2001a) the success of European LCCs is based upon 'the provision of easily accessible scheduled short-haul services at very low unrestricted fares close to those of charter airlines but with "no frills", that is, without many of the traditional product features of either scheduled or charter services'.

Initially, the European market for LCCs was accounted for by five companies, namely Ryanair, easyJet, Virgin Express, Go-fly and Buzz. While most of the low-cost carriers are independent, a few differ from this and are owned by major airlines or travel agencies. As the following Table 11.2 depicts, KLM, Lufthansa and TUI have a foot in the 'no-frills market' door by owning majorities of certain European LCCs. Furthermore, the table summarizes information about the corresponding fleet size as well as the LCCs hub location and destinations.

Company Profiles of Major European LCCs

The aim of the following subsections is to characterize the two major European markets for LCCs, i.e. Great Britain and Germany. General emphasis is placed on each of the countries' biggest market players. With respect to Great Britain these are Ryanair and easyJet whereas for Germany respectively, Germanwings, Hapag-Lloyd-Express, Air Berlin and Germania.

Ryanair Ryanair was the first European airline company that can be assigned to the Low-Fare segment. The company was founded in 1985 by the Irish businessman Tony Ryan with the intention to tear down the duopoly market power of Aer Lingus and British Airways. Initially, Ryanair operated like a traditional airline, offering full service during their flights, yet with low fares. However, unclear strategic positioning, lack of cost control and a fleet consisting of four different plane types culminated in a price war with Air Lingus that ended shortly after in a 10 million pound deficit.[14]

This situation called for a turnaround in strategy. Consequently, the thirty-year-old Michael O'Leary was put in charge as the new CEO of Ryanair in January 1991. Because O'Leary was very familiar with the strategy of the US LCC Southwest Airlines, he decided to adopt this kind of business strategy and started to imitate the concept.[15] The route network was dramatically decreased from 19 to 5 routes and the fleet was concentrated on a single type, the Boeing 737-200. In addition, free on-board service was no longer offered and the emphasis was put on attracting the market segment of Visiting Friends and Relatives[16] (VFR) between Ireland and Great Britain.[17] Consequently, Ryanair's market share on this route rose to 40% and the overall financial situation sustainably improved.

---

[14] See Rothfischer 2002, p. 58.

[15] See Schmitt 2001, p. 4.

[16] The VFR segment is a subcategory of the spare-time traveller. The other segment belongs to the vacation traveller.

[17] Another target group were business men travelling between both countries. See Mason et al. 2000, p. 33.

With the end of liberalization, Ryanair expanded not only its fleet continuously but also its route network. At the end of February 2001, the company opened up its first base on continental European ground. Similar to British and Irish airports, the Bruxelle-Charleroi airport is of secondary type equipped with ramp handling times of 25 minutes allowing for more take-offs a day than is possible at major airports like, for example London-Heathrow.

Currently Ryanair possesses five continental European hubs and four on the British and Irish isles. In sum, Ryanair operates on 82 routes serving 58 different destinations. With a total of 1700 employees, Ryanair's net income in March 2003 was 230 million euro. Approximately 13 million passengers travelled with Ryanair and the company expects 27 million more in 2010.[18] In order to be able to handle this fast growing demand, Ryanair ordered 22 more Boeing 737-800 with an option to buy a further 78 airplanes. The competitive situation was further improved by the acquisition of Ryanair's rival Buzz in February 2003. It is now able to serve 17 million passengers a year and to compete directly with its major competitor easyJet. In order to further economise, Ryanair has just announced its plan to make drastic cuts with respect to each passenger's accompanied luggage weight. The long-term strategy is to transport only those passengers who use carry-on luggage. In this manner, costly security checks can be avoided, thereby reducing employment costs. It is estimated that such a strategy will save one-third of the overall ground airport costs.

easyJet As opposed to Ryanair, easyJet followed the business strategy of a LCC right from the start. Consequently, its founder, Stelios Haji-Ioannou copied the strategy profile from Southwest Airlines; two Boeing 737 were leased and the operation was launched from London's Luton airport. Two routes were served: From Luton to Glasgow and to Edinburgh. Aggressive marketing campaigns, e.g. 'Fly to Scotland for the price of a pair of jeans', helped the company to establish itself very quickly in the market.[19]

In 1998, three years after easyJet was founded, the company owned six Boeing 737-300 airplanes and operated on twelve routes between five European countries. At the same time, the company decided to expand into the Swiss market, which was at that time untouched by the European air-transport market liberalization. Forty percent of the Swiss charter airline TEA Basel AG was acquired and in 1999 the company was renamed 'easyJet Switzerland'.[20]

In August 2002, the 374 million pound purchase of Go-fly, an affiliate company of British Airways, put the company in the position of being Europe's largest LCC. This deal greatly enlarged easyJet's fleet. The company now owns 64 Boeing 737 planes and serves 83 routes between 35 destinations. The geographical spread of easyJet covers London's airports, i.e. Stansted, Luton and Gatwick, two Parisian airports, i.e. Orly and Charles de Gaulle, as well as the major airport of Genf and Amsterdam. Thus, easyJet differs from Ryanair in the fact that the fleet is stationed at primary airports and not on secondary ones. Intended is the direct

[18] See Sparaco 2002b, p. 62.
[19] See Sull 1999, p. 23.
[20] See Lawton 2002, p. 114 con't.

competition with regular airlines and additionally to target the segment of business travellers.[21]

In June 2002, easyJet transported a total of 14 million passengers.[22] Since their beginning of operation, the company grew continuously, a fact that is also reflected by its financial data. EBIT (earnings before interest and tax) increased from 9.1 million pounds in 1999 up to 71.58 million pounds in 2002. In order to sustain such a growth, the company recently ordered 120 new Airbus A319 planes over the next five years and furthermore holds an option to buy 120 additional airplanes of the same type until 2012.

Germanwings On October 27th, 2002 the German LCC Germanwings started their airline business from the airport Cologne-Bonn. Germanwings is an affiliate company of Eurowings, one of Germany's most successful regional airlines. In turn, Lufthansa currently holds 24.9% of Eurowings and has the option to enlarge its share in the company up to 49%. With respect to the financial year 2003, Germanwings was able to attract 2.4 million customers, which represents 50% of the total number of passengers transported by the Eurowings group.

Germanwings' business concept is adopted from other major LCCs like e.g. Ryanair or easyJet. The customer can only make use of call centres or the Internet in order to place their booking and there are no on-board services during the flights. Consequently, over 93% of the flights are booked over the company's Internet platform. The fleet counts 12 airplanes and consists of two types (Airbus A319 and A320) which depart from first category major German airports. With respect to the location of Germanwings' aircraft fleet, two major German regions are targeted: North-Rhine Westfalia (Airport Cologne-Bonn) and Baden-Wuerttemberg (Airport Stuttgart).[23]

Hapag-Lloyd-Express Hapag-Lloyd-Express (HLX) was founded in 2000 and is owned by the German TUI AG, one of the world's largest travel and tourist agencies. Currently HLX offers 17 destinations that are served from the Rhine-Ruhr airport in Cologne-Bonn. This airport serves as HLX homebase, however, the company has opened up another base in Hannover in early 2003. The company owns eight Boeing 737-700 planes thus adopting the same single-plane strategy as its opponents.

In contrast to other LCCs, HLX sells its tickets not only via the Internet or through call centres. It also takes advandage of TUIs travel agencies. In addition, strategic alliances have been formed with the European car-rental company Europcar and the insurance company Elvia. One of the distinctive differences, however, is that HLX only administers sales and organizational functions. Aircraft, crew and maintenance are under the control of Germania. To date, HLX has sold some 500,000 tickets and an increase up to 1.3 million transactions in 2003 calls for further fleet expansion. Consequently, the company has recently enlarged its fleet by a further two Boeing 737s.

---

21 See Lawton 2002, p. 119.
22 See easyJet 2002, as of 30.09.2002.
23 See Germanwings 2004, as of 22.07.2004.

Recently, HLX has also won a public bid for the transportation of German public officials, the so-called 'Beamtenshuttle' ('Civil Servant Shuttle') on the routes Cologne-Bonn and Berlin. Monthly, some 12,000 public officials commute on these routes. HLX has accordingly increased their daily flight frequency from six to seven flights a day. Such future public bids represent a new source of innovative growth potential for LCCs.

Air Berlin After the Second World War, only aircrafts from the Allied powers were allowed to land in Berlin. For this reason, the 'Air Berlin Inc.' was founded in 1978 in Oregon, USA. Air Berlin commenced its services in 1979 with a leased Boeing 707 on a flight from Berlin to Palma de Mallorca, and thereafter with flights to the USA. Today Palma de Mallorca is still the prime destination of Air Berlin's route network. The 'Majorca Shuttle' operates daily flights from 12 German airports. The total number of weekly flights to Majorca has increased from 140 (summer 2001) to over 160 (summer 2002). In the summer of 2003 there were as many as 193 flights to choose from.

Air Berlin was the first holiday airline to operate from 'fringe airports' such as Muenster/Osnabrueck or Paderborn. Air Berlin aimed to provide 'blanket coverage' to be closer to its passengers. Originally flights were only available to package tour operators. Once the route network was extensive enough and the flights were sufficiently frequent, 'flight-only' tickets were sold directly to the public. This applied to the Majorca Shuttle, for instance, offering daily flights from 12 German airports. Flight-only tickets are now available on all Air Berlin routes – and at prices that represent exceptional value.[24]

In September 2004 the director of Air Berlin forecasted a 30% increase in passenger numbers this year, contributing to a 20% rise in turnover. Achim Hunold said that he expected turnover to reach 1.07 billion euros (1.29 billion dollars), compared with 894 million euros last year. He also forecasts that 12.5 million passengers would fly with Air Berlin, the country's second largest no-frills airline, this year, compared with 9.6 million last year.[25]

Germania Originally located in Cologne-Bonn, 'SAT-Special Air Transport' was set up as a holiday charter airline in 1976 with turboprop Fokker F-27 Friendships operating out of German regional airports. Charter services from Germany to the Mediterranean, North Africa and the Canaries were in flight. Throughout the 1970s and 1980s it slowly grew as it purchased three ex-LTU Caravelle 10R jets and two Boeing 727s. In 1986 the airline changed its name to Germania.[26] Germania was founded by the German millionaire Heinrich Bischoff, who according to press announcements owns some 80% of the capital stock. The remaining 20% are held by the Preussag/TUI company.[27]

---

[24] Air Berlin Homepage (www.airberlin.com) as of 15.09.2004.

[25] Pott 2004.

[26] Airline History Homepage (http://airlines.afriqonline.com/airlines/467.htm) as of 16.09.2004.

[27] N.A. 2002c.

Towards the end of the 1990s Germania was flying sub-charters for other carriers and for holiday companies. In 1995 the airline was taken over by the Hetzel travel company that, however, went insolvent. Several aircrafts had been leased to Condor but flew with Germania crews. Subsequently Germania Airlines and Berlin's SAT-GmbH bought the Stuttgart travel agent Hetzel out. According to management, Hetzel would not be drawn on the purchase price, and they emphasized that nothing would change for their customers. The change of ownership would not affect travelling and it would be business as usual.[28]

Air traffic to Berlin continued to increase and Germania shifted their main focus to this route. At the same time, they also enhanced their activities on the Cologne, Bonn, Leipzig, Erfurt, Dresden, Muenster, Stuttgart and Nuremberg routes. In 1995 Germania ordered 12 new Boeing B737-700 airplanes and started an intensive collaboration with TUI, the largest travel agency in Europe. They also collaborated with other travel agencies such as LTU, Olimar, Meier's Weltreisen, 1.2-Fly, Tjaereborg, and Jahn Reisen. In 2000, Germania further entered into a mutual agreement with the enterprise Siemens and advertised on their airplanes the Siemens slogan 'Be inspired'.

Heinrich Bishoff recently purchased 19 Fokker F-100 airplanes from the US Airways bankruptcy assets for 2.6 million US-dollars per plane and has an option on a further 100 airplanes. These ten year old planes, inclusive the so-called C-Check (general overhauling), have comparative costs of less than one-tenth of respective new Boeing 737 airplanes, which it is argued will give Germania a significant cost advantage. Even though the Fokker F-100 planes are smaller, with 100 seats as compared to 140 seats of the competition, Germania contends that for flights up to two and a half hours time they possess a valuable cost advantage compared to using the relatively unprofitable larger Boeings for short distances. In July 2003 Germania Express had already achieved a 67% capacity load, with a small profit emerging.[29]

On the 12th of November 2001 Germania scheduled the first flight number ST 1061 from Frankfurt am Main to Berlin-Tegel for only 99 euro. This was the beginning of their reorientation away from charter market business towards a 'No-Frill-Business' strategy. Germania offered four flights in each direction per working day. A round-trip ticket costed 198 euro.[30]

Lufthansa, the incumbent monopolist on this route reacted aggressively to this encroachment of a discount carrier on its domestic market. At the time, Lufthansa offered 14 daily flights (in both directions) at a price of 485 euro for the full flexible round-trip ticket in the economy class. Lufthansa responded by reducing ticket prices down to 200 euro round-trip, which reflects a reduction of approximately 58%. Germania counteracted this move with a price reduction from 99 euro to 55 euro.[31]

---

[28] German News (English Edition), Fr, 05.07.1996, http://www.germnews.de/archive/dn/1996/07/05.html#9 as of 16.09.2004, www.vvtel.com/vvtravels/003_bycompany/01_airlines/germania-gmi/profile.html–36k, as of 16.09.2004.

[29] Flottau 2003.

[30] Bundeskartellamt 2002, p. 2-3, Hueschelrath 2003.

[31] Hueschelrath 2003.

As well documented by Hueschelrath (2003), the Lufthansa-Germania case and resulting regulatory German Federal Cartel Office decision constitutes a milestone in predation antitrust economics. The predator (Lufthansa) was convicted and the Federal Cartel Office had to solve the sanctioning problem. On the 18th of February 2002 the German Federal Cartel Office passed a landmark decision along the lines of Section 19 of the 'Act Against Restraints of Competition'. Applying a minimum-price-distance rule Lufthansa was prohibited from setting a price (including passenger fees) for a one-way ticket per passenger on the Frankfurt–Berlin route, which was not at least 35 euro above Germania's price. The ruling remains in place for two years, under the assumption that during this time span Germania should be able to gain enough reputation capital and establish a clientele base.[32] The Germania-Lufthansa case has made it quite evident that a variety of antitrust questions still remain open regarding how to maintain and promote competition in aviation markets.

4. Opportunities and Threats in the European LCC Market

At the European market level, LCCs are growing thus far into a senescent segment of the aviation industry, just as was observed in the American market in the years following airline deregulation policies. Several new discount airlines were established within the last 10 years. Nevertheless, one fact seems to be absolutely clear: with more entrants showing up on the low-fare stage, the more fierce competition will become. A further augmentation, i.e. an increase of supply on already existing routes, will bring fares further down. This leads to an even more intense struggle. It will not be a fight for market share only, ultimately this will be a battle for survival.

This section scrutinizes the threats and opportunities facing low-cost airlines. Various elements will be subject to discussion. Firstly, general aspects of sustainability relevant to low-cost air transportation will be analysed. Secondly, the influence of different growth tactics on the success of LCCs will be analysed, followed by a closer look at the growth aspects of Europe's major players. Thirdly, we analyse the situation of an LCC in relation to their competitors (section 4.3) and to their clients (section 4.4). The last two parts of this section will bring into focus the threats from potential alternative suppliers and safety issues for LCCs, illustrating why especially discounters have to deal most sensitively with this topic.

Base Competences

We forecast that the number of low-end competitors after the upcoming price war will eventually be shrunk to a maximum of two or three survivors: the mighty ones with the biggest market penetration and those who realize the lowest unit costs in the long run. This implies that continued success for a low-cost airline will crucially depend on its ability to transfer its competitive advantage, thus its base

[32] Hueschelrath 2003, p. 7.

competences, into the future. Hereby LCCs' base competences are a combination of lower input costs, cheaper product design and cheaper process design.

Input cost As cost of labour is a large proportion of an airline's input costs, LCCs' sustainability especially depends on the development of this cost category. In this context it has to be considered that the more a company matures, the more the employees' seniority will rise. Rising seniority commonly implies rising payments per employee. Furthermore, in times of general economic improvement low-cost airlines could lose staff when traditional airlines headhunt for qualified personnel, thus bringing low-cost airlines under pressure to increase payments for their existing workforce. Further, better economic times may stimulate employees to demand higher wages, consequently the influence of unions might increase. Strikes might be a demonstration medium, therefore some or all flights might have to be cancelled, meaning an absolute cost explosion and a certain loss of autonomy for the company.

Countries such as Germany, France and Belgium face environments with well-organized strong unions that may tend to cause problems in general for LCCs. Further, these countries are often characterized by extensive restrictive social legislation policies. This is regularly reflected in the existing labour market restrictions and resulting inflexibilites observed in these countries as compared to the more flexible policies maintained e.g. in Great Britain.[33] Illustrative of such developments and union power was the recent failure to reach a compromising salary tariff bargaining solution between the German pilot organization 'Pilotenvereinigung Cockpit' and the Deutsche BA (DBA), as well as the unsuccessful acquisition attempt of DBA by easyJet in March 2003.

All these variables are likely to melt away the LCC-advantage of lower wages in the long run. Recently, e.g. easyJet faced union problems when in September last year their staff voted to strike, even though easyJet spokeswoman Samantha Day said: 'We are a long way from any strike action.'[34]

A similar ominous threat may materialize given the bargaining situation between airport facilities, dispatchers, shipping agents and LCCs. Initially, the Start-Ups of the low-cost airlines were in a rather favourable position to bargain for concessionaires and contract terms with the minor airports, dispatchers and shipping agents. However, as competition in tendering for concessions ensues, their bargaining position will be weakened and conditions of contract renewal will become more costly as the LCCs experience higher costs of airport and slot fees.[35]

In addition to the above uncertainties, the procurement and purchasing of airplane kerosine presents another high risk factor. Due to the global interdependencies in the world fuel markets it is almost impossible to accurately forecast future fuel prices. Consequently, for low cost carriers an additional problem emerges as they now face a situation of potential fuel cost price increases which will become progressively more difficult not only to calculate in advance, but also more difficult to pass on to passengers ex post. This will inherently decrease the future profit margins of the low cost carriers. In addition, European

[33] See Lawton 2002, p. 65 con't.
[34] N.A. 2002a.
[35] See Pompl 2002, p. 122.

airlines must also take account of the fact that global fuel transactions are commonly payable in US dollars. Thus, the necessity of securing an optimal hedging-policy due to the uncertainty of exchange rate changes will also boost cost elements.

Finally, since the terrorist attack of September 11, 2001, the world aviation industry faces higher insurance costs. Extra cost premiums of up to 2%, over and above the normal invoiced one-and-a-half to 3% calculated, are now commonly demanded by the insurance companies to cover the future risk of war and terrorist activities. However, given the current emphasis on short circuit European flights of European low cost carriers, the probability and necessity of war insurance coverage is minimal.

Product and process design Whether the anticipated success of LCCs' lean process design will still be realizable in future, depends intrinsically on the prospective sustainability of their low-cost product service design. Should European discount airlines expand their routing to more distant destinations such as, for example, transatlantic flights, the advantage of the short turnaround times could fade away. Long-haul airplanes commonly transport up to 400 people. Passenger boarding will require more time, as well as fuelling the aircraft for long distance flights. However LCC's may still have an advantage vis-à-vis their competitors.

Furthermore, passengers travelling on long haul flights mostly carry luggage. The on- and off-baggage loading procedures may require thirty minutes and more. Even though customer surveys commonly indicate that customers would consider flying long haul on a low-cost airline as well, it may still be maintained that passengers travelling on long haul flights will continue to demand higher standards in in-flight catering, in-flight entertainment and seat comfort. The extremely narrow seat pitches on short haul airline carriers will most likely not be useable on long haul flights, not only for reasons of comfort but for the risk of venous thrombolism, the so called 'economy class syndrome', or deep vain thrombosis (DVT). Therefore, unit cost advantages cannot be achieved as easily as on short haul flights with an average sector length of 500-800 km.

On short haul flights the competitive advantage of low-cost product and process design has strong potential to be a sustainable factor of success compared to traditional airlines. However, the advantages arising from the pure low-cost structure cannot be simply copied for long haul flights. Several attempts from large traditional players to replicate the low-cost design have failed.[36] Qantas' subsidiary, Australian Airlines, however appears to be successful. It is difficult to make the customer understand that a brand that so far has offered 'all-inclusive services' now offers 'no-frills' instead. Therefore passenger expectations associated with names such as British Airways or Lufthansa still reflect upper-class air travel combined with rich in-flight catering and in-flight entertainment.

[36] See e.g. Calder 2002, p. 36.

Growth Strategies

Since low-cost airlines are the most rapidly growing sector in aviation industry, LCCs' expansion tactics have a significant influence on the upcoming success of the individual company. Thus two key questions emerge: By how much and how fast can low-cost airlines grow and at the same time continue their success? In order to adequately answer these questions, the advantages and impediments of various expansion tactics will be highlighted in the following.

Organic growth Organic growth displays augmentation achieved by expansion within the company. However, rapid organic growth faces pertinent obstacles. Southwest Airlines, without any doubt the most successful peanut-player in the history of 'cheap seats', has grown organically in a quite conservative manner over more than three decades. Former President and CEO Herb Kelleher and the Board have always decided in favour of very modest growth rates. It took Southwest nearly twelve years to increase their assets to a fleet of fifty aircraft, 'new capacity was used to add frequencies to existing routes and only a small handful of new routes were added each year'.[37] The management of Southwest Airlines chose a strategy of slow-but-steady growth, and their success proves them right.

One drawback of very rapid growth arises from the fact that large amounts of fresh capital are needed for expansion. This consequently could result in unfavourable terms for urgently required loans, e.g. high interest rates or unpropitious repayment periods.

Once bought and hired, assets and new staff have to be integrated into the production process. This causes a significantly higher cost weight if a large number of variables have to be changed at once. Later on, the communication within the firm has to be guaranteed, leading to a more complex system of administration. The advantage of a lean and simple controlling cycle therefore might fade away, since its fast implementation strongly limits the effect of the learning curve.

Another problem due to rapid expansion is, for example, the effect of over-expansion. This portrays a situation of expanding too fast compared with market demand, creating strong pressure on an airline to bring down ticket prices to fill up the newly created capacity. If new routes into lower traffic markets were added, it would be increasingly difficult for the operator to attract enough passengers in order to achieve the desired load factors required by yield management. An alternative attractive option is conceivable through collaboration. For example, in order to increase the passenger load of their airplanes, Hapag-Lloyd Flug (HLF) and the German LCC Air Berlin recently agreed on stronger cooperation. The unique feature of this venture is that both companies will coordinate their flight schedules and routes. Consequently, HLF will serve the long-haul routes while Air Berlin will concentrate on the short-haul routes.

Mergers and acquisitions Expansion can also be achieved by means of mergers and acquisitions. In contrast to organic growth, a merger describes the situation in which one company is being absorbed by another. An acquisition too is a type of a

---

[37] Doganis 2001, p. 129.

merger, but is generally considered a more hostile action of integrating another company.

Merger activities can be carried out in various ways of which the three most important types are:

- the horizontal merger
- the vertical merger
- the mixed merger

A *horizontal merger* reflects a union of competitors in the same line of business, or at one level, e.g. a fusion of Lufthansa and British Airways. *Vertical merger* activities reflect the integration of supplier and customer in the chain of production, e.g. the amalgamation of Italian brake specialist Brembo and the car producer Porsche. *Mixed mergers* represent the combination of firms that were neither customers, suppliers, nor direct competitors of one another. For example, JetBlue Airways recently acquired satellite TV operator DIRECTV®, in order to offer the passenger an attractive in-flight entertainment programme of 24 live TV channels free of charge.[38]

After long years of steady organic growth in America and Europe, horizontal mergers have become a common medium of expansion among airlines. Large operators often swallow small regional carriers. The aim is to achieve a wider spread of routes with a minimum of effort and to reduce competition. But the desired turnout is not always achieved. To give an example, in 1985 Southwest Airlines bought Transtar and in 1993 it acquired Morris Air, both being interesting operations at first glance. But soon after its takeover, Transtar had to be closed down[39] due to poor performance.

Growth Strategies of European LCCs The two major players in the European discount-sky are Irish Ryanair and British easyJet. Interestingly, both airlines follow very different growth strategies. Ryanair increases its size slowly and steadily with focus on maintaining the lowest competitive cost position. With this, a clear point of convergence is the maintained preference for secondary airports, thus avoiding head on competition with other LCCs. Since competition has increased sharply, Ryanair has cut back its plans to expand into the German market. Thus Ryanair maintains overall low costs and thereby sustains its ability to offer the lowest average fares on the market.

easyJet, as its main competitor, has decided in favour of growth by acquisition. Apparently, organic development alone does not sufficiently fit the strategic plans of its Board. Centre of attention is the time sensitive traveller, which leads to the necessity of more airport slots and higher frequencies. To achieve this, easyJet completed on August 1st 2002 the acquisition of Go-fly by paying a sum of 374 million pounds.[40] easyJet plans to operate both airlines as a single unit, thus Go-fly has been completely integrated into easyJet operations.

---

[38] See JetBlue Airways 2002, as of 17.11.2002.
[39] See Calder 2002, p. 36.
[40] See easyJet 2002b, as of 17.11.2002.

The emerging opportunities from this deal are the prompt increase in frequency on existing routes as well as the proliferation of the easyJet network. Additionally, the new airline will possibly achieve cost savings, e.g. in maintenance cost (Go-fly has the same airplane type) and in marketing expenses. Furthermore, easyJet receives a second operating base close to London. Departures will now be offered from London-Luton as well as from London-Stansted.

Despite the advantages of these strategic moves, impediments must be noted as well. Go-fly, for example, operated from larger airports and had assigned seating, hence its cost basis was not purely designed on a 'low-cost' model. And even though the homepage of Go-fly stated that the normal Go-fly service will be continued and that seat allocation will be kept,[41] the integration process to a single entity will surely be carried out, adapting the easyJet-no-frills-model in order to bring down the cost basis. Merrion Stockbrockers, an Irish financial group, estimates that the completion of the easyJet-Go-fly merger will likely take approximately two years and cost about 25.4 million pounds.

Another step underlines the 'bang-boom' growth strategy of easyJet. In May 2002 it bought an option to acquire the German subsidiary of British Airways (BA), Deutsche BA (DBA). At the time the option was worth € 5 million, the deal itself would bring BA an estimated €30-39 million plus about €0.6 million additional monthly payments until the option is being exercised. All in all, the purchase of DBA could easily cause total acquisition costs of up to 52 million euro. Advantages of this deal come from the fact that easyJet is already allowed to start restructuring DBA into a pure low-cost operator. An additional benefit is the Deutsche BA network with a strong emphasis on the German market, offering slots e.g. at Cologne-Bonn and Frankfurt/Main. This at the same time brings in obstacles by driving up costs, since fees at those airports are much higher than they are at secondary airfields. All in all, the integration of DBA seems much more risky than the implementation of Go-fly, since DBA has not been profitable since its foundation in 1992.

The Competitve Situation

Market Entry and First-Mover-Advantages In the recent past, the European LCCs Ryanair and easyJet have benefited the most from their activities, making them the strongest and most successful providers to date. This is largely due to their First-Mover advantage. Attracted by low prices, the demand for air travel increased substantially.[42] Creating feasible no-frills low cost air travel for new low-budget passengers and market segments enhanced growth rates rapidly without initially distracting loyal passengers of other airlines to any great extent. Basically, a new market segment has evolved which did not exist in this form before. Travellers who generally used alternative means of cheaper transportation than flying now began to consider the possibility of travel by air. For example, commencing 1995 to 1998, the LCCs succeeded in increasing the number of passengers travelling by

[41] See Go-fly 2002, as of 17.11.2002.

[42] See Doganis 2001a, p. 139.

air on the route London to Barcelona from half a million to roughly one million passengers.

The LCCs' market share for this route rose to approximately 25%. Similar success stories can be observed for other European routes. The annual passenger growth rate for new low cost airline circuit routes is commonly above 20%, which is significantly higher than traditional growth rates. The LCC incumbents benefit the most from these new developments. Indeed, the consulting firm McKinsey estimates that within this 'winner-takes-all dynamic' only Ryanair (the original LCC) will be capable to sustain high average operating margins. They forecast that Ryanair's competitors will face difficulties in achieving full passenger flights. The main reason identified being the exclusive and primary access to the regional and national resources (e.g. airport, flight slots, landing rights, brand power, etc.), which imitators cannot appropriate to the same degree.[43]

For example, a successful LCC incumbent has not only established a loyal clientele, but also favourable conditions with airports and it possesses valuable flight slot concessions and enjoys a strong reputation (Brand Power). Commonly, the incumbent additionally benefits from an easy access to national infrastructures and may even be supported by subvention policies of the airport itself or the local transportation ministry. Such advantages were consequently implemented by Ryanair and easyJet. Both have been highly successful in marketing their low cost flights into the United Kingdom, Germany, France and Holland, thereby stimulating flight demand while achieving a strong brand name position.[44] Simultaneously, Ryanair has found an ideal airport partner in 'Frankfurt Hahn Airport', located in the middle of Germany allowing extensive access to the German and continental European market. Given that the Frankfurt Hahn Airport is a relatively small airport with limited capacity, as compared e.g. to the larger Frankfurt Main Airport, its resources are not adequate to provide for new carriers, which keeps potential competitive market entry at bay.

Market Entry of New Competitors One reason why the number of European LCCs has recently increased is due to the low market entry barriers faced by new carriers. Historically aviation regulation laws were significant entry barriers, however deregulation of the airline market now entails much less government intervention. If legislative barriers to entry do exist, they are frequently either totally or partially circumvented. easyJet, for example, operated officially until October of 1997 under a licence authorization for Air Foyle and GB Airways until they obtained their own permits.[45]

Significant barriers to entry exist for new incumbents with regard to airport capacity limitations, especially with respect to the practice of obtaining airport take-off or landing slots. It is now almost impossible for new LCCs to get slots for main airports during the peak load flight times (early morning and evening flights). Partly responsible for this difficult situation is the method of the slot distribution system applied. For example, according to the present system if an airline carrier has slot rights from the previous time period, which they used with a minimum of

---

[43] See Grant 1998, p. 184.
[44] See Binggeli and Pomeo 2002, p. 5.
[45] See Schmitt 2001, p. 20.

80% fulfillment, they will be scheduled with the future slot position. Such problems, in addition to comparatively high fees demanded by the major airports, as well as the slot dilemma for LCCs, are an important reason why LCCs such as Ryanair consequently only fly to secondary airport locations.

Other often asserted barriers to entry, such as computer reservation systems, procurement provision systems and frequent flyer programmes may not necessarily represent great obstacles since due to cost they are essentially irrelevant innovations from the LCCs' perspective, or the LCCs already have their own systems in place.[46]

Rivalry among existing competitors In the European market the low-cost no-frills budget airlines compete not only with the major conventional scheduled airline carriers, but also with the charter (non-scheduled) airlines. However, competition is increasing also among the budget airlines themselves. This is illustrated by the fact that whereas during the late nineties only five European budget airlines existed, currently twelve budget airlines are competing. Accordingly, alternative competitive strategies abound. Ryanair, for example, only services secondary airports specifically targeting leisure vacation passengers. In contrast, easyJet and Virgin Express mainly service major airports, thereby targeting potential business class passengers.[47]

In addition, the market leaders Ryanair and easyJet apply different growth strategies. Whereas Ryanair follows a hub-and-spoke network expansion strategy, easyJet has attempted to enhance their network competitive position by triangulation.[48] By this means many European flight routes are serviced by only one budget airline. This is, however, not necessarily the case for the more lucrative routes such as London–Barcelona, where there is room for more than one LCC.[49] It has been observed that up to the middle of 2002 only 17 low budget airline routes involved direct competition. In contrast, 111 European routes were serviced by only one LCC carrier.[50] This situation however, may rapidly change as market entry attempts persist.

Since the services of the low budget airlines are commonly quite similar, competition is mainly along the lines of price competition. The new market incumbents attempt to gain market shares by very low pricing strategies, which often leads to high competitive intensities on particular European routes. Such predatory pricing conduct may even lead to a price race to the bottom. Along these lines, Ryanair has been especially aggressive.[51] The weaker the market incumbents are, the more they will be naturally confronted with a highly risky business environment. First of all, they commonly do not have the necessary financial reserves to support large marketing campaigns and the high fixed costs of maintaining daily business at the same time, while also fighting a price war with competitors. Secondly, their brand name is normally not well known enough and

[46] See Wiezorek 1998, pp. 175-181.
[47] See Rae 2001, p. 334.
[48] See Lawton 2002, p. 118 con't.
[49] See Sull 1999, p. 26.
[50] See Binggeli and Pompeo 2002, p. 5.
[51] See Lawton 2002, p. 109.

recognized by passengers compared to the LCC incumbent's brand name. Empirically, such elements often lead to market withdrawal of the weaker competitor. For instance, Virgin Express withdrew from the route London–Brussels, as it was unable to compete with the more favourable Ryanair prices. Similarly, the low budget carrier Go-fly had to discontinue their Dublin–Edinburgh connection.[52]

Simultaneously with the increasing price competition, a recent phase of consolidation within the European low budget airline market segment is observable. Illustrative of this tendency is easyJet's acquisition of Go-fly and Ryanair's acquisition of Buzz. Being the major players, they both have enhanced hereby their market positions as well as diminished the number of competitors in the market.

Table 11.3 Charter airlines' cost advantages compared to low-cost scheduled carriers

Advantages	Details
Use larger more economical aircrafts	180 to 350 seats compared to 130-160 seats
Higher daily aircraft utilization	Fly through the night
Capacity offered closely matches demand (despite higher seasonality)	High daily frequencies not important
	No need to offer flights in low season
Higher passenger load factors (85-90%, not 65-75%)	Flights pre-sold to holiday tour operators
	Poor flight cancelled or consolidated
Very low sales or advertising expenditures	Most capacity sold to a few large tour operators
	Vertical integration with largest tour operators
BUT – some frills and more expensive in-flight service	

Source: Doganis, 2001a, p. 159.

Today's no-frills low budget airlines face even further competition from the airline charter industry, which have been practising a successful low-cost business model in Europe since the nineteen-sixties. Approximately 13% of the international passenger traffic today, in terms of passenger kilometers, is generated by charter flights.[53] The charter airlines are generally part of vertically integrated business enterprises that usually also consist of tour guide and travelling agency

[52] See Sparaco 2002a, pp. 59-61.
[53] See Doganis 2002, p. 152.

services. Naturally, even though the low cost business model of the LCCs and the airline charter industry vary in many ways, they have one goal in common: that is offering their services at particularly low fares. Since the charter airlines regularly fly to the same destinations in Europe as the LCCs they compete directly for customers.[54] Despite intensive competition, some LCCs have maintained strong positions. For example, easyJet started its route London Luton–Palma in 1999, thereby successfully pushing the British charter airline Monarch out of the market. Further similar price battles between LCCs and charter airlines are now also taking place on routes such as Luton–Malaga.[55]

Table 11.3 illustrates the most important reasons for cost advantages of charter airlines compared to low cost carriers.

Predatory Pricing and other reactions As mentioned above, the competition among the European low budget carrier airline group itself is becoming progressively more intense. Furthermore, the LCCs are also directly attacking the direct short circuit routes of the traditional flag carriers who throughout the decades have often enjoyed a comfortable domestic monopoly or duopoly situation. The following reactions of the flag carriers could be possible.

Confronted with a new LCC entrant in their market, flag carriers in general often first react with ticket price reductions on the affected routes. One promising strategy that is applied frequently is predatory pricing, i.e. the flag carriers offer passengers tickets under the real costs of supply in the hope of crowding out the new market entrant.[56] Predatory pricing means pricing so low that competitors quit rather than compete, permitting the predator to raise prices in the long run. One of the main causes of predatory pricing in the aviation market is the increased competition and efficiency resulting from the deregulation of this sector and the privatization of state-owned airlines.

Predatory pricing is subject to the competition laws and policies of most industrialized countries, but there has been a lively controversy over what standards should be applied. This is especially the case in the aviation industries since more and more low cost carriers have entered the market over the years. This begs the question if political intervention is really necessary to counteract a problem that so far has not been truly proven? Some theories about predatory pricing go so far as to urge that any rule against predation will do more harm than good by depriving consumers of the benefits of vigorous price competition. For example, when has it ever been possible for consumers to fly from Frankfurt to London for less than 50 euro? On the other hand concerns arise that the major airlines, which are often dominant in their markets, will seek to unfairly protect their positions against new entrants.[57]

The arguments that predation exists came under attack by McGee (1958, 1980). He argues that predation is more costly to the predator than the victim, given the predator's larger market share, and that as the campaign and market share increase, so do the costs. Against these large and certain costs are future profits

[54] See e.g. Baker 2000, pp. 76-81; Williams 2001, pp. 279-283.
[55] See e.g. Gallacher 1999, p. 70.
[56] See Cabral 1999, p. 37.
[57] See OECD 1989, p. 5 and Milgrom 1987, p. 937.

which must be discounted in two ways, once to reduce them to present value terms and then again to reflect the uncertainty that they will arise. He then gives reasons why the future monopoly profits might not be realized: the prey might enter into long-term contracts with customers (who would not want to see a competing supplier disappear), find financing to ride out the price cutting or shut down and wait for prices to rise. This last argument can be applied especially in the aviation market each time the major airlines have the possibility to start with a low cost offer. They do not face large sunk costs because they already have the infrastructure at their disposal. Even if the victim does go out of business, new entry is possible in the post-predation period. Most importantly, however, is McGee's point that the threat of predation is not credible because it would not pay to carry it out, the argument being that the dominant firm would lose more by predating than by coexisting with a rival. He further contends that better monopolizing strategies exist. For example a merger leading to a monopoly situation, where legal, would make more sense to the dominant firm, as the large losses incurred by a predatory campaign could be avoided. Given these losses, McGee doubts that a predatory campaign could sufficiently lower acquisition costs and believes that the losses to the predator will be greater than the generated savings.[58]

Besides predatory pricing, other strategies exist, which can be used by major airlines in order to impair existing LCCs. TUI, Germany's largest travel agency and majority owner of Hapag-Lloyd Flug (HLF) and the LCC Hapag-Lloyd Express, has threatened the Hannover airport with the withdrawal of its fleet if the airport allows the British rival easyJet to set up a German base in Hannover. In reaction to this, easyJet withrew its plans to start from Hannover and switched to the regional airport in Dortmund. Moreover, it is possible for the traditional carriers to offer additional attractive passenger benefits such as frequent-flyer-programme, i.e. Lufthansa's 'Miles and More'. Furthermore, their membership in global strategic aviation alliances such as 'Star Alliance' spans an advantageous network permitting global passenger servicing. Thirdly, the traditional scheduled fight carriers have established in the meantime their own LCCs in an attempt to secure their market shares. In 1997, British Airways was the first scheduled carrier to establish a no-frills low budget supplier Go-fly. KLM followed in January 2000 by establishing Buzz. It is, however, to some extent questionable whether or not such attempts have been successful. Go-fly, for example, has directly competed with BA on their own routes ('market cannibalization' effect). The image of BA was damaged as passengers' perceptions of the mother affiliate were distorted due to an unclear branding of the new carrier.[59]

Are European LCCs possible victims of predation? As stated before, predatory pricing in the airline industry is assumed, where an incumbent significantly lowers its prices in response to the market entrance of a new rival. An illustrative example has been the price war between American Airlines and Vanguard Airlines regarding the intra-American connecting traffic between Dallas and Kansas City.

---

[58] See McGee 1980, p. 295 con't. See further Hueschelrath (2003).
[59] See e.g. Calder 2002, p. 140, Reed 1998, p. 46.

Before the market entry of Vanguard Airlines, American Airline sold its one-way ticket for US-$108. When Vanguard Airlines entered the route in 1995, American Airlines reacted to Vanguard Airlines' price offer and lowered the price on this route down to US-$80. In addition, American Airlines increased its number of daily flights by almost 100% that is up to 14 flights. As a consequence, Vanguard Airlines was squeezed out of the market in December 1995. Upon success, American Airlines increased its prices to US-$147 and scaled back the number of flights.

Since the liberalization of the European air traffic, the European Commission and national cartel offices were called several times to pass judgement over alleged acts of predatory pricing. In 2000, Ryanair filed a formal complaint against the German airline Deutsche Lufthansa A.G. Ryanair accused Lufthansa of having behaved in a predatory manner on the routes between London and Germany. Although already serving Frankfurt–London/Heathrow and Frankfurt–London/Gatwick, Lufthansa had indeed immediately announced the operation on the route Frankfurt–London/Stansted following Ryanair's entrance on that same route. In addition, Lufthansa lowered its fares on this route by approximately 60%. Fares on the other two routes however remained unaffected by this price cut. Lufthansa made use of a similar price strategy while competing with Go-fly and Deutsche BA on the route London/Stansted–Munich. Right after both LCCs abandoned their operation on these routes Lufthansa increased its prices again.[60]

In 2002, as briefly mentioned in Section 3.1.6, Lufthansa found itself again accused of predatory pricing. This time, the German Bundeskartellamt ('German federal cartel agency') interpreted Lufthansa's price cuts on the route Berlin–Frankfurt as a motive to put the regional carrier Germania out of business. On November 12, 2001 Germania started operation between Berlin and Frankfurt a.M. at a price of €99 for a single flight ticket. In reaction, Lufthansa lowered their return ticket from initially €486 to €200, a roughly 60% decrease. Although providing frequent flyer points, on-board service and catering as well as higher flight frequency between Berlin and Frankfurt, Lufhansa's offer is equally priced with respect to Germania's offer. Based upon these facts, the German Bundeskartellamt saw strong indications for price predation. Consequently the case was brought to court, where it was decided that Lufthansa must charge at least €35 more on the one-way ticket than Germania for the next two years in order to fairly account for the extra service provided.[61]

Recently, the British low-cost carrier Virgin Express has made similar allegations against the Italian airline Alitalia. The routes between Brussels – which serves as the main European hub for Virgin Express – to Milan and Rome are particularly affected. Alitalia has been accused of offering below costs fares on these routes. Similar allegations were made against the Spanish airline Iberia, which Virgin Express accused of predatory pricing on the routes from Brussel to Barcelona and Madrid. It remains to be seen how these instances evolve in the near future.

---

60 See Steen and Soregard 2002, p. 217; Machatschke 2000, p. 52.

61 See e.g. Bundeskartellamt 2002, p. 2 con't., Steen and Soregard 2002, p. 217 con't., Capell et al. 2002, p. 17, Böge 2002, p. 1473 con't., Hueschelrath 2003 and Wiezorek 1998, pp. 182-186.

In another recent case of predation, Norway's competition watchdog came to the conclusion that the Scandinavian airline SAS did not act in a predatory manner while operating in the Norwegian market. The commission found no evidence for squeezing out SAS's domestic rival Norwegian Air Shuttle (NAS). Late 2002 NAS had begun competing on the four main Norwegian domestic routes with the major airline, which had caused SAS to reduce its ticket prices.

As these examples illustrate predatory pricing is still a major issue in the European airline industry. Steen and Soregard (2002) reason that the following lessons can be drawn from the persistent predatory pricing behaviour of major airlines. First, reactions by the authorities need to be enforced quickly and second, ex post regulation might not be as efficient as ex ante regulation in the case of predation.[62] The recent years show evidence of a market consolidation in the European low-cost airline sector. Last year, Ryanair bought the LCC rival Buzz Stansted Ltd. from the Dutch airline KLM for aproximately US-$25.8 million. While KLM itself failed to acquire the British LCC Go-fly from the British BA, easyJet succeeded in merging with Go-fly for roughly £374 million only a few months later. This transaction makes easyJet Europe's number one low-cost airline. So far, none of the LCCs has been accused of predatory behaviour. Nonetheless, network economies of scale available to large scale LCCs and other altering circumstances may eventually increase further market consolidation and speed up the likelihood that a highly concentrated LCC market will begin to protect itself by acts of predation.

Customer Profile

Customer Segmentation The global airline industry can be differentiated into two major customer segments; the price inelastic business travellers and the price elastic leisure and business travellers. The elastic segment is commonly further differentiated in vacationers and passengers visiting friends and relatives ('VFRs') and business travellers paying their own fares. Whereas the business traveller's ticket (inelastic segment) is paid by the passenger's affiliated enterprise, the leisure passenger is individually responsible for the financing of the entire trip.[63] Given the observed variations of the price elasticity of demand in the different market segments, passengers react to price changes differently reflecting the market segment they belong to.

As Doganis (2001a), among others, points out, the passenger mix will vary from market route to market route. For example, it is commonly observed that flight desintinations such as to the southern part of Spain, France and Italy are mainly dominated by leisure passengers, whereas flight routes such as London–Paris, London–Frankfurt, or London–Brussels are primarily utilized by business travellers. Which category a passenger belongs to generally determines the respective expected service. For example, business passengers expect flight-schedules to be on time, as well as short check-in times, numerous scheduled flights, good board service, frequent flyer programmes, etc.[64] In contrast, leisure

[62] Steen and Soregard 2002, p. 218. See further e.g. Hueschelrath 2003.
[63] See e.g. Doganis 2002, p. 183.
[64] See Mason 2000, p. 110.

passengers are less demanding. They essentially value low cost safe transportation. In the following the potential development aspects of LCCs, as reflected by servicing the two respective passenger market segments, will be surveyed.

Leisure passengers Leisure passenger expenditures are generally restricted by personal incomes and savings, subsequently the individual willingness to pay for airline tickets is largely constrained by the traveller's budget. Hence the price elasticity of demand of leisure passengers is highly elastic. In contrast, the price elasticity of demand for business class travellers is relatively inelastic. The low budget airlines' business model concept takes this point into account. Moreover, economic downturns increase passengers' cost consciousness further making the services offered by the LCCs even more attractive. From a demographic perspective, the growing number of single and double income households with higher disposable incomes has also influenced the observed tendency of increased flight demand.[65] The LCCs are in part responsible for such increases as airline tickets become more accessible to a wider public. No frills low budget airlines have not only benefited from an increasing propensity of passengers to fly, but also from their own ability to offer low cost seats enabling them to capture greater market shares.

Complementing leisure passengers, LCCs have also concentrated on flight routes that are attractive to VFRs, i.e. visiting friends and relatives. Illustrative of such a European VFR-market is the case of Ireland and England. The carrier Ryanair initially serviced this market and the route Dublin–London rapidly developed into one of the most trafficked routes.[66]

Given the enhanced flexibility of low budget flight scheduling and online booking procedures, the propensity of passengers to book additional weekend trips to southern European locations has increased. Moreover, the cheaper flights were welcomed by northern Europeans with dwellings in southern Europe. In the past, they were often dependent upon the offered charter flights.[67] It must also be mentioned that the LCCs have given city tourism a new stimulus. Rather than simply booking an all inclusive city bus tour, it is now possible to book a low budget flight and hotel separately at lower rates than alternative common package deals on offer.

Business passengers Meanwhile, business passengers are also increasingly turning to the LCCs. Since the 1990s, the number of business passengers on low budget flights has more than doubled. Indeed, today some 40% or more of passengers on LCC flights are business passengers. On some of its routes easyJet services over 50% are business travellers. In an interview study conducted by Mason,[68] some 74% of business passengers at Heathrow Airport (a major scheduled airline airport facility) confirmed the desire to fly with a LCC in the future.

---

[65] See Pompl 2002, p. 191 con't.
[66] See Mason et al. 2000, p. 7.
[67] See Williams 2001, p. 284.
[68] See Mason 2001, p. 103 and p. 107.

Short route business passengers are also becoming increasingly cost conscious. Commonly, the passengers are entrepreneurs of small and medium sized enterprises who are highly cost sensitive. The willingness to forgo comfortable on board service is compensated by low ticket prices. To date business passengers from larger enterprises have been more reserved in booking LCC flights, however, management is increasingly aware of the potential savings on offer.

However, obstacles still remain before the business clientele of large enterprises may be successfully reached by the LCCs. In order for the low budget airlines to gain market shares in this passenger segment of the market they will need to offer first of all more flights to major airports, thereby decreasing total passenger travel time. Secondly, LCCs will have to convince large enterprises that are used to the convenience of booking through travel agencies to consider direct online reservations. This implies also a reconsideration of certain business passenger service elements such as re-routing and refunding non-used flight tickets.

Strong relationships with customers The airline industry may be further differentiated into two major customer demand groups. First, travel agencies are large institutional demanders. Secondly, individual passengers in the final instance are the ultimate demanders. Travel agencies, being large institutional airline customers, such as Carlson Wagonlit, Neckermann and Thomas Cook have strong bargaining positions. In part, their bargaining position is founded upon the lack of specific brand preferences of the ultimate passenger. This enables them to influence the choice rationality of passengers when they book their flights. Travel agencies thus demand high commissions from airlines in promoting their tickets. It may be argued that LCCs will not be significantly affected by the transactions of the travel agencies as such since LCCs sell their tickets directly online.

However, the direct distribution channel of the Internet and direct online ticket sales may enhance the bargaining position of the individual passenger as comparative ticket price arbitrage potentials are taken into account. LCCs offer a general homogeneous service of similar quality, permitting easy identification of the cheapest supplier. As already mentioned above, passengers in this market segment react strongly to price differences. In addition, the low budget segment of the airline industry is characterized by low level 'switching costs', i.e. of changing to another airline. Since most LCCs offer no 'Frequent Flyer Programmes', the passenger propensity to change airlines in this market segment has no additional lock-in costs of changing suppliers.

Competition by Substitutes

Naturally, alternative means of transportation such as the car or inter-city high-speed train connections are a serious substitute to be considered. However, to what extent a real source of competitive danger exists for LCCs is highly variable within the individual European borders.

In general, Europe is characterized by an excellent infrastructure with regard to train networks and international highway networks. Automobile mobility, for example, often offers the greatest amount of individual transportation flexibility, along with direct locational point-to-point transportation. For small groups of individuals collective transportation by car is indeed often cheaper than by train or plane.[69]

Of special competitive interest are of course the intercity high-speed train connections. Especially France with its TGV trains and Germany with its ICE trains have well developed high-speed intercity connections. As train-travelling times are further reduced in the near future LCCs will face further competition. The additional advantage of train travelling may be seen in the light of possible last minute or on-board ticketing aspects, as well as the fact that lengthy boarding times and check-ins are alleviated. Uncomplicated train ticket re-scheduling and a high daily frequency of intercity train connections are another further advantage. Currently, at least in Germany, the high price schemes of the Deutsche Bundesbahn, along with frequent scheduling delays and long travel times are an important reason why low budget airlines have rapidly gained customers. For example, the route Berlin–Cologne experienced large drops in profits due to the fact that a normal train ticket was priced at 72.40 euro, which was on average 20 euro more expensive than a flight ticket from a LCC.

Future strategies and innovations of high-speed train transportation providers may prove to be a real threat to low budget airlines. In the near future, for example, the French TGV group is planning to implement a double-decker train with sufficient room for more than 1000 passengers. They also plan to lower their ticket prices and sell tickets over the Internet. Similar reactive strategic planning is also being considered by the Deutsche Bundesbahn, along with the completion of further new high-speed intercity routes such as Cologne–Frankfurt.

Great Britain faces a somewhat different situation. Due to the fact that it is an island passengers were traditionally dependent upon air transportation to and from Europe. The duration of alternative sea transportation is relatively longer and often inconvenient and laborious. Meanwhile, the Channel Tunnel Rail Link (CTRL) and Eurostar service approximately 60% of the London–Paris route and some 45% of the London–Brussels route.

Safety Issues

Since the European market is fiercely contested, the discussion about low budget airlines and safety issues is always prominent. The question emerges: does low-cost equal low-safety? First of all: Southwest Airlines as 'the mother of low-cost air transport' has been operating now for more than 30 years without a fatal crash. Their excellent flight record is unbeaten in the history of civil (short haul) aviation and only the Australian carrier Qantas has achieved such excellence thus far. Since the most dangerous phases in a flight are the take off and the landing phase, this flight record is even more astonishing. It has to be considered that Southwest Airlines has an average of 2,547 daily departures and its average aircraft stage length is 540 miles with an average flight time of 1 hour and 36 minutes.

European LCCs face the same problems as all other LCCs worldwide. They are aware of being at the centre of public attention. For example when a traditional

---

[69] See Pompl 2002, pp. 208-211.

carrier has a fatal crash, this can easily result in economic trouble for the company. However the economic and public consequences would be even worse if a low-cost carrier came down. Not only would the media thankfully centralize the topic 'low-cost-low-safety', but traditional airlines as well as cockpit and cabin unions would also point at low-cost operations, critically assessing LCCs' safety standards. Eventually this could result in bringing down operations, just as it happened to ValuJet after its crash in 1996.

Since Southwest proved to be one of the safest carriers, customer perceptions take this into account. Today's LCCs predominantly operate very young fleets, thereby using latest technology aircraft. Their standard operating procedures (SOPs) are directly adapted from the aircraft manufacturers, thus minimizing risks. Additionally, most of the times they benefit from the fact that the cabin crew of LCCs provides a safety advantage as well. All cabin attendants are extensively trained on a single aircraft-type, thus reducing the danger of confusion in case of an emergency. LCCs know how thin the line is between low operating costs and cost savings in safety aspects. Ryanair's CEO Mr. O'Leary says: '*There is a challenge every day in this business not to do something stupid.*'

To sum up, it can be stated that today's low-cost players do not operate unsafely compared to their 'full-service' competitors. A cost analysis would clearly show that the cost savings achieved by LCCs are not based on reductions in safety standards or in crew training. It could rather be proven that easyJet, Ryanair, Germanwings or Buzz are flying new aircraft, thus operating at very high standards. They know that a crash would hit them much more intensely than the traditional carriers. But since they expand rather quickly these days, a lot of monitoring must be carried out in order to maintain those high standards.

5. Summary

The analysis has largely focused on the current and upcoming developments in the European low-cost aviation segment. The first part briefly summarized the liberalization process which made no-frills airlines possible. Furthermore, an overview about major European LCCs as well as additional stylized facts were given. It was demonstrated in section 4 that continued success for the LCCs will depend on their future ability to maintain their competitive advantage. As this advantage arises from cost efficiency based on input costs and the product and process design, it could be shown that continuing cost advantages over traditional and new competitors will significantly depend on the growth strategy a LCC will take. As Berster and Wilken (2004, p. 2) point out, 'simplicity of all processes is the key success factor'. In addition, organic growth has proven to be an important strategy to increase size with a minimum of risk. Nevertheless, especially in Europe the low-cost market has been rapidly changing over the last five years. With the maturing of this industry, mergers and acquisitions will most probably increase. The opportunities and the threats arising from this type of growth were also demonstrated. The long-term perspective of the industry is thus uncertain. Ryanair CEO Michael O'Leary has the vision that in the near future tickets in the LCC segment can be given to the customers at no costs. Instead, the LCCs will earn money by bringing the customers to the destination's hotels, to the corresponding airports and their shops as well as to the car rental companies who will pay at the end for this service. If this is the case, O'Leary speculates that the market will further consolidate and only very few LCCs will survive within Europe over the next five years.

References

AEA Jahrbuch (1998), Association of European Airlines, Brüssel, July 1998.

AEA Jahrbuch (1999), Association of European Airlines, Brüssel, July 1999.

Air Berlin (2004), Homepage, http://www.airberlin.com (Accessed: 15.09.2004).

Airline History (2004), Homepage, http://airlines.afriqonline.com/airlines/467.htm (Accessed: 16.09.2004).

Baker, Colin (2000), War of independents, *Airline Business*, No. 10, pp. 76-81.

Berster, P. and Wilken, D. (2004), *Market Penetration and Demand Generation of Low Cost Carriers in Germany*, Working Paper presented at the ATRS Conference, July 2004, Istanbul, Turkey (contact: Dieter.Wilken@dlr.de).

Binggeli, Urs and Pompeo, Lucio (2002), Hyped hopes for Europe's low-cost airlines, in: *The McKinsey Quarterly No. 4*, http://www.mckinseyquarterly.com/article_page.asp?ar=1231&L2=23 (Accessed: 19.01.2003).

Böge, Ulf (2002), Missbrauchskontrolle bei Predatory Pricing, in: *Wisu – Das Wirtschaftsstudium*, Vol. 12, pp. 1473-1474.

Borghetto, Martin, Berthelot, Boniface and Gibbons, Owen (2002), European Low-Cost Airlines: This Is a Buying Opportunity!, in: *Morgan Stanley-Studie vom 19.09.2002*, http://www.ryanair.com/download/morgansep.pdf (Accessed: 30.01.2003).

Bundeskartelamt (2002), *Beschluss in dem Verwaltungsverfahren gegen Deutsche Lufthansa AG*, Köln, B 9-144/01, Bonn.

Cabral, Luis (1999), Case Study – easyJet's $500 Million Gamble: Commentary 4, in: *European Management Journal*, Vol. 17, No. 1, pp. 37-38.

Calder, Simon (2002), *No Frills. The Truth Behind The Low-Cost Revolution In The Skies*, London: Virgin Books.

Capell, Kerry, Matlack, Carol and Tierney, Christine (2002), Day Of The Discounters: Short-haul carriers imperil the giants, in: *Business Week Europe* vom 11.02.2002, pp. 16-18.

Council Directive 83/416/EEC concerning the authorisation of scheduled inter-regional air services for the transport of passengers, mail and cargo between Member States; 25.07.1983.

Doganis, Rigas (2001), *Airline Business in the twenty-first century*, Routledge, New York.

Doganis, Rigas (2001a), *The airline business in the 21st century*, London/New York: Routledge.

Doganis, Rigas (2002), Flying Off Course, *The Economics of International Airlines*, London/New York: Routledge.

easyJet (2002), *Annual Reports and Accounts*, 30.09.2002, http://www.easyjet.com/en/about/2001-2002FullYearResults.pdf (Accessed: 07.04.2003).

easyJet (2002a), *easyJet acquires option to buy DBA*, 2002-11-17, http://easyjet.com/en/about/20020814_01.html (Accessed: 17.11.2002).

easyJet (2002b), *Our merger with GO*, 2002-11-15, http://easyjet.com/en/about/gomerger.html (Accessed: 15.11.2002).

Flottau, Jens (2003), Nur verdienen ist schöner, Die deutschen Billig-Fluglinien expandieren wie wild, *Die Zeit*, Nr. 33, http://www.zeit.de/2003/33/Billigflieger (Accessed: 17.09.04).

Gallacher, Jackie (1999), New challenge to charter, in: *Airline Business*, No. 10, pp. 68-70.

Germanwings (2004), Press statement, 22.07.2004, http://www10.germanwings.com/de/50/1302_6596.html (Accessed: 22.07.04).

Gilbert, David, Child, David and Bennett, Marion (2001), A qualitative study of the current practices of 'no-frills' airlines operating in the UK, in: *Journal of Vacation Marketing*, Vol. 7, No. 4, pp. 302-315.

Go-fly (2002), Will I still have an allocated seat?, 2002-11-17, http://www.go-fly.com/templates/FAQItem/DisplayFAQItem.jsp?text_id=FAQItem1361566&language=EN (Accessed: 17.11.2002).

Grant, Robert M. (1998), *Contemporary Strategy Analysis. Concepts, Techniques, Applications*, Malden/Oxford: Blackwell.

Grundmann, Silvia (1998), *Marktöffnung im Luftverkehr. Hoheitliche Eintrittsbarrieren in den USA und in der EG*, Baden-Baden: Nomos Verlagsgesellschaft.

Hueschelrath, Kai (2003), *How to Prey on Predators? The Lufthansa-Germania Case*, WHU Graduate School of Management, Institute for Industrial Organization, Vallendar, Germany, Working Paper, March, http://www.econ.ku.dk/cie/workshops/Norio%20IV/pdf%20-%20papers/Hueschelrath-Predation.PDF (Accessed: 16.09.2004).

JetBlue Airways (2002), Factsheet, 2002-11-17, http://www.jetblue.com/learnmore/factsheet.html (Accessed: 17.11.2002).

Lawton, Thomas C. (2002), *Cleared for Take-Off: Structure and strategy in the low fare airline business*, Hants/Burlington: Ashgate.

Lewis, Claire Simone (1992), EC Air Transport Policy: Third Package Proposals for Liberalisation; in: *EIU European Trends*, No. 2, p. 42.

Machatschke, Michael (2000), Luftgefechte, in: *Manager Magazin*, No. 4, p. 52.

Mason, Keith J. (2000), The propensity of business travellers to use low cost airlines, in: *Journal of Transport Geography*, Vol. 8, pp. 107-119.

Mason, Keith J., Whelan, Conor and Williams, George (2000), Europe's Low Cost Airlines: An analysis of the economics and operating characteristics of Europe's charter and low cost scheduled carriers, in: *Air Transport Group Research Report*, No. 7.

Mason, Keith J. (2001), Marketing low-cost airline services to business travellers, in: *Journal of Air Transport Management*, Vol. 7, pp. 103-109.

Maurer, Peter (2002), Luftverkehrsmanagement. Basiswissen, München: Oldenbourg.

McGee, John (1958), Predatory Price Cutting: The Standard Oil Case, in: *Journal of Law and Economics*, 1, pp. 137-169.

McGee, John (1980), Predatory Pricing Revisited, in: *Journal of Law and Economics*, 23, pp. 289-330.

Milgrom, Paul (1987), Predatory Pricing, in: *The New Palgrave: A Dictionary of Economics*, Vol. 3, pp. 937-938.

N.A. (2002a), *easyJet staff vote to strike*, 2002-09-04, http://news.bbc.co.uk/2/hi/business/2236479.stm (Accessed: 04.09.2002).

N.A. (2002b), Fly me, I'm cheap, 2002-09-09, http://www.economist.com/background/PrinterFriendly.cfm?Story_ID=1280855 (Accessed: 09.09.2002).

N.A. (2002c), Lufthansa droht neue Konkurrenz. Positive Zahlen von Robinson-Tochter, *Die Welt*, June 24, http://www.ig-oekoflughafen.de/Verschiedenes_98.htm.

Niejahr, Michael (1999), Regulation and Liberalisation of International Air Transport – Prospects for Reform from the Perspective of the European Commission, in: Pfähler; Wilhelm; Niemeier, Hans-Martin; Mayer, Otto G. (ed.): *Airports and Air Traffic. Regulation, Privatisation and Competition*, Frankfurt: Lang, pp. 57-67.

OECD (1989), *Predatory Pricing*, OECD, Paris.

Pompl, Wilhelm (2002): *Luftverkehr: Eine ökonomische und politische Einführung*, Berlin/Heidelberg/New York: Springer.

Pott, Wolfgang (2004), Wenig Arbeit, viel Umsatz in der Franchise-Branche; in: *Welt am Sonntag*, 5.9.2004.

Rae, David (2001), easyJet: a case of entrepreneurial management?, in: *Strategic Change*, Vol. 10, pp. 325-336.

Reed, Arthur (1998), Up in BA's Blue Sky, in: *Air Transport World*, No. 1, pp. 46-47.

Rothfischer, Brigitte (2002), Ryanair: Irische Invasion, in: *Aero International*, No. 4, pp. 58-59.

Schmitt, Stefan (2001), No-frills-Anbieter im europäischen Luftverkehrsmarkt – Potentielle oder tatsächliche Konkurrenz?, in: *Dortmunder Diskussionsbeiträge zur Wirtschaftspolitik*, No. 110, pp. 1-31.

Sparaco, Pierre (2002a), Low-Cost Carriers Steal The European Show, in: *Aviation Week and Space Technology*, Vol. 157, No. 21, pp. 59-61.

Sparaco, Pierre (2002b), Ireland's Ryanair Seeks European Supremacy, in: *Aviation Week and Space Technology*, Vol. 157, No. 21, p. 62.

Steen, F. and Sorgard, L. (2002), From regulated duopoly to a private monopoly: The deregulation of the Norwegian airline industry, in: *Swedish Economic Policy Review*, Volume 9, pp. 191-221.

Sterzenbach, Rüdiger (1999), *Luftverkehr. Betriebswirtschaftliches Lehr- und Handbuch*, München: Oldenbourg.

Sull, Don (1999), Case Study – easyJet's $500 Million Gamble, in: *European Management Journal*, Vol. 17, No.1, pp. 20-38.

Wiezorek, Bernhard (1998), *Strategien europäischer Fluggesellschaften in einem liberalisierten Weltluftverkehr*, Frankfurt: Lang.

Williams, George (2001), Will Europe's charter carriers be replaced by 'no-frills' scheduled airlines?, in: *Journal of Air Transport Management*, Vol. 7, pp. 277-286.

Part E:
Theoretical and Policy Aspects of Predatory Pricing and Competition

Chapter 12

A Case of Beer and Pretzels

William G. Morrison

1. Introduction

Predatory pricing is a notoriously slippery concept due to the apparent existence of a fine line between healthy competition (in which falling prices and increased productive efficiency benefit consumers over the long term) and the possibility of unhealthy competition (in which consumers enjoy lower prices initially only to be faced with the consequences of market power in the future). The dynamic evolutionary nature of market structures lies at the heart of the debate: how can we tell over a given period of time, whether observed competition in a market is likely to continue into the future given the possibilities of either naturally occurring or strategic entry and exit? The answer to this question depends on our ability to detect and measure behaviour that is harmful to sustained competition and consequently on our ability to distinguish naturally occurring from coerced market exit. In this paper a fictitious example is constructed to explore some aspects of predation that are especially relevant to air travel markets. In particular, the example is contrived to emphasize the importance of capacity utilization, product differentiation and product complementarities. The example is viewed through the lens of a stylized investigation algorithm developed in Morrison (2003) that captures common elements in predatory pricing rules across various jurisdictions. The algorithm highlights key decision nodes in the investigation process and the findings necessary to proceed towards conviction.

Section 2 outlines a stylized description of predatory pricing investigations. In section 3, a fictitious case is developed in which strategic interaction between two pretzel producers leads to allegations of predatory pricing. Section 4 discusses the merits and issues raised by the competition authority's case and by the accused firm's defence. Section 5 relates the issues in the case to competition between full service and value-based airlines in air travel markets.

2. Defining and Determining Predatory Pricing

Morrison (2004) draws from an international comparison of predatory pricing legislation and procedures to create a composite stylized algorithm for the decision process underlying the investigation of predatory pricing allegations. The algorithm incorporates commonly observed elements in predatory pricing institutions around the world. Suppressing some elements and emphasizing others in the algorithm allow us to match it to particular cases or jurisdictions. Moreover the algorithm serves to highlight key differences across jurisdictions by reference to the relative emphasis placed on the various elements.

As illustrated in Figure 12.1, the algorithm has three main decision nodes, relating to the accused firm's market power and behaviour before, during and after the period of alleged predation. The first question to be asked is whether the accused firm held a significant degree of market power prior to the alleged predatory conduct. Secondly, during the period when predatory pricing is alleged to have occurred, what was the defendant's price in relation to costs of production? More specifically what was the defendant's price (P) in relation to total average cost (ATC) and average avoidable cost (AAC).This yields three possibilities:

i. $P > ATC$
ii. $AAC > P$
iii. $ATC > P > AAV$

Thirdly following the period of alleged predation, can the accused firm reasonably expect to recoup profits that were foregone during the period of the alleged predation?

The first decision node requires that the accused firm have sufficient market power to abuse a dominant position. In many academic discussions of predatory pricing, the defendant is characterized as a monopolist, who is faced with the entry of a competitor, however in practice, market shares of 35% or more are used to define a threshold of sufficient market power.

That is, the firm must have some ability to affect the prices of competitors. The second decision in the process relates the pricing policies of the accused firm to its costs to determine whether or not the firm engaged in activities that resulted in avoidable losses. Here, the traditional standard – the Agreeda-Turner test – of prices relative to average variable cost, is often not practical given multi-product firms. Consequently, the implementation of a cost-based standard for predatory pricing has graduated towards comparing price to the accused firm's *average avoidable cost*, which includes product specific fixed costs and variable costs, but not sunk costs.

The second decision in the process relates the pricing policies of the accused firm to its costs to determine whether or not the firm engaged in activities that resulted in avoidable losses. Here, the traditional standard – the Agreeda-Turner test – of prices relative to average variable cost, is often not practical given multi-product firms. Consequently, the implementation of a cost-based standard for predatory pricing has graduated towards comparing price to the accused firm's *average avoidable cost*, which includes product specific fixed costs and variable costs, but not sunk costs.

Considering the three possibilities in the algorithm, clearly if price is found to exceed average total costs, then the firm cannot be found to have set unreasonably low prices and an allegation of predatory pricing will most likely be dismissed at this stage. On the other hand, if price is found to be below average avoidable cost, then the accused firm can be said to have sustained unnecessary losses during the period. As such, price below AAC is at least a necessary condition for a finding of 'per se' predation and in some jurisdictions might be a sufficient condition.

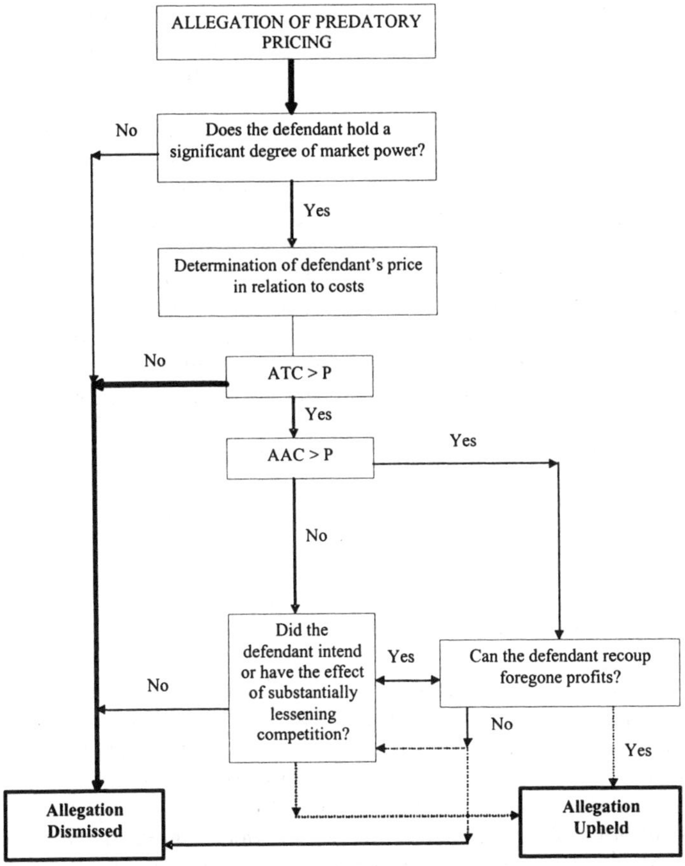

Figure 12.1 A stylized predatory pricing investigation algorithm[1]

The trickier potential finding is a price that falls in the range between average total cost and average avoidable cost. Such a price does not define avoidable losses, yet a price below ATC nevertheless means that losses were sustained, which might still be construed as predation, particularly if prices were above ATC in the recent past. Pricing between ATC and AAC thus constitutes a 'grey area' that suffers from both conceptual and measurement problems. In such situations the competition authority may not abandon prosecution, choosing instead to look for other measures or mitigating factors, such as the intent of the accused firm's pricing policies, the relative efficiency of the alleged victim, the possibility of reputation effects resulting from the accused firm's actions and the actual or perceived extent of technical or strategic barriers to entry in the future.

In the grey area, it becomes more difficult to relate the exit of a competitor to a lessening of competition. For example Button's (2003) argument concerning the

[1] Source: Morrison (2004).

possibility of an empty core in the market could be used to provide a plausible alternative explanation of events thought to be consistent with predatory pricing.[2] These measurement problems are likely to place a great deal of weight on the ability of any investigation to show pricing below average avoidable cost.

The third node in the decision algorithm asks whether the accused firm has or can recoup profits forgone during the predatory pricing period. If the defendant set its price below average avoidable cost but cannot recoup forgone profits, then in some jurisdictions (the US in particular), the allegation of predatory pricing will likely not be upheld. In other jurisdictions, relatively more weight might be placed on whether the accused firm's actions had the effect of substantially lessening competition. That is, even if recoupment is found not to be possible, reputation effects or other barriers to future entry could be used to argue that a substantial lessening of competition occurred. The dotted lines in Figure 12.1 indicate that several decision paths are possible here, depending upon the legislative focus.

In the next section I develop a fictitious market interaction between two firms who produce pretzels. This example deliberately de-emphasizes the pre-predation dominance and post-predation recoupment questions in order to focus upon the measurement and inference of price in relation to costs during an alleged period of predation.

3. A Case of Beer and Pretzels

Consider the following scenario involving the interactions of two firms. Firm X produces two products: X-beer and X-pretzels, which are gross complements. The company's market research (and independent industry research) shows that some of the customers that buy its pretzels also buy its beer, however the beer and pretzels are sold at different stores. Significant capital requirements in the production of both of these products are vats in which ingredients are mixed. Firm X owns two large vats and one smaller one. These vats are *internally* flexible in that they can be deployed in the production of either beer or pretzels and can be switched if required. However the vats are difficult and costly to dispose of externally, and as such the vats are regarded by firm X as sunk costs in relation to its overall operation.

---

[2] Button argues that market structures in passenger aviation are devoid of a stable market structure. This 'empty core' property is found in theoretical location models of the type developed by Hotelling in 1929. In such models, a stable (Nash Equilibrium) market structure may exist with two firms, such that each firm makes positive economic profits. However the existence of this equilibrium evaporates if the model is used to analyse competition between three firms because one firm always ends up located in between the other two, with no market share. Once the unprofitable firm exits the two remaining firms will make positive profits until the entry of a new competitor squeezes one of the existing incumbents out of the market. The market structure may thus be plagued by continual entry and exit. Of course this reasoning relies on low entry barriers and limited foresight by potential entrants, who do not realize that once they become incumbents, they themselves will be squeezed out of the market by subsequent entry.

Chronology of Events: February

In February, firm X was a monopolist in the pretzel market, had a significant market share in the beer market and was profitable overall (total revenues exceed total costs). Each bag of X-pretzels sold in a re-sealable bag for $1. As part of an ongoing sales promotion strategy, firm X included a coupon ('pretzel cents') inside each bag that provided a 10 cent discount on a future purchase of X-pretzels. At this time, firm X was deploying its two large vats in beer production and its smaller vat in the production of pretzels. The average avoidable cost per bag of pretzels can be calculated in two ways: with or without the inclusion of costs relating to vats. With this deployment of its vats, firm X's average avoidable cost per bag of pretzels was 40 cents if vat costs were omitted, and 65 cents if they were included. The capital costs associated with packaging the pretzels in resealable bags are sunk and are not included in this avoidable cost calculation.[3]

Chronology of Events: March

In March, a rival firm (Y) entered this market with a competing pretzel product. Firm Y did not compete in the beer market, but distributed its pretzels through the same snack food store as firm X. Firm Y offered fewer pretzels (half the amount offered by X) in a plain, non-resealable bag and priced its pretzels at an 'everyday low price' of 55 cents per bag with no discount coupons. Firm Y used a small vat in production and had an average avoidable cost (including vat costs) of 40 cents per bag.

Chronology of Events: April

In April, firm X reduced its price per bag of pretzels to 55 cents (matching Y's price). In addition, firm X redeployed its vats, moving one large vat into pretzel production and switching the smaller vat into beer production. As a result of this switch, average avoidable costs (including vat costs) fell to 60 cents per bag in pretzel production, while in beer production, average avoidable costs (including vat costs) increased slightly. Firm X remained profitable overall (total revenues exceeded total costs).

Chronology of Events: August/September/October

In August, Firm Y exited the pretzel market and in September, firm X redeployed its vats in the old configuration: the smaller vat was once again dedicated to pretzel production and the larger vat to beer production. Firm X raised the price of pretzels back to its old February price of $1 per bag. In October, Firm Y formally filed a complaint against X with the competition authority, alleging that X engaged in predatory pricing.

---

[3] The issue of whether the vats constitute variable or avoidable costs will be discussed later in the paper.

Some Details of the Investigation

In deciding to pursue an investigation of the alleged predatory pricing, the competition authority issues a statement stating that in its opinion:

a) X was in a dominant position prior to Y's entry
b) firm X engaged in a programme of selling its pretzels at a price below average avoidable cost and that
c) After the exit of firm Y, firm X raised its price to pre-entry levels with the intent and likelihood of recouping the losses it sustained during the period of alleged predation.

Applying the 'facts' of our fictitious case to the investigation algorithm, the competition authority begins by asking if firm X had sufficient market power in the relevant product market, prior to entry by Y. A broad market definition that includes other snack foods would reduce the measure of X's market power, but for now let us assume that pretzels are viewed as the relevant product market. Since firm X was the only producer, prior to entry by firm Y, we can regard X as an ex-ante monopolist and in a position of dominance. Also, the market structure returned to a monopoly after firm Y's exit and firm X's price returned its original level, therefore there is a strong argument that X will recoup any profits forgone during the period of alleged predation. With the ex-ante and ex-post criteria largely met, the focus becomes X's pricing policy in relation to its costs. The authority cites evidence that firm X's price (55 cents) was below its avoidable costs of production (60 cents). Further, the authority argues that firm X's conduct substantially lessened competition by creating a credible reputation for predatory conduct that serves as a strategic barrier to future entry.

Firm X's Defence

In response to the charges, Firm X submits a defence, with five components:

i. The authority has incorrectly neglected the beer market, which must be considered in conjunction with the pretzel market, due to the complementary nature of the products. In this context, the price cut for pretzels should be viewed as a legitimate 'promotion' strategy to encourage beer sales.
ii. Language found in internal memos and emails during the period of Y's tenure in the market (such as the comment 'wiping Y off the face of the earth') simply reflect the language of aggressive competition between two market rivals. No evidence exists of communications that show an unambiguous intent to deliberately sustain losses in order to cause or hasten Y's exit, with the goal of recouping lost profits afterwards.
iii. Firm X's response to Y's entry in the pretzel market was competitive but not predatory. Its pricing policy simply involved matching the prices of a rival firm and since it lowered its price to do this, it had to increase capacity to keep up with demand, hence the switching of vats.

iv. The vats should not be counted in the calculation of average avoidable costs. These costs should be treated as sunk costs because firm X could not easily or without substantial cost, sell or otherwise dispose of these assets. Consequently it is firm X's opinion that its price of 55 cents was above its average avoidable cost of 40 cents.

v. Even if vat costs are included, the relevant cost test for predation should be to compare firm X's price to firm Y's average avoidable costs. Quoting from Baumol (1996), firm X states that predation has to do with a dominant firm causing the exit of an equally or more efficient rival. Since firm Y had a lower cost structure than firm X, it would only be forced to exit by firm X's pricing behaviour if firm X's price was lower than firm Y's average avoidable cost. Firm X's price of 55 cents per bag exceeded firm Y's average avoidable costs of 40 cents (which is an overestimate due to the inclusion of vat costs).

4. How Strong is the Case against Firm X?

Some element of X's defence can be dealt with more easily than others. Part (v) of the defence which argues that the proper price cost test is to compare X's price to Y's costs, is clearly fallacious. Baumol's (1996) arguments are only appropriate when the products concerned are homogeneous. In this case the products are differentiated, with firm X offering more quality (vertical differentiation) relative to firm Y. When offered at the same price, more pretzels in a resealable bag will be unambiguously preferred to a smaller number of pretzels in a plain bag. Thus it can easily be determined that price matching by X was de facto price undercutting after adjusting for product differentiation.

However beyond determining this, an additional problem is the calculation of an appropriate price premium to define 'quality adjusted' price matching. Since there is the potential for horizontal as well as vertical differentiation, the competition authority needs to uncover the distribution of consumer preferences over all the relevant attributes of each product in order to calculate the price premium. This is likely to be costly and difficult. The competition authority might nonetheless elect to formulate a more practical solution using a rougher cost-based measure of the premium. Such a premium could be calculated as a normal rate of return over the average cost of packaging X-pretzels in resealable bags plus the value of the 'pretzel cents' coupon (10 cents). There is however a possibility of underestimating the premium since the particular bundle of attributes offered by X could be valued more highly than the sum of values for each attribute added independently.

As an alternative to the calculation of a price premium, the competition authority could simply require that X's price be compared to its avoidable costs, which would include product specific fixed and variable costs.[4] However, this will also bias this decision by underestimating AAC to the extent that capital costs associated with the resealable packaging are sunk and therefore not included in this calculation.

---

[4] Those jurisdictions where emphasis is placed on comparing lost profits to ex-post recouped profits would also prefer to compare the defendant's price and own average avoidable cost.

Product Complementarities

The complementarities between X-pretzels and X-beer is being used to justify price cutting by X as form of promotion that increased beer revenues and thus maximized joint profits from both markets. In response to this, the authority could argue that X's price cut was excessive (X has never engaged in such a drastic price cut before) and that the timing which coincided with Y's entry indicates predatory intent. Secondly and more importantly, if the primary purpose of X's pricing programme was to promote beer sales, then why did X not bundle the two products together more directly? For example, X could have offered beer coupons in each bag of pretzels or could have attached free promotional bags of pretzels to its cases of beer. Firm X could have used the pretzel market to promote beer sales in ways that would not necessitate lowering the price of pretzels *across the board*. Notice that here, it is important to separate the intent from the effects of firm X's pricing programme. Even if it was not a conscious attempt to force Y to exit the market, there was a negative external effect of the pricing policy that was avoidable, give its stated intent as a promotion.

Avoidable Costs

If firm X's price is to be compared to its own avoidable cost, then the issue of how avoidable costs are calculated becomes very important. If costs associated with the vats used by firm X are excluded from the calculation, then price will not be below average avoidable costs, but will be judged to be in the grey area between AAC and ATC.

If price is in the range between AAC and ATC, then the authority's case for upholding the predation charge is significantly weakened. The authority may still attempt to show that there was intent and that the effect of the pricing by X was to substantially lessen competition. However, given the complicating factor of product market complementarities, and product differentiation as sources of reasonable doubt, the charges are more likely to be dismissed if there are no concrete measures that relate X's actions to a lessening of competition rather than to a harming a particular rival.

In contrast, if the authority is able to convince the judging body that price is below average avoidable cost, then the assessment is more likely to lean more towards 'per se' predation. This is because a price below average avoidable cost can be more clearly linked to a lessening of competition rather than a competitor. In particular such a finding suggests that firm X willingly incurred *avoidable* losses during the period of alleged predation. This also strengthens the credibility and magnitude of reputation effects that could deter future entry by Y or other firms.

Should the vat be included in the calculation of avoidable costs? In arguing that they should not be included, X is pointing to the fact that disposal of these assets is not easy or costless, however the real issue here is that capacity was *redeployed* from one market (beer) to another (pretzels). Therefore, even though complete disposal of the vats indicates that they represent sunk costs overall to firm X, the competition authority can argue that since the firm can (and did) redeploy vats across markets, the large vat which was redeployed as part of the pretzel pricing strategy is a legitimate avoidable cost. Furthermore, the threat of redeploying capacity in this way provides a source of credibility to limit pricing as an entry deterring strategy that mitigates the usual critique of the Bain-Sylos postulate. That is, the incumbent firm has a credible means of increasing capacity temporarily until the threat of entry has passed. Then by redeploying capacity elsewhere in the system, the incumbent can reduce capacity and return to profitability.[5]

Note however that this strategy need not be costless. In this example we need to assess whether there are forgone profits in the beer market resulting from the redeployment of the vats.

When a large vat was switched with the small one, average costs increased in the beer market and these foregone profits represent opportunity costs of using the large vat in the pretzel market. The competition authority can therefore argue that these opportunity costs should be included in the average avoidable cost calculation. Note that since firm X redeployed its capital in a way that resulted in lost profits in the beer market, this strengthens the arguments for including the vat costs as part of average avoidable costs because an alternative redeployment would have been more profitable. Further, the redeployment to the old configuration in the presence of these opportunity costs is consistent with recoupment.

In summary, the 'facts' of the beer and pretzel case have been manufactured to create an environment that mimics conditions that are key to understanding predatory pricing issues in air travel markets. The example highlights how product differentiation, product market complementarities and redeployments of capacity complicate and impact a competition authority's ability to uncover and demonstrate the existence of predatory pricing. The stylized algorithm employed to represent the decision path of an investigation indicates that these elements of the problem primarily effect the determination of price in relation to cost and the degree to which firm X's conduct caused a substantial lessening of competition. In particular, if redeployed capacity is not included in the calculation of avoidable costs, and revenues beyond the pretzel market are included in profit calculations, a finding against firm X is quite unlikely, even though ex-ante and ex-post market dominance can be demonstrated.

5. Beer, Pretzels and Air Travel Markets

The extent to which the reader finds the example excessively contrived is perhaps a rough estimate of how the particular characteristics of air travel markets deserve to be given separate guidelines or legislation in competition policy.

The beer and pretzel case was developed to provide direct analogies to air travel markets. The appropriate analogies between the example and competition in air travel are outlined in Table 12.1.

---

[5] See Bain (1956) and Sylos-Labini (1962) for the original limit output argument. See Spence (1977) for a critique and extension of the limit-output model that focuses on the credibility of sunk capacity investments.

Table 12.1 Analogies between air travel markets and the beer and pretzels example

Competition and competitors in air travel markets	Competition and competitors in the beer and pretzels example
A full service (hub-and-spoke network) airline (FSA)	Firm X
A value-based (point-to-point) airline (VBA)	Firm Y
A city pair, where the FSA's hub airport, is located in the destination city	Pretzel market
A city pair where the origin is the FSA's hub airport	Beer market
In flight services (meals, seat design and spacing, seat pre-booking) and airport services (covered bridges, preferred locations in terminals, large number of ticket desks and auto-check-in centres)	Resealable bags
Customer loyalty programmes offered by the FSA (frequent flier miles)	Pretzel cents coupon
Smaller aircraft (lower seating capacity and range)	Small vat
Larger aircraft (larger seating capacity and range)	Large vat

The beer and pretzel case illustrates how the characteristics of product differentiation, demand complementary internally flexible capacity within a network of city pairs combine to create special circumstances for competition and predation in air travel markets.

The issues highlighted by the beer and pretzels example suggest that competition policy legislation relating to predatory pricing in air travel should stipulate

a) That average avoidable cost calculations should include aircraft costs when aircraft are redeployed in support of price cuts and that net profits are foregone as a result of the redeployment.
b) That product differentiation must be taken into account when assessing competition between airlines. In particular an FSA should not expect that price matching can be used as a defence of the rival firm is a VBA. However the difficulties in uncovering preferences suggests that focusing upon average avoidable costs, is more pragmatic, even though such a measure will probably underestimate the appropriate price differential between airlines offering different service bundles.

c) That demand complementarities should not be allowed as a 'beyond revenues' defence against price cuts, then the pricing policy does not specifically target those travellers who will travel beyond the city pair in question. Note that this is not a question of intent. It might be the case that an FSA has innocently designed its business operations to offer low fares to all travellers between a city pair a means of enhancing beyond revenues. But here the criterion must be the potential or actual harm to competition.

While questions still remain concerning whether recoupment should be a necessary condition for a finding of predatory pricing, the above changes in approaching price/cost decision node in the process would represent a move that more clearly defines the 'rules of the game', in a way that is both measurable, transparent and relies less upon subjective evidence of intent. Clearly defined regulatory rules would reduce the level of regulatory uncertainty which would provide a positive stimulus to aviation market investment and decision making and provide consistency of decision making by competition authorities and the associated resource costs currently being employed in seeking to either prove or dispute the existence of predatory pricing. These resources could then be reallocated to more productive uses [...] including perhaps, the production and consumption of beer and pretzels.

Acknowledgements

I wish thank participants at the 'Predatory Behaviour in Aviation' Workshop, sponsored by the Hamburg Institute of International Economics, for comments on an earlier draft.

References

Bain, J. (1956), *Barriers to new competition*, Cambridge, MA, Harvard University Press.

Baumol, W.J. (1996), Predation and the logic of the average variable cost test, *Journal of Law and Economics*, 39(1), 49-72.

Button, K. (2003), Does the theory of the 'core' explain why airlines fail to cover their long-run costs of capital?, *Journal of Air Transportation Management*, 9(1), 5-14.

Competition Bureau (Industry Canada) (2001), *Enforcement Guidelines on the Abuse of Dominance in the Airline Industry*, 19-20.

Morrison, W.G. (2004), Dimensions of predatory pricing in air travel markets, *Journal of Air Transport Management* (10), 87-95.

Spence, A.M. (1977), Entry, capacity, investment and oligopolistic pricing, *Bell Journal of Economics*, 8, 534-544.

Sylos-Labini, P. (1962), Oligopoly and technical progress, Cambridge, MA, Harvard University Press.

Chapter 13

Investigating Airline Managers' Perception of Route Entry Barriers: A Questionnaire-Based Approach

Mirko C.A. Schnell

1. Route Entry Barriers and Deregulation

In the following paper a recently completed research project applying a new investigative approach to the area of route entry barriers in airline markets will be presented. The main research question under consideration was: 'How do airline managers perceive barriers to route entry?' Taking into account that only *perceived* entry barriers will prevent actual entry, managers and analysts need to know which economic agent notices which entry difficulties and to what extent these obstacles deter entry in practice. Hence, the present research aims at shedding further light on the issue of predation in aviation markets by directly analysing airline managers' perception of the effectiveness of predatory strategies in the aviation industry.

Although theoretical models and empirical studies have greatly contributed to the understanding of what drives entry the overall knowledge in this area remains rather limited. This lack of knowledge is largely due to the lack of data, the disregard of an intra-firm perspective and too restrictive behavioural assumptions respectively. However a profound knowledge of entry barriers is of fundamental importance for the formulation of adequate policy and business strategy alike, given that market characteristics such as entry barriers restrict airlines' playing field per definition. Hence the existence of entry obstacles severely limits competition in a market; whereas deregulation intends to intensify it.

Over the past couple of decades airline markets around the globe have been deregulated. Regulators not only removed institutional entry barriers in domestic markets such as Canada or the USA, but also lifted restrictions that were formerly imposed on operations by foreign carriers in multi-country regions like Australia, New Zealand and the European Economic Area.[1] Governments largely expected that these pro-active moves would automatically increase competition, which in turn would enhance the industry's efficiency as well as social welfare in general.

[1] The European Economic Area comprises all countries of the European Union and, additionally, Iceland and Norway. Liberalization became effective in this region.

In the following, the paper first discusses the value of a positive definition of entry barriers for assessing the effect of predatory measures on entry. Second, it identifies knowledge gaps regarding entry barriers. Third, it describes a methodology that may address some of these gaps. Finally, the findings of two research projects using this new methodology will be presented. The two projects investigated entry difficulties in the aviation markets in Australia/New Zealand, Canada, the European Union and the USA. The findings show that entry into a market is rendered more difficult by predatory behaviour compared to a scenario in which only a single incumbent serves the route. Respondents also ranked them among the most effective entry deterring conditions on route-level. Further it appears that respondents routinely expect predatory measures by the incumbent on at least some routes considered for entry by their airline.

2. Defining Route Entry Barriers

'A contestable market is one into which entry is absolutely free and exit is absolutely costless' (Baumol, 1982, p. 3). Free entry, however, does not mean that entrants face no costs in starting the business, but that cost discrimination against entrants is impossible (Baumol, 1982). This represents a normative definition of entry barriers, which above all is relevant to competition policy.

Entry barriers do not only appear in contestability theory, but are also relevant in many other economic theories (e.g. Connor, 1991; Dixit, 1980; Spence, 1977). To the extent that these theories or models generally assume 'rationality', actual barriers become the same as perceived ones; that is, theories or models take a normative view of the issue of barriers to entry.

However by using a normative definition of entry barriers, economic theories of market entry may risk overlooking some entry barriers and hence may overestimate the threat of potential competition. This is especially the case when considering that airline managers may in fact not act in line with the assumption of 'rationality'. Moreover, such a normative definition cannot inform business strategy as it fails to acknowledge impediments that may deter airlines, such as brand name or superior quality. Consequently, although economic theory aims at investigating the important issue of barriers to route entry, it risks playing down just this very concern.

Given the above and past and present attempt at explaining real competition among airlines, questions concerning what *does* (i.e. positive perspective) rather than what *should* (i.e. normative perspective) deter entry according to 'rational' behaviour are of crucial importance. Hence, a distinct definition is chosen for the purpose of the research project presented: here a barrier to route entry is one which does actually prevent an airline from operating flights on a particular route.[2]

The usefulness of this positive definition stems from its broad perspective in two respects. First, it enables economists to gain a better understanding of competition. In fact as long as additional obstacles remain in place, rivalry among airlines might not develop even where all conventional entry barriers (i.e.

[2] In the following, entry barriers refer to entry on route-level.

according to normative definitions, cf. Demsetz, 1982) have been removed. However, when choosing a positive definition – caution needs to be exercised about drawing immediate conclusions for regulatory policy. Not every entry existing obstacle as defined by the above positive characterization should trigger actions by regulators. Second, the positive definition used here informs business strategy as it reveals which (strategic) actions may keep potential entrants from entering an incumbent's market. Conventional definitions of entry barriers largely only determine which of these measures should prompt regulatory interference. Hence, those deterrence devices not banned by the conventional perspective represent opportunities to airlines by which they can shelter their markets from competition without infringing on competition law.

Regardless of the way in which entry barriers are defined, a clear distinction between natural and strategic entry barriers is largely discernible. The former are inherent features of a market, that is, they exist independently of any incumbent's behaviour. The latter describes obstacles either unintentionally or purposely erected due to the conduct of the incumbent(s). As is the case with many categorizations, the resulting groups tend to overlap: for example, advertising expenses to inform potential passengers of a new service (cf. natural) may increase where there is an incumbent on the route in question (cf. strategic). Similarly, as suggested by the S-curve relationship (Fruhan, 1972), the scale of route entry required for a profitable service depends on the offered capacity compared to rivals.[3] Hence, incumbents can reinforce the effects of natural barriers. Since it is difficult to distinguish the natural effect from an artificial effect imposed by the incumbent(s), such reinforcing measures may represent a subtle way for airlines to restrict competition.

3. What do we know about Route Entry Barriers?

Earlier studies of airline markets have greatly enhanced the knowledge of how airlines compete, how airlines' conduct affects social welfare and the extent to which this is in line with economic theory. In exploring airline competition, research found that fares may depend, for example, on measures of concentration, the number of incumbents and potential entrants (Graham et al., 1983; Strassmann, 1990; Peteraf/Reed, 1994; Gimeno, 1999).[4] Moreover, findings show that prices tend to fall when entry occurs (Joskow et al., 1994). Overall it appears that most results largely question the contestability of airline markets.

Despite their merits, previous economic studies following the traditional approach have some shortcomings: they only observe the result of non-contestability rather than its cause(s), they neglect airline characteristics and the perspective of airline managers, they treat only some entry obstacles while overlooking others and they mainly use US-data.

[3] The S-curve relationship refers to the phenomenon that an airline that offers more than 50% of route capacity has a market share above its capacity share, while an airline that offers less than 50% of route capacity has a market share below its capacity share.

[4] An overview is available from the author upon request.

Although previous analyses do detect relationships among variables contradicting contestability theory, their insights are less systematic when it comes to why this contradiction may be the case. More precisely, these studies say rather little on the following questions: Which assumption(s) of contestability theory is/are precisely infringed in real markets? For example, could it be that incumbent airlines are in the position to alter their prices more quickly than assumed by theory? If this was the case then an entrant's hit-and-run strategy would not be feasible. Furthermore do entry barriers exist, which render entry and exit costly? Indeed it can be reasonably argued that airline markets exhibit features that qualify as entry difficulties (e.g. Doganis, 1994) and some empirical studies support such reasoning (e.g. Sinclair, 1995; Gimeno, 1999). However, there are two potential problems with the application of such logic. First, there appears to be a tendency to find plausible explanations (e.g. entry barriers) for phenomena (e.g. non-contestability) that, however, may not correspond to the perception of airline managers and hence may not influence their behaviour in the predicted way. Second, airline managers may well perceive a certain condition denoted as a difficulty by logic, but may then find ways to overcome the obstacle. Consequently, the obstacle in question would not deter entry, which is in contrast to the expectations formed by logic. Hence it can be argued that pure logic may in some cases result in misleading conclusions. Therefore, the relevant question needs to read: How do airline managers view their competitive environment? Again, previous evidence does not systematically inform on this issue as the necessary data is not readily obtainable on which a satisfactory answer could be based. To the best of the author's knowledge, Levine (1987) published the only comprehensive paper that exhibits the perspective of an airline manager on the issue of barriers to entry.

Likewise, what conclusions could be drawn from the lack of statistically significant findings with respect to a specific entry barrier? It either shows that the characteristic in question is too rarely observed (or does not exist at all) to exert considerable influence on competition or that it is indeed often present, but without having the supposed negative effect on competition. In any case, the lack of statistical evidence however does not rule out that a specific entry barrier never occurs (Lieberman, 1987).

Furthermore, all these studies implicitly assume that industry participants share the same perception about entry difficulties. Such implicit assumption however overlooks two important facts. First, market environments are not unambiguous realities as confirmed by research in management science (Stubbart/Ramaprasad, 1988; Fombrun/Zajac, 1987) and economics (Aiginger et al., 2001). Hence, perception of entry barriers is likely to vary among airline managers. Second, this insight is not trivial, as a large body of theory and evidence shows that managers' perception of environmental conditions matters (Penrose, 1959; Duncan, 1972; Snow, 1976; Paine/Anderson, 1977; Lefebvre et al., 1997). Similarly, differences among (groups of) carriers have been so far hardly addressed. As opposed to economic theory, airlines do not form a homogeneous group of firms. Consequently, a specific market characteristic may have very different effects on the entry behaviour of two distinct carriers.

It needs to be noted that most of the above objections are due to the lack of data and not to the respective methodology applied to analyse the available information. The only modest reward by the economics profession for the effort to collect primary data (Schmalensee, 1988) may explain why the dearth of data has persisted. Yet, as such data would yield valuable insights for the society in general, this lack appears to be a special case of market failure.

The lack of data results in two additional knowledge gaps. First, it prevented so far to determine the importance of strategic measures, such as frequent-flyer schemes for competitive advantage (Borenstein, 1991). Therefore, the absolute significance of certain route characteristics as an entry barrier remains unclear. Second, the data problem has made researchers focus their empirical analyses on only a few barriers to entry. Systematic research dealing with the long list of entry barriers suggested by theory and statements of airline managers is not available. Hence, a comprehensive comparison of the effectiveness of entry obstacles is so far not possible. However it appears unlikely that all entry barriers would inhibit entry to the same extent. To ensure that policy measures are effective and efficient, they need to tackle the severest problems first. That again requires knowledge of which impediments are most effective in preventing entry.

To circumvent the restrictions set by the lack of data, researchers have applied proxies, which aim at capturing the phenomenon in question. However, such substitutes portray the actual phenomenon only partly. Besides, while researchers naturally try to derive proxies by using logical reasoning, the relationship between the actual phenomenon and its proxy may not correspond to what airline managers perceive. For example, Peteraf (1995) uses a vector of sensible proxies to account for sunk costs. Yet, the proxies may not tally with the way airline managers calculate non-recoverable costs, provided they take account of such costs at all. In this respect it needs to be noted, that only what managers *do* consider (i.e. positive view) affects their behaviour, and not what they *should* consider (i.e. normative view). Hence given that the concept of 'rational behaviour' does not hold in reality, the discrepancy between the positive and the normative view does exist.

An additional noteworthy problem is that the insights at hand are almost exclusively based on data from US-markets, reflecting the availability of data. The 10%-ticket-sample, which US-airlines have to provide by law, offers a unique and rich data set for research. To the best of the author's knowledge, a comparable source of data is not accessible for air transport markets in any other country at affordable cost. Consequently, only few studies based on systematic data focus on markets outside the USA. For example, Schwegmann (1998) analyses fares on intra-European routes by using data sold by the Reed Travel Group.

Such shortage of data leads to general policy recommendations based mainly on US-evidence. However, this general approach neglects the question of whether a specific entry difficulty is equally effective in all deregulated airline markets. It appears rational to argue that for some reason an entry barrier may be more effective in one region than in another at any one point in time. This may be due to the fact that airline markets, once they are deregulated, go through different phases regarding the intensity of competition. In fact, Joesch/Zick (1994) and Fawcett/Farris (1989) provide findings that shortly after deregulation the US-airline markets were contestable but become non-contestable due to measures applied by incumbent airlines. Additionally, based on the assumption that a carrier's characteristics influence the effectiveness of entry impediments, the composition of the industry with respect to the type of carriers within one region may affect barriers' effectiveness. Moreover, a carrier's strategies and the barriers' height may interact. In this context, new entrants in the European airline markets report a management emphasis which differs from the one of their US-counterparts (Gudmundsson, 1997). This difference may be (partly) due to the different perception regarding route entry .

In conclusion, many questions relating to entry barriers still await answers. Considering that only perceived entry barriers prevent entry, it is of interest which entry difficulties are noticed by whom and to what extent they prevent entry in practice. Such knowledge would largely contribute to explaining systematic differences in entry behaviour. Moreover, it would provide some substantial information needed to develop policy measures that aim at fostering competition. It would also represent a valuable base for developing and assessing airlines' strategies.

It needs to be stressed that the above remarks about the limitations of existing evidence should not be taken as a negative assessment of it in general. It goes without saying that no methodology can answer all possible research questions. However some approaches may be more appropriate to answer a certain question than others and that different methodologies can generate distinct valuable insights into the same issue. Moreover, the above review reveals that many shortcomings result from the mere lack of data rather than the methods applied to analyse the data.

4. A Proposal how to Enhance the Knowledge about Route Entry Barriers

Following the above argument, a research methodology different from the ones used so far is advocated. The chosen approach subsequently should address the knowledge gaps discussed above, which appear largely caused by the lack of readily available data. To tackle this objective, the research design has to meet the following requirements: First, it has to generate a new data set to alleviate the problems caused by insufficient data. Second, the data should portray an intra-firm (or even intra-manager) perspective – a point recently made by some economists (Mueller, 1992; Helper, 2000). Third, the data collection procedure should be capable of handling a comprehensive list of entry barriers. Apart from these conditions, the results should comply with the need of generalizability and therefore conclusions should be based on more than a handful of airlines. Moreover, the approach should be affordable and manageable with respect to time constraints. In view of this set of requirements, the use of a questionnaire-based survey seems appropriate.

Such a survey-based method provides intra-firm views and data not otherwise available. It has the potential to investigate how airlines' strategies and resources influence the perception of entry barriers. Furthermore, it can take the effect of additional characteristics of a carrier (e.g. size, scope or age) into account when assessing the way obstacles are being perceived. Moreover, with respect to respondents, the affiliation to a certain department within a specific airline might also have an impact on the assessment of entry difficulties. In summary, this method does not only consider hypotheses advanced by Industrial Organizations-theory, but also by the resource-based view (e.g. Penrose, 1959; Conner, 1991; Peteraf, 1993), which has recently attracted increasing attention in the field of strategic management. Thereby, it does not only inform policy, but also airlines' strategies: It helps explain entry behaviour and not only suggests effective deterrence devices but also means to overcome them. Eventually, it may shed new light on why some airlines fail.

From the perspective of long-term competition policy, the results spawned by this methodology may be a first step to question some of the conventional wisdom about competition on monopolistic routes. Suppose such a survey finds the following: some potential obstacles (e.g. monopolist's strong brand name, superior quality or sufficient supply of capacity) which, by normative definition, are not regarded as barriers to entry prove to be indeed quite effective. In this respect, one can think of obstacles such as the monopolist's brand name, superior quality or sufficient supply of capacity, for example. If airline managers report that these route conditions do prevent entry, this would mean in turn that first of all actual competition does not exist because entry does not occur and consequently the monopoly is sustained. However, as suggested by evidence, actual competition is most relevant to control an incumbent's behaviour. Second, if the monopolist in question is aware of both possessing such an advantage and the advantage's entry deterring effect, potential competition does not exert a threat since the monopolist does not fear entry. Consequently, neither potential nor actual competition restrains the monopolistic incumbent.[5] To render things worse, competition policy does not recommend taking measures, when applying normative definitions of entry barriers. However, before deciding how to deal with this challenge, the evidence needs to be viewed, as this issue may in fact be irrelevant.

Despite its potential, this survey-based methodology is relatively new to the field of entry barriers. In cross-sectional studies, Smiley (1988) and Bunch/Smiley (1992) explore, from an intra-firm perspective, how frequently certain strategic entry deterrence measures are applied. Investigating the marketing executives' perspective, Karakaya/Stahl (1989) provide empirical evidence on the importance of innocent and strategic entry barriers in consumer and industrial goods markets. Brewer (1996) analyses which industry conditions might prevent potential entrants from a commitment to the British rail freight industry. Although all these studies have made important contributions to the knowledge of entry difficulties, only one of these investigates how industry entry barriers are being perceived by firms whose entry might be prevented. Moreover, none of these studies have concentrated on the airline industry. However, air carriers apply some peculiar measures (e.g. code-sharing, hub-and-spoke system) when serving a market. Hence, a conclusion about the effect on competition cannot be readily drawn by mere inference from evidence available for other industries. This is because comparable means hardly exist in other industries, and airlines may overcome

---

[5] Of course, for this reasoning to hold, competitors must not serve the relevant city-pair by offering connecting services, as this would introduce at least some competitive pressure.

industry-specific obstacles by strategies (e.g. frequent-flyer programmes) unique to the airline industry.

It goes without saying that the presented method has some shortcomings, which should be mentioned. First, answers by respondents may be biased, for example respondents may have incentives to deliberately under- or overstate specific obstacles in order to accomplish a certain objective (e.g. influencing regulatory policy). Distortion resulting from biased answers represents a problem that is inherent to most surveys and that can be alleviated (e.g. by ensuring confidentiality) but can never be fully controlled for. Yet, there are methods to estimate the extent of biased answers. For one, the respondent's answers to other sensitive questions can be used to assess the extent of bias. If these answers seem to be honest in that it would be disadvantageous for the respondent if the answer were made public, then it can be reasonably assumed that the remaining answers correspond fairly closely to facts. Furthermore the stated severity of a problem can be compared with actual conditions on recently entered routes. Follow-up interviews with some respondents also offer the opportunity to verify answers.

Second, a usual objection to survey data refers to the subjectivity of the stated answers. However, in the given context, subjectivity does not constitute a problem because 'barriers to entry are effective if they are perceived to exist' (Borenstein, 1989, p. 345). The issue of subjectivity therefore relates to the distinction between a normative and positive definition of barriers to entry.

Third, the survey may not ask the 'right' questions. To avoid such bias, the phrasing of route entry conditions were based on information available in academic and industry journals and the resulting questionnaire was further discussed in a preliminary stage with aviation experts.

Finally, the survey was not sent to representatives of those airlines that did not enter the aviation market at all because the barriers were too high. The survey rather focuses on barriers to route (not industry) entry, hence such a selection bias is unlikely to distort the results. In conclusion, the proposed method serves as a complement, rather than a substitute, for the traditional approach in testing for entry barriers.

5. Application and Results of the New Research Approach

The survey-based methodology was applied in two research projects whose results are presented below. There are two reasons why both projects mainly considered characteristics relevant at the route rather than the industry level. While competitive pressure can stem from start-up as well as established airlines, only entry impediments at the route level are relevant to both groups of carriers. First, a start-up airline, for which barriers at the industry level are also of importance, has to enter at least one route since an airline without a single route serves no market. Second, entry difficulties at the route level play a role for established carriers' considerations about expansion. Evidence on barriers to route entry offers sufficient insight for regulatory policy makers and airline management alike – though additional substantiation of barriers at the industry level are not worthless.

First Project in 1998

The first survey dealt with entry barriers in Europe's airline markets one year after full deregulation came into effect in 1997 (Schnell, 1998).[6] Data from 39 airlines based in the European Economic Area was collected in spring 1998.[7] Some of the results were subsequently published in Pitelis/Schnell (2002). Table 13.1 provides an overview of the considered entry difficulties and their severity as assessed by European airline managers. The sample endorses the suppositions set above, that is, (1) entry barriers differ in the extent to which they impede entry, (2) airline managers have distinct perceptions with regard to a certain obstacle, (3) at least some airline managers' stated perceptions cannot be reconciled with normative theory on barriers to entry (e.g. regarding sunk costs).

Above, it was suggested that the data gathered via a questionnaire-based survey may help to enhance the knowledge of why some carriers fail. Four years after the responses were received, they can be further analysed to ascertain whether this hypothesis is supported. In fact, the sample largely corroborates the assumption that barriers to entry exert some influence on an airline's ability to survive. Airline managers that participated in the survey in 1998 and whose company does not exist any more in March 2002 (i.e. due to bankruptcy, take-over or merger) stated a higher effectiveness of many strategic barriers as compared to airlines still operating in March 2002. Differences are significant for code-sharing and limited amount of discount tickets (10%), frequent-flyer programme and post-entry expansion of capacity by incumbent (5%), as well as the mere existence of an incumbent and counter-actions on other routes (1%).[8] Such a striking disparity between both groups is not observable with respect to natural barriers, as only airport fees are significantly more effective (at the 5%-level) from the perspective of the disappeared airlines.

Second Project in 2001/2002

A new survey with an extended scope was carried out in late 2001. First, it does not only cover airlines within the European Economic Area, but also includes carriers from other deregulated markets, i.e. Australia/New Zealand, Canada and the USA.

Second, with regard to entry obstacles, the questionnaire distinguishes between the mere existence of an impediment and its effect. First, the survey asked whether a certain route condition that could potentially qualify as an entry barrier exists at all.

---

[6] A copy of the Master thesis can be requested from the author.

[7] The European Economic Area comprises all countries of the European Union and, additionally, Iceland and Norway. Liberalization became effective in this region.

[8] The Mann-Whitney-U-test was applied and numbers refer to the one-tailed significance level.

Table 13.1 Route entry difficulties as assessed by managers of airlines in the European Economic Area

Difficulties	Perceived as a difficulty	in-effective (-1)	neither effective nor in-effective (0)	effective (+1)	very effective (+2)	absolutely effective (+3)	rank
Market research about new route is very costly	43%	16%	16%	16%	8%	0%	13.9
New route is not part of your competitor's hub-and-spoke system but yet served by your competitor	22%	11%	38%	24%	5%	0%	12.7
Competitors offer superior service to customers compared with your company	37%	8%	18%	21%	13%	3%	11.4
Competitors segment customer groups and offer differentiated services to them	26%	8%	32%	21%	8%	5%	10.9
Some of the route entering costs cannot be recovered if you give up this route	24%	11%	24%	21%	13%	8%	10.3
Competitors advertise heavily	19%	8%	27%	30%	14%	3%	10.2
Additional aircraft are required to serve new route	27%	5%	19%	22%	22%	5%	9.7
Heavy advertising expenses are required to inform potential passengers on new route about your company's flights	21%	0%	29%	24%	21%	5%	9.2
Competitors offer limited amount of deeply discounted tickets before and after entry of your company	11%	11%	24%	37%	16%	3%	9.2
Competitors may undertake counter-actions against your company on routes other than the entered one	5%	0%	42%	32%	18%	3%	8.7

Your company's total average costs on the new route are higher than the presumed total average costs of your competitor(s)	24%	5%	18%	16%	24%	13%	8.6
Fees for using airport facilities (e.g. slots, ground facilities)	11%	11%	13%	34%	24%	8%	8.1
Competitors share flight codes with other competitors on new route	16%	3%	19%	30%	24%	8%	7.9
Introduction of new route requires vast amount of capital compared to its turnover	16%	3%	22%	27%	30%	3%	7.8
New route represents spoke of your competitor's hub-and-spoke system	16%	5%	11%	27%	35%	5%	7.2
Competitors offer frequent-flyer programmes	16%	8%	13%	21%	32%	11%	7.1
Certain scale of route entry is required to achieve a profitable average costs level	16%	0%	14%	30%	35%	5%	7.1
Before entry of your company, competitors set fares at a low unattractive level for your company	8%	3%	24%	34%	18%	13%	6.9
After entry of your company, competitors may reduce fares to a low unattractive level for your company	3%	3%	18%	45%	29%	3%	6.1
New route connects two hub airports of your competitor	11%	0%	22%	22%	25%	19%	5.9
Competitors may expand their number of flights after entry of your company	5%	3%	8%	34%	42%	8%	4.9
Attractive slots (regarding daytime) are not available as your competitors own them	5%	0%	3%	11%	24%	57%	1.8

Source: Pitelis/Schnell, 2002.

Second, for the route conditions existing according to the respondent's perception, he/she has to indicate the degree of influence exerted by the condition in question in preventing entry (i.e. 'never', 'unlikely', 'probably', 'likely' or 'always').[9] The distinction between a phenomenon's existence and its effect appears meaningful for the following reasons. In view of the intended comparison of different regulatory regions, the mere existence of a certain characteristic appears to be of interest. Moreover, only existing difficulties influence an airline's conduct and thus help explain the current stage of competition. Therefore, competition policy should primarily focus on real issues and not hypothetical effects.

Yet, this distinction comes at a cost. From the perspective of a business strategist, it is worth knowing whether a certain route condition could potentially prevent entry and would consequently shelter an airline's markets from competition. Of course, this shortcoming could have been avoided by a slightly different design of the answer section. However, there are a practical and a findings-related reason for not having done so. First, this would have increased the time to complete the questionnaire, which is considered to be already at the upper limit regarding length. Moreover, airlines not confronted with a certain route characteristic (and therefore having no experience in the feature's effect) may assess this condition wrongly and hence provide misleading advice to rivals.

The database of addressees was mainly formed by details available in *World Airlines 2001* and complemented by information out of industry journals as well as airlines' websites. Questionnaires were sent to airlines' top-management and line managers in charge of route entry decisions and other market-related strategies. To ensure a high response rate, mailings were personally addressed to the managers of concern. To accomplish a reasonable response rate and a sufficiently large sample in absolute terms, additional endeavours were taken: respondents were offered a feedback report covering the results of the survey in return for participation. Moreover, strict confidentiality on the respondent- and airline-level was promised. Furthermore, a more detailed description of the research project was offered on the author's website. To further increase the response rate on the airline-level, more than one questionnaire was mailed to a single airline, provided the names of potential respondents were known.

1180 questionnaires were sent in late 2001 to airline managers working at 306 different carriers with their home base in one of the regions mentioned above. A reminder letter was sent three weeks after the initial mailing. A further four weeks later, the questionnaire was emailed to the managers to further increase the response rate. A final email reminder was sent another three weeks later. Ultimately, 239 completed questionnaires from 156 different airlines were received. This equals an overall response rate of 51.0% at the airline level and 20.3% at the manager level. Given the industry state after the terrorists' attacks of 11th September 2001, the response rate appears to be surprisingly high. The results presented in Table 13.2 focus on entry barriers related to predatory behaviour.

---

9 The layout of the questionnaire is designed in the following way: the respondent has either to indicate that a certain phenomenon never exists or that it exists while specifying the extent of impeding entry according to the scale provided.

The aim has been to provide insights useful not only to researchers, interested in determining whether a phenomenon has the supposed impact, but also to policy makers and the aviation industry in general. In this respect an argument advanced earlier needs to be considered: a phenomenon can only deter entry when it is perceived to exist. Therefore, efficient competition policy should primarily focus on real issues and not hypothetical effects. Thus, in order to fulfil the needs of researchers and practitioners alike it is crucial to ascertain the existence of the conditions first – and that from the perspective of managers. Hence, it needs to known whether airline managers perceive the conditions at all, before analysing conditions' effectiveness. Consequently, the respondents were first asked whether the respective condition exists on at least some routes or not.[10]

Table 13.2 Existence of route entry conditions relating to predation[11]

Conditions	Does not exist on any route	Exists at least on some routes
Competitors may expand capacity after your entry	0.0	100.0
Competitors may (closely) match your departure times	0.0	100.0
Competitors may undertake counter-actions against your company on routes other than the entered one	0.0	100.0
After your entry, competitors may reduce fares to a low unattractive level for your company	0.5	99.5
Competitors have a reputation for being tough	1.5	98.5

Except for the threat of price reduction post-entry, retaliation measures, which are by definition not observable pre-entry, are expected by all respondents on at least some routes. To the extent that such an expectation results from experience, this suggests that entrants believe that incumbents do not generally accommodate entrants. In conclusion, the considered conditions challenge airline managers on at least some routes. Therefore, the conditions can be assumed realistic.

Having ascertained the existence of the conditions, the conditions' effectiveness in deterring entry is being scrutinized. Those respondents who reported a specific condition to exist were asked to indicate to what extent this

[10] Of course, a question that asks for the percentage of routes on which a condition exists would have generated additional information, especially for practitioners. However, for the sack of low complexity and thus time required to fill in the questionnaire we renounced this aspect. Moreover, we believe that the question of existence is generally less affected by subjectivity than is effectiveness and thus can more easily be answered by traditional data sources.

[11] In percentage of respondents.

condition deters entry of their airline. They had to specify the likelihood of deterrence on a five-point Likert-scale.[12]

The following refers to the conditions' effectiveness rather than the conditions' likelihood by which entry is prevented. Concerning the aspects dealt with in this study, effectiveness and likelihood of deterrence are the same: the less easily a condition can be overcome, that is, the more effectively it prevents entry, the more likely it prevents entry. The height of entry-barriers is not considered for the following reason: height may not affect effectiveness and hence they are not the same.[13] Moreover, it is effectiveness, not height, that ultimately matters for conduct (i.e. entry) and thus for strategy and policy implications.

Table 13.3 presents the distribution of answers. Respondents facing a certain condition do not assess it uniformly regarding its effectiveness. With respect to all investigated conditions, answers are distributed along the full scale. Thus the sample suggests that a particular condition hinders different airlines to different degrees. Again, this emphasizes the fact that managers (and thus firms) within a certain industry face different market conditions.

Table 13.3 Extent of the effectiveness of route entry conditions related to predation[14]

Conditions	Never	Unlikely	Probably	Likely	Always	Mean	S.D.
Competitors may expand capacity after your entry	7.4	30.0	43.8	14.8	3.9	1.78	0.93
Competitors have a reputation for being tough	13.4	31.2	29.7	20.3	5.4	1.73	1.10
Competitors may (closely) match your departure times	10.4	32.2	34.7	20.3	2.5	1.72	0.98
After your entry, competitors may reduce fares to a low unattractive level for your company	10.8	29.6	38.9	18.7	2.0	1.71	0.96
Competitors may undertake counter-actions against your company on routes other than the entered one	9.8	35.1	33.7	19.0	2.4	1.69	0.97

[12] The categories were labelled as follows: 0....'Never'; +1....'Unlikely'; +2....'Probably'; +3....'Likely'; +4....'Always'.

[13] Consider in this respect that the height of a mountain does not necessarily affect the likelihood of its conquest and thus the effectiveness of deterring its conquest.

[14] Likelihood of deterrence in percentage of respondents who perceive the respective condition. Percentage numbers may not add up to 100% due to rounding errors.

All the anticipated retaliation measures considered in the survey are being classified among the most effective strategic conditions[15] and represent long-standing issues in economic theory and policy. They have a similar average effectiveness, which is somewhat higher than the effectiveness of the presence of two competitors on a route (Mean 1.59).

These small differences imply that on a barrier-by-barrier basis, the predation measure may be overstated in the discussion about obstacles to competition in airline markets. However, in reality, an entrant is likely to be confronted with a bundle of these conditions at any one point in time. Therefore, as several of the above conditions may be simultaneously at work in reality, the findings need to be qualified when drawing conclusions on the extent of deterred entry.

92.9% of respondents believe that no single predatory barrier is effective in deterring entry at all times. A largely different picture emerges though, once the category 'likely' is also taken into account: 58% of respondents report that no single predatory condition is either likely to deter entry or always deters entry. This finding emphasizes the interpretation given above, namely that the effect of predatory measures on entry may, on average, be overrated.

So far, the analysis has largely focused on the severity of entry barriers in absolute terms, that is on the extent to which a specific condition inhibits entry. However, in order to better assess the effects of a certain condition the relative effectiveness of this condition (i.e. how effective is it compared to other conditions) needs to be known. Such knowledge would inform policy makers and industrialists alike. It would enable policy makers to concentrate on the most pressing issues, while business leaders would be in the position to shelter their companies from competition more effectively. Therefore, in order to verify the tentative conclusions drawn above, it will be analysed in the following whether the differences in perception on which the conclusions were based are indeed significant.

First, via the application of a t-test, it is assessed whether the respondents perceive the sole presence of a competitor as less effective than other conditions related to predation (1-tailed hypothesis) (cf. Table 13.4). Table 13.4 shows: the mean response per condition related to predation, the mean of the benchmark scenario and the significance level of the difference of the two means as calculated by the t-test. For example, capacity expansion gets an average effectiveness of 1.79, while the benchmark gets an average of 1.42. With a t-value of 5.216 the difference is significant at the 0.1%-level.

Overall, the analysis reveals that all predatory measures considered impede entry significantly more effectively than the mere presence of a rival. This finding clearly renews the case of a long-standing issue in competition policy (i.e. predation).

For exploratory purposes the next section investigates whether the predatory measures differ in effectiveness, that is, whether their means are significantly different from each other. Again, a t-test was applied and its results are listed in Table 13.5. The table shows the mean effectiveness of condition A and B and the t-value for the difference of the respective means. For example, post-entry fare reduction has an average effectiveness of 1.72 in comparison with the mean of

[15] A hub-to-hub route is the most effective strategic barrier (Mean 1.92).

post-entry capacity expansion of 1.75 – which is not a significant difference according to the t-test.

Table 13.4 Comparison of barriers' effectiveness using existence of an incumbent as benchmark[16]

Conditions	Mean (condition)	Mean (benchmark)[17]	|t-value|[18]
Competitors may expand capacity after your entry	1.79	1.42	5.216****
Competitors have a reputation for being tough	1.77	1.43	4.779****
Competitors may (closely) match your departure times	1.73	1.42	4.507****
After your entry, competitors may reduce fares to a low unattractive level for your company	1.72	1.42	4.088****
Competitors may undertake counter-actions against your company on routes other than the entered one	1.71	1.43	3.770****

Furthermore post-entry capacity expansion has a mean of 1.79 in comparison with the mean of departure time matching of 1.73 – which does not represent a significant difference, either.

In fact it appears that the retaliation measures under consideration do not differ in their effectiveness. Thus, the threat of post-entry price reduction seems to be as effective as a hypothetical post-entry capacity expansion, for example.

6. Summary and conclusion

The present paper began with a brief discussion of the different definitions of barriers to entry and the rationale for replacing the usually employed normative definition with a positive definition of barriers to entry was explained. Subsequently, as it was argued that only perceived entry barriers matter for airlines' entry behaviour, knowledge gaps in the current academic literature were identified, especially the perception-based perspective which has largely been neglected by previous studies, and which was the major impetus for the present research programme.

---

[16] The scale was labelled as follows: 0....'Never'; +1....'Unlikely'; +2....'Probably'; +3....'Likely'; +4....'Always'.

[17] Mean of benchmark varies due to missing values.

[18] 1-tailed significance-level: *....10%, **....5%, ***....1%, ****....0.1%.

Table 13.5 Comparison of the effectiveness of predatory measures[19]

Condition A	Condition B	A	B	|t-value|[20]
After your entry, competitors may reduce fares to a low unattractive level for your company	Competitors may expand capacity after your entry	1.72	1.75	0.525
	Competitors may undertake counter-actions against your company on routes other than the entered one	1.71	1.67	0.616
	Competitors have a reputation for being tough	1.72	1.72	0.000
	Competitors may (closely) match your departure times	1.71	1.70	0.151
Competitors may expand capacity after your entry	Competitors may (closely) match your departure times	1.79	1.73	1.142
	Competitors have a reputation for being tough	1.79	1.74	0.610
	Competitors may undertake counter-actions against your company on routes other than the entered one	1.78	1.69	1.334
Competitors may (closely) match your departure times	Competitors may undertake counter-actions against your company on routes other than the entered one	1.73	1.69	0.589
	Competitors have a reputation for being tough	1.74	1.74	0.000
Competitors may undertake counter-actions against your company on routes other than the entered one	Competitors have a reputation for being tough	1.70	1.73	0.492

In the second part of the paper some selected results of the author's research in this field were presented. It appears that predatory measures render entry more difficult compared to a scenario in which only a single incumbent serves the route (cf. Table 13.4). However, on average the analysed samples imply that the role of predation in reducing competition may be largely overrated. The majority of respondents reported that it is unlikely that entry is mainly prevented by threats of

[19] The value of the mean of a specific condition may not be the same for all comparisons due to missing values.
[20] 2-tailed significance-level: *....10%, **....5%, ***....1%, ****....0.1%.

predatory moves of incumbents. However, it needs to be noted that airlines, which failed because of predation, were not included in the sample and may have articulated a different assessment of the effectiveness of predation in deterring entry.

In conclusion, it was shown that an effort to gather primary data exhibiting an intra-firm perspective contributes to improving the knowledge of barriers to entry. The insights from such an analysis do not only inform competition policy, but also airlines' business strategies.

References

Aiginger, K., McCabe, M., Mueller, D.C. and Weiss, C. (2001), Do American and European Industrial Organization Economists differ? In: *Review of industrial organization*, 19, 383-405.

Baumol, W.J. (1982), Contestable markets: an uprising in the theory of industry structure. In: *American economic review*, 72(1), pp. 1-15.

Borenstein, S. (1989), Hubs and high fares: dominance and market power in the U.S. airline industry. In: *RAND journal of economics*, 20(3), pp. 344-365.

Borenstein, S. (1991), The dominant-firm advantage in multi-product industries: evidence from the U.S. airlines. In: *Quarterly journal of economics*, 106(4), pp. 1237-1266.

Brewer, P.R. (1996), Contestability in UK rail freight – The economics of open access. In: *Transport policy*, 3(3), pp. 91-98.

Bunch, D.S. and Smiley, R. (1992), Who deters entry? Evidence on the use of strategic entry deterrents. In: *Review of economics and statistics*, 74(3), 509-521.

Conner, K.R. (1991), A historical comparison of resource-based theory and five schools of thought within industrial organization economics: do we have a new theory of the firm? In: *Journal of management*, 17(1), pp. 121-154.

Demsetz, H. (1982), Barriers to entry. In: *American economic review*, 72(1), pp. 47-57.

Dixit, A. (1980), The role of investment in entry-deterrence. In: *Economic journal*, 90(357), 95-106.

Doganis, R. (1994), The impact of liberalisation on European airline strategies and operations. In: *Journal of air transport management*, 1(1), pp. 15-25.

Duncan, R. (1972), Characteristics of organizational environments and perceived environmental uncertainty. In: *Administrative science quarterly*, 17(4), pp. 313-327.

Fawcett, S.E. and Farris M.T. (1989), Contestable markets and airline adaptability under deregulation. In: *Transportation journal*, 29(1), pp. 12-24.

Fombrun, C.J. and Zajac, E.J. (1987), Structural and perceptual influences on intra industry stratification. In: *Academy of management journal*, 30(1), pp. 33-50.

Fruhan, W. (1972), *The fight for competitive advantage – A study of the United States domestic trunk carrier*, Harvard University, Boston.

Gimeno, J. (1999), Reciprocal threats in multimarket rivalry: staking out 'spheres of influence' in the U.S. airline industry. In: *Strategic management journal*, 20(2), pp. 101-128.

Graham, D.R., Kaplan, D.P. and Sibley, D.S. (1983), Efficiency and competition in the airline industry. In: *Bell journal of economics*, 14(1), pp. 118-138.

Gudmundsson, S.V. (1997), The difference between European and US airline management practice: the case of new-entrant airlines. In: *Journal of air transport management*, 3(2), pp. 75-82.

Helper, S. (2000), Economist and field research: 'You can observe a lot just by watching'. In: *American economic review*, 90(2), pp. 228-232.

Joesch, J.M. and Zick, C.D. (1994), Evidence of changing contestability in commercial airline markets during the 1980s. In: *Journal of consumer affairs*, 28(1), pp. 1-24.

Joskow, A.S., Werden, G.J. and Johnson, R.L. (1994), Entry, exit, and performance in airline markets. In: *International journal of industrial organization*, 12(4), pp. 457-471.

Karakaya, F. and Stahl, M.J. (1989), Barriers to entry and market entry decisions in consumer and industrial goods markets. In: *Journal of marketing*, 53(2), 80-91.

Lefebvre, L., Mason, R. and Lefebvre, E. (1997), The influence prism in SMEs: the power of CEO's perceptions on technology policy and its organizational impacts. In: *Management science*, 43, pp. 856-878.

Levine, M.E. (1987), Airline competition in deregulated markets: theory, firm strategy, and public policy. In: *Yale journal on regulation*, 4(2), pp. 393-494.

Lieberman, M.B. (1987), Excess capacity as a barrier to entry: an empirical appraisal. In: *Journal of industrial economics*, 35(4), pp. 607-627.

Mueller, D.C. (1992), The corporation and the economist. In: *International journal of industrial organization*, 10, pp. 147-170.

Paine, F.T. and Anderson, C.R. (1977), Contingencies effecting strategy formulation and effectiveness: an empirical study. In: *Journal of management studies*, 14(2), pp. 147-158.

Penrose, E. (1959), *The theory of the growth of the firm*, Oxford.

Peteraf, M.A. (1993), The cornerstones of competitive advantage: a resource-based view. In: *Strategic management journal*, 14, pp. 179-191.

Peteraf, M.A. (1995), Sunk costs, contestability and airline monopoly power. In: *Review of industrial organization*, 10(3), pp. 289-306.

Peteraf, M.A. and Reed, R. (1994), Pricing and performance in monopoly airline markets. In: *Journal of law and economics*, 37(1), pp. 193-213.

Pitelis, C.N. and Schnell, M.C.A. (2002), Barriers to Mobility in Europe's civil aviation markets: Theory and new evidence. In: *Review of industrial organization*, 20(2), pp. 127-150.

Schmalensee, R. (1988), Industrial economics: an overview. In: *Economic journal*, 98, pp. 643-681.

Schnell, M.C.A. (1998), *Entry barriers into the European civil aviation markets: Theory and new evidence*, M.Phil. thesis, University of Cambridge.

Schwegmann, V. (1998), *Preisstrategien im europäischen Linienluftverkehr*, Köln.

Shepherd, W.G. (1984), 'Contestability' vs. competition. In: *American economic review*, 74(4), pp. 572-587.

Sinclair, R.A. (1995), An empirical model of entry and exit in airline markets. In: *Review of industrial organization*, 10(5), pp. 541-557.

Smiley, R. (1988), Empirical evidence on strategic entry deterrence. In: *International journal of industrial organization*, 6(2), 167-180.

Snow, C. (1976), The role of managerial perceptions in organizational adaptation: an exploratory study. In: *Academy of management proceedings*, pp. 249-255.

Spence, M. (1977), Entry, capacity, investment and oligopolistic pricing. In: *Bell Journal of economics*, 8(2), 534-544.

Strassmann, D.L. (1990), Potential competition in the deregulated airlines. In: *Review of economics and statistics*, 72(4), pp. 696-702.

Stubbart, C.I. and Ramaprasad, A. (1988), Probing two chief executives schematic knowledge of the U.S. steel industry using cognitive maps. In: *Advances of strategic management*, 5, pp. 139-164.

Chapter 14

Why do Low Cost Carriers Arise and How can they Survive the Competitive Responses of Established Airlines?: A Theoretical Explanation

Markus Reisinger

1. Introduction

After the deregulation of the European airline market in the 1990s, there was one striking development. This was the entry of low cost carriers. Starting in summer 1995 with Ryanair there are now almost 50 low cost airlines in Europe which fly on many different routes.[1] These developments dramatically changed the European airline market.[2] The business strategy of these low cost airlines to save costs is to offer no frills, have fast turn-rounds, use only a single-type fleet of planes, and to fly mainly to cheap secondary airports like Stanstead instead of Heathrow in London or Frankfurt-Hahn instead of Frankfurt. They also save a huge amount of administration costs by only accepting credit card payments and thus reducing the paperwork to a minimum. With this strategy they are able to offer low fares and attract many customers.

The market share of low cost airlines in 2004 is now 10%.[3] This is in terms of passenger counts. In the European Transit Report it was estimated that this share will grow up to 24% in the year 2010. Looking at past data the market share of low cost airlines in 2004 is almost fifteen times higher than in 1995. This provides evidence that the market segment of the low cost airlines was heavily growing in the last years and will also do in the future.[4]

But the development in Europe was only a successor of the development in the USA where after entry of the first low cost carrier, Southwest Airlines, the market changed drastically. Southwest Airlines was founded before the deregulation phase in the USA. After its market entry it had to compete against traditional airlines which initiated price wars and established own low cost carriers

---

[1] See Economist, May 9th 2004.

[2] For a detailed discussion of Europe's low cost airlines see the chapter of Gilroy, Lukas and Volpert (2004) in this book.

[3] See European Transit Report 2004.

[4] In the US the market share in terms of passengers was in 2003 already 20%. See Januszewski and Lederman (2003).

to induce Southwest Airlines to exit the market. But after Southwest Airlines survived these hard times it grew steadily and is now one of the safest and most profitable airlines in the USA. Despite its growth Southwest Airlines has not deviated from its business strategy which is no frills, low costs, and low fares. The market share increased from 0% in 1970 to around 10% in 2001 and its number of passengers is now 60 times higher than 30 years ago.[5]

Dresner, Lin and Windle (1996) tested this empirically. They found that after entry of Southwest Airlines and other low cost airlines prices decreased heavily and passenger counts increased. This was not only the case in routes in which the low cost airlines entered but also in competing routes and on neighbouring airports.

This paper tries to answer the questions why low cost airlines arise and why they have been more successful than low cost carriers of established airlines. I want to answer these questions in a theoretical framework and analyse a annibal model which tries to capture some main aspects of the airline market.

Since the established airlines have operated in the market for a long time, they have already chosen the market segment which is most profitable. This segment consists of the business travellers. To avoid fierce competition a new entrant has to choose another segment, the leisure travellers, from whom he could demand only lower prices but could offer lower quality which results in lower costs.

As fares decline, the profits of the flag carrier also decrease. Thus this carrier might want to take some actions to induce the rival to leave the market again. One idea is to create a low cost carrier on its own. This was done by British Airways with the foundation of the low cost carrier GO in 1998 but it turned out that this was not profitable.[6] In 2001 British Airways sold GO to its management for 110 million pounds, which again resold it to easyJet in 2002 for 354 million pounds.[7] The question arises why British Airways set up GO initially. The reason might be that British Airways established it for predatory reasons. If costs of the new competitors like Ryanair and easyJet are high enough these airlines must exit the market because they are unable to recover their costs and predation has been successful. But if costs have been overestimated by the traditional airline, predation is unsuccessful and it is better to withdraw its own low cost carrier. The reason is that with fierce competition in the low cost segment prices for leisure travellers are very low. But this also affects the fares of the flag carrier. Thus the established airline annibalises itself and does better in withdrawing its low cost carrier.

The remainder of the paper is organized as follows. The next section sets out the model and provides a discussion of the assumptions. The equilibrium of the model is derived and analysed in section 3. In section 4 I consider the predation strategy of building an own low cost carrier. Section 5 presents three possible extensions of the model and its policy implications. The conclusions are contained in section 6.

---

[5] For an overview of the story of Southwest Airlines see the chapter of Knorr and Arndt (2004) in this book.

[6] According to CNN money, January 11th 2001, the losses of GO in the financial year 1998/1999 were 15.5 million pounds and 15.2 million pounds in 1999/2000.

[7] See BBC News, May 16th 2002.

2. The Model

Consider a model in which there are two firms i=1,2. The firms decide sequentially about the quality q_i they want to produce. Firm 1 decides about q_1 in stage 1 and firm 2 decides q_2 in stage 2.[8] Firm 1 is considered the incumbent and firm 2 the entrant. In an airline market firm 1 is an established airline while firm 2 is a new airline which can enter the routes of firm 1. The variable q_i includes all the things that make up flight quality; for example the flexibility of booking time, if the airport of an airline is a main airport or a secondary airport, if a meal is served during the flight and so on.[9] In stage 3 both firms simultaneously choose their prices after observing the quality choice of the other firm. This multi-stage game captures the notion that firms can change their prices almost instantaneously but a change in quality takes some amount of time because it might involve new production techniques or new computer programming.

Passengers have the utility function

$$U(\theta) = \theta q_i - p_i, \qquad i=1,2$$

where q_i is the quality of a flight with firm i and p_i its price. If a consumer does not fly she gets a utility which is normalized to 0. θ is uniformly distributed between [0,1] and represents a passenger's preference for quality. Thus a high θ stands for a business traveller because for her it is very important that she has lots of flexibility in booking time and is landing at a central airport. A low θ represents a leisure traveller. The mass of consumers is normalized to 1. The resulting demand function, if only one firm would be active in the market, is then

$$D(p) = \begin{cases} 1-\dfrac{p}{q_i} & \text{if} \quad q_i \geq p \\ 0 & \text{if} \quad q_i < p \end{cases}$$

The revenue of firm i is the respective price p_i times the number of its passengers. Each airline bears its operating costs C_j with j=H,L and $C_H > C_L$. So the firm with the higher quality has to bear higher operation costs. To make the model tractable marginal costs are assumed to be zero for both firms. With positive marginal costs the formulae would be much more complicated and the results are hard to interpret but the basic conclusions would remain the same.

---

8 The model is a variation of Shaked and Sutton's (1982) model of product differentiation. In their model there are n firms which choose their qualities simultaneously. In equilibrium only two firms are active in the market and choose different qualities. But it is not possible in their model to predict which firm will choose the higher quality. As will become clear in the next section this is possible with a sequential quality decision and the modelling fits airline competition very well.

9 The main reason for this assumption is tractability. Models with multidimensional quality choice are usually very hard to analyse. Some progress in handling them has been made in recent years. See e.g. Rochet and Chonet (1998) or Rochet and Stole (2003).

Both airlines have no capacity constraint but it is assumed that there is an upper bound on quality $\overline{q}$ which can be produced. To obtain clear results this upper bound is fixed at $\overline{q} = 1/2$. So both airlines can produce quality between 0 and 1/2.[10]

Discussion of Assumptions

First the model has a classical leader follower structure where one firm initially has a monopoly and then entry occurs. This structure describes what is going on in the European airline market after deregulation especially on national routes where many established airlines were near monopolists during times of regulation. The market was dominated by national carriers such as Lufthansa, British Airways or Air France for almost 50 years.

Secondly, the model is a Bertrand game with vertical product differentiation. This is different to many airline competition models which assume Cournot competition.[11] The reason is that this paper tries to endogenously explain quality differences between airlines. In standard Cournot models firms produce homogenous goods and thus are facing the same price after their quantity decision. In the model here, as well as in real world, the quality of a flight is different with different aviation companies and thus the Cournot model is not directly applicable. After firms have chosen their respective qualities they compete for passengers in prices. It is therefore assumed that there are no capacity constraints. This assumption is a strong one but it is clear that in equilibrium firms set prices in such a way that not the whole population demands a flight. Thus no firm needs unlimited capacity.[12]

Third, another key assumption is that there are no network differences between firms. Usually the established airline is equipped with a network of connecting flights due to its long presence in the market. In contrast the entrant does not have such a network. This feature can be incorporated in the model by assuming that some fraction of consumers will only buy from the incumbent because this fraction needs a connecting flight. Of course, this network difference is then an exogenous one and is not explained within the model. But we observe several established airlines setting up low cost carriers and their key objective is not to attract people with connecting flights. Rather they focus on people who want to travel cheaply from A to B. The main aim of this paper is to analyse the circumstances under which an established airline will set up such a carrier.

---

[10] This assumption is made for simplicity. Relaxing it would also not change the qualitative results.

[11] In a recent paper Goolsbee and Syverson (2004) found only weak evidence for capacity competition of an established airline facing the threat of entry. Instead they found strong evidence that the incumbent is dropping its fares.

[12] Despite this fact firms earn positive profits in equilibrium because both exert market power. So prices are above quality adjusted marginal costs.

3. Equilibrium: The Entrant Offers Lower Quality

In this section the equilibrium is derived. Since the game is dynamic (firms move sequentially) the equilibrium concept is subgame perfect Nash equilibrium.[13] The idea is that each player has to play a best response after each possible action of the other player and not only after those which were actually played. This concept rules out incredible threats.[14] The solution technique is backward induction.

We first look at the third stage of the game. If both airlines offer the same quality $q_1=q_2=q$, this would lead to standard Bertrand competition. Since both firms offer the same product if one firm slightly undercuts its competitor's price it would win all passengers from the other airline which leads to an equilibrium price of zero. The profit of each firm would be negative because of the operating costs. So assume both airlines produce different qualities, namely q_H and q_L with $q_H > q_L$.[15] To get some consumers, the firm with the lower quality must offer a lower price $p_L < p_H$. This low quality firm gets consumers with a lower preference for quality, that means with a low θ. Consumers with a high preference for quality fly with the high quality airline. The marginal consumer θ' who is indifferent between both airlines is characterized by

$$\theta' q_H - p_H = \theta' q_L - p_L$$

which gives

$$\theta' = \frac{p_H - p_L}{q_H - q_L}$$

Thus all consumers between θ' and 1 are flying with the high quality airline.

A passenger who is indifferent between flying with the low quality airline and not flying is characterized by

$$\theta'' q_L - p_L = 0$$

or

$$\theta'' = \frac{p_L}{q_L}$$

All passengers between θ'' and θ' fly with the low quality airline and all passengers between 0 and θ have such low valuations that they decide not to fly at all.

---

[13] This concept was developed by Selten (1965).

[14] For an introduction into dynamic games of perfect information see Gibbons (1995).

[15] Later it will be analysed which firm produces q_H and which one q_L.

Thus the profits of the airlines are given by

$$\Pi_H = p_H\left(1-\frac{p_H - p_L}{q_H - q_L}\right) - C_H$$

and

$$\Pi_L = p_L\left(\frac{p_H - p_L}{q_H - q_L} - \frac{p_L}{q_L}\right) - C_L$$

Differentiating Π_H with respect to p_H and Π_L with respect to p_L and setting the resulting formulae equal zero yields the first order conditions. From them we can calculate the reaction functions of both airlines

$$p_H(p_L) = \frac{q_H - q_L + p_L}{2}$$

and

$$p_L(p_H) = \frac{p_H q_L}{2q_H}$$

Solving these two equations yields equilibrium prices (dependent on qualities) of

$$p_H = 2q_H\left(\frac{q_H - q_L}{4q_H - q_L}\right)$$

and

$$p_L = q_L\left(\frac{q_H - q_L}{4q_H - q_L}\right)$$

Having solved the equilibrium in stage 3 we move backwards to stage 2. In stage 2 we have two different cases to consider. Either firm 2, the entrant, is the low quality seller or firm 2 is the high quality seller. First analyse the case in which firm 2 is the low quality seller. In this case firm 2's profit function after plugging in p_H and p_L can be written as

$$\Pi_2 = q_L\left(\frac{q_H - q_L}{4q_H - q_L}\right)\left(\frac{2q_H - q_L}{4q_H - q_L} - \frac{q_H - q_L}{4q_H - q_L}\right) - C_L \qquad (1)$$

Simplifying equation (1) yields

$$\Pi_2 = \left(\frac{q_H - q_L}{(4q_H - q_L)^2}\right) q_H q_L - C_L \tag{2}$$

Differentiating (2) with respect to q_L and setting the result equal to zero yields the reaction function

$$q_L(q_H) = \frac{4}{7} q_H \tag{3}$$

which is dependent of q_H because firm 1 has already produced q_H in stage 1 and this could be observed by firm 2.

Firm 1 in stage 1 now takes into account that firm 2 will react to q_l according to equation (3). After some calculations firm 1's profit is given by

$$\Pi_1 = \frac{7}{48} q_H - C_H$$

which is strictly increasing in q_H. Thus it is optimal to set q_H at its upper bound, $\overline{q} = 1/2$. Firm 2 therefore chooses $q_L = 1/2$. Calculating prices and profits we get

$$p_1 = \frac{21}{148},\ p_2 = \frac{3}{74},\ \Pi_1 = \frac{7}{96} - C_H, \text{ and } \Pi_2 = \frac{147}{10952} - C_L$$

Before having a closer look at this equilibrium and providing some intuition for the results, consider the second possible case in which firm 2 offers higher quality than firm 1. The profit function of firm 2 in this case is

$$\Pi_H = (q_H - q_L)\left(\frac{2q_H}{4q_H - q_L}\right)^2 - C_H$$

Building the first order condition yields

$$4q_H^2 - 3q_H q_L + 2q_L^2 = 0$$

The potential solution of q_H can be obtained from the quadratic formula and is given by

$$q_{H,a}, q_{H,b} = \frac{3q_L \pm \sqrt{-23}q_L}{8}$$

which is not possible. This shows that it cannot be the case that firm 2 offers q_H.

Thus in the unique subgame perfect equilibrium firm 1, the incumbent, offers

$$q_1 = \frac{1}{2}$$

and earns profits of

$$\Pi_1 = \frac{7}{96} - C_H$$

and firm 2 offers

$$q_2 = \frac{2}{7}$$

and earns profits of

$$\Pi_2 = \frac{147}{10952} - C_L.$$ [16]

Interpretation of the Results

It is shown that by the mere fact that firm 2 enters the market after firm 1, firm 2 offers a lower quality. What is the intuition behind this result? Since firm 1 moves first in this game it has a first mover advantage. Firm 1 chooses that market segment, namely the business travellers, which guarantees highest profits. Why is firm 2 not doing the same after its entry? Because this would result in cut throat competition since both airlines would offer almost the same product. Thus firm 2 must find another market segment, the leisure travellers, which guarantees some profits and lessens competition with firm 1. This can explain what was going on in the airline market after deregulation.

This line of reasoning is also confirmed if we compare the duopoly situation with the situation before entry in which firm 1 is monopolist. Calculating quality, prices, and profits in the monopoly situation we get $q^{mon} = ½$, $p^{mon} = ¼$, and $\Pi mon = 1/8 - C_H$. Only passengers with a θ between ½ and 1 are served. In comparison with duopoly the quality of firm 1 stays the same ($q_1 = 1/2$) but p_1 and therefore Π_1 decreases a lot because of the additional competition effect. This is also what we observe in real world. Established airlines have not substantially reduced the quality of their traditional flights but have lowered the price in reaction to entry of competitors.

[16] It should be mentioned that these numbers look very small but this is only an implication of the assumptions. For example, consumer mass is normalized to unity. If instead we would assume a consumer mass of e.g. 10 the profits would be 10 times higher.

Looking at the passenger numbers in duopoly we see that θ' is at 35/74 and θ'' is at 21/148. This means passengers with a θ between 21/148 and 1 fly with either one of both airlines. So entry has increased the passenger base significantly, namely from ½ to 127/148, and some passengers have switched from the established airline to the low cost carrier. Because of the decreased prices all passengers gain while the incumbent clearly loses from entry.

If one would incorporate that there are network effects on the side of the incumbent the results would persist up to some degree. In this case passengers needing a connecting flight cannot use the low cost airline because it has no network. This holds true independent of the quality taste parameter. Thus the entrant cannot gain so many new passengers. This in turn lowers competition because some passengers must choose firm 1. Prices would not fall by such a large amount as in the model. Thus network effects would dampen the results of the model but the effects would still be present.

Introducing different marginal costs because of quality differences would not change the effects of the results either. If one assumes that firm 1 has to bear higher marginal costs because of higher quality production, then the quality differences would perhaps not be as high as in the preceding analysis. But still firm 1 would choose to be the high quality firm if the business traveller segment is the more profitable one. As mentioned above here I abstain from assuming positive marginal costs to keep the model tractable.

4. Predation by Setting Up an own Low Cost Carrier

In the previous section it was shown that established airlines have a first mover advantage compared to a new entrant. If its costs are not very high they should earn higher profits. Yet, their profits are much smaller than in the monopoly case because of price competition. Therefore the incumbent might think of entry deterrence strategies or, after entry has already occurred, strategies that drive the low cost carrier in the loss zone so that it must exit the market. A popular strategy pursued by some incumbents to achieve this, is to establish a low cost airline on its own. Examples are the low cost carrier GO which was set up by British Airways[17] or Germanwings set up by Lufthansa.[18] But, as mentioned in the introduction, this strategy was sometimes not successful.

This section attempts to find a theoretical explanation why it may be rational for an established airline to set up an own low cost carrier but withdraw it from the market in later periods. The reason is that the own low cost carrier turns out to be too much of a competitor to the business of the parent firm. Of course, this result does depend on the assumption that the low cost subsidiary is flying on the same routes as the incumbent airline. But this is also what is observed in the market. For example, Germanwings has its hub in Cologne-Bonn, yet, Lufthansa is still flying regularly to this airport.

[17] For a more detailed analysis of GO, see Lawton (2002).

[18] Also in the US many of the established airlines built low cost carriers in reaction to the entry of South West Airlines. See the chapter of Knorr and Arndt (2004) in this book.

To formalize this idea the model of the previous section is extended. We keep the first three stages of the game. Both firms have already chosen their respective qualities

$$q_1 = \frac{1}{2} \text{ and } q_2 = \frac{2}{7}$$

and earned profits in stage 3 of

$$\Pi_1 = \frac{7}{96} - C_H \text{ and } \Pi_2 = \frac{147}{10952} - C_L$$

There are now two additional stages. In stage 4 firm 1 can decide if it wants to build its own low cost carrier. The quality of this low cost carrier is assumed to be the same as the quality of firm 2, $q_1^{LCC} = q_2 = 2/7$.[19] If it does so it has to bear additional operating costs of C_1^{LCC}. These costs C_1^{LCC} need not be the same as the 'true' low cost airline's cost C_L. Both firms choose optimal prices and earn profits accordingly in stage 4. If firm 1 does not establish a low cost carrier the outcome is the same as in stage 3.

After stage 4 the sum of firm 2's profits in period 3 and period 4 might be above or below zero. If it is below zero the firm is bankrupt and must exit the market. Firm 1 is monopolist in stage 5 and sets prices for its traditional airline and its low cost carrier in the optimal monopolistic way. If firm 2's profit is above zero it is still active in stage 5. Before competition in stage 5, firm 1 has the option to withdraw its low cost carrier. If it does so both firms compete as in stage 3. If it does not withdraw the firms compete as in stage 4. The time line of the model is described in Figure 14.1.

Stage 1	Stage 2	Stage 3	Stage 4	Stage 5
Firm 1 chooses q_1	Firm 2 chooses q_2	Firms set prices, payoffs are realized	Firm 1 chooses to build a LCC or not, firms set prices, payoffs are realized	Firm 2 goes bankrupt or not, firm 1 can withdraw its LCC, firms set prices, payoffs are realized

Figure 14.1 Time line of the game

[19] This is made for simplicity, letting firm 1 decide about the optimal quality of its low cost carrier would not change the qualitative results but would only complicate the analysis. But in the real world there is certainly a difference between the two qualities. For example, frequent flyer programmes or interlining would render the quality of the incumbent's low cost airline higher.

One last component is needed to make the model work. It is assumed that firm 1 does not know the operation costs of firm 2 with certainty and has only an expectation $E[C_L]$ about C_L. Also no discounting takes place, so if there were a discount factor δ, it would be equal to 1.

Analysis of the Game

The game is solved by backward induction as in the previous section. Starting at stage 5 I begin the analysis with the case that firm 1 has built a low cost carrier in stage 4. First look at the case if firm 2 is not bankrupt. Firm 1 can now decide if it wants to withdraw its low cost airline or not. If it withdraws we are in the same situation as in stage 3 of the previous section. Only the traditional airline of firm 1 and the competitor's low cost carrier are in the market. Thus prices and profits in stage 5 are the same as in stage 3.

If firm 1 does not withdraw its low cost carrier the situation is different. Both low cost airlines are offering the same quality. By the standard Bertrand argument price competition would lead to zero prices because each low cost airline can win all passengers from the other one by slightly undercutting the competitor's price. Thus $p_2 = p_1^{LCC} = 0$ and $\Pi 2 = -C_L$. The marginal passenger who is indifferent between flying with the traditional airline of firm 1 and with one of the low cost airlines is characterized by

$$\frac{1}{2}\cdot\theta' - p_1 = \frac{2}{7}\cdot\theta'$$

which yields

$$\theta' = \frac{14p_1}{3}$$

The optimal p_1 can be calculated by differentiating

$$\Pi_1 = p_1\left(1 - \frac{14p_1}{3}\right) - C_H - C_1^{LCC}$$

with respect to p_1 and building the first order condition. This yields

$$p_1 = \frac{3}{28}$$

Firm 1's profit in period 5 is therefore

$$\Pi_1(5) = \frac{3}{56} - C_H - C_1^{LCC}$$

θ' is given in this case by 0.5. Because the established airline lowers its fare p_1 so heavily, it gets a demand of 1/2 while both low cost carriers share the rest of the market and get a consumer mass of 1/4 each.

But if firm 1 had withdrawn its low cost airline its profit would have been

$$\Pi_1(5) = \frac{7}{96} - C_H$$

which is higher than the profit without withdrawing. Therefore if firm 2 is not bankrupt it is optimal for firm 1's low cost airline to exit the market.

If firm 2 has gone bankrupt firm 1 is a monopolist in stage 5 and sets optimal prices for its traditional airline and its low cost carrier. It can be shown that firm 1 would also withdraw its low cost carrier from the market if it is a monopolist.[20] Profits are therefore

$$\Pi_1(5) = \frac{1}{8} - C_H$$

Now look at stage 4 where firm 1 has two possibilities namely to set up an own low cost carrier or not. The analysis of stage 4 is similar to that in stage 5. If firm 1 has to build a low cost airline in stage 4 both low cost airlines prices are zero and profits are

$$\Pi_1(4) = \frac{3}{56} - C_H - C_1^{LCC} \text{ and } \Pi_2 = -C_L$$

If firm 1 does not build a low cost airline profits are as in stage 3,

$$\Pi_1(4) = \frac{7}{96} - C_H \text{ and } \Pi_2(4) = \frac{147}{10952} - C_L$$

Under which circumstances does firm 2 goes bankrupt? This is only the case if the sum of stage 3 and stage 4 profits are below zero. If firm 1 sets up a low cost airline these profits are

$$\Pi_2(3) + \Pi_2(4) = \frac{147}{10952} - C_L - C_L$$

[20] This result is only an artefact of the model parameters and must not be true in general. It also depends on the assumption that withdrawing the low cost carrier is possible without any costs.

which is below zero only if $C_L > \dfrac{147}{21904}$

If firm 1 does not build a low cost airline firm 2's profits are simply

$$\Pi_2(3) + \Pi_2(4) = \frac{147}{5476} - 2C_L$$

which is positive because otherwise firm 2 would not have entered the market.

Now look at firm 1's profits. If it does not build a low cost airline in stage 4 its profits in period 4 and 5 are

$$\Pi_1(4) + \Pi_1(5) = 2(\frac{7}{96} - C_H) = \frac{7}{48} - 2C_H$$

If it builds a low cost airline these profits depend on the probability of bankruptcy of firm 2. Denote this probability $p = p(E[C_2])$ where p is obviously increasing in $E[C_2]$. Thus profits are given by

$$\Pi_1(4) + \Pi_1(5) = \frac{3}{56} \text{-} C_H - C_1^{LCC}$$

$$+ p[\frac{1}{8} - C_H] + (1-p)[\frac{7}{96} - C_H]$$

$$= \frac{85}{672} + \frac{5}{96}p - 2C_H - C_1^{LCC}$$

Comparing this profit with the profit without building a low cost airline yields that building a low cost airline is profitable if

$$\frac{85}{672} + \frac{5}{96}p - 2C_H - C_1^{LCC} > \frac{7}{48} - 2C_H$$

or

$$p > \frac{96 \cdot C_1^{LCC}}{5} + \frac{13}{35} \tag{4}$$

Since p is increasing in $E[C_2]$ this shows that the higher $E[C_2]$ the higher is the probability that firm 2 goes bankrupt and the more firm 1 is willing to build a low cost carrier.

This shows that if (4) is satisfied firm 1 sets up a low cost airline in stage 4. But despite this if firm 2 has not gone bankrupt it is optimal for firm 1 to withdraw its low cost airline in stage 5. There are two reasons for this seemingly inconsistent behaviour. First firm 1 can save the operating costs C_1^{LCC}. Second, the competition of the low cost airlines drives profits of the traditional business of the established airline down.[21] Thus firm 1 in some sense cannibalizes itself. To see this look at the stage revenues in case of existence of a low cost airline of firm 1 and in case of non existence. If firm 1 builds a low cost airline the revenue is 3/56 while without a low cost airline it is 7/96. Since the low cost airline earns zero revenue because of Bertrand competition the revenue of the traditional airline has decreased. The reason is that, because of competition between the low cost carriers, their prices are so low that many passengers want to switch to the low cost airlines. Firm 1 has to react and lower the prices of its traditional airline which drives its profits down. Thus firm 1 does better in withdrawing its low cost airline if firm 2 is still active.

This model can explain why an incumbent airline establishes a low cost carrier and withdraws it in later periods. The low cost carrier is used to induce its rival to leave the market. However, if the rival's costs have been overestimated and the rival survived the predation it is optimal to withdraw the low cost airline later.

As an example for this look at the case of GO. It was founded in 1998 by British Airways, one year after the low cost carrier easyJet was founded. The formation of GO can probably be seen as an attempt to compete with easyJet because both airlines do fly on almost the same routes. A goal of British Airways could be to inflict losses on easyJet and force it to exit the market. In August 2002 easyJet purchased GO from its management which made easyJet the largest low cost carrier in Europe. Thus the story of GO was not successful for British Airways. Despite this it was rational for British Airways to set up GO from the perspective of 1998. At this time easyJet was not large and it was uncertain how the airline market would evolve in the next years. It could likely have been that easyJet did not attract many passengers, or that the low cost carriers did not manage to build up reputation because people think that they are not safe enough or because of other reasons. But market conditions turned out well for low cost carriers and easyJet's earnings grew continuously. Therefore it was no longer reasonable for British Airways to keep GO because this only strengthened competition and lowered profits of British Airways' traditional business. Thus the best decision was to sell GO.

Another example is the US market where in the years after its foundation Southwest Airlines was not successful. One of the reasons was that established airlines like Delta and American decreased their prices significantly and established own low cost carriers. But Southwest Airlines survived these difficult years. After that, established airlines returned to their usual strategy of higher prices and stopped predation. Their low cost carriers folded and the success story of Southwest Airlines began.

---

[21] This argument was first developed by Judd (1985) in a model with qualitatively similar but differentiated products (apples and oranges). As is shown here this also holds for vertical product differentiation.

5. Extensions

This section considers three possible extensions, namely competition between incumbents, the introduction of a minimum quality standard and price discrimination of airlines.

Competition between Incumbents

Up to now we have only considered cases where one incumbent, the established airline, competes with an entrant, the low cost carrier. Donnenfeld and Weber (1995) analysed a model with competition between multiple incumbents. In their model there are two incumbents which can produce different qualities and face the threat of a potential entrant.

They first analyse what the outcome will be if the incumbents do not collude. The result depends on the value of the fixed costs of the entrant. If these costs are low the firms can do nothing against entry and will accommodate it. They will select extreme qualities. This means that one firm produces high quality while the other one produces low quality. They do this to differentiate their goods from that of the entrant. The entrant chooses a quality in the middle to soften price competition. If the entrant's costs are in some middle range the opposite is true. The incumbents produce similar qualities which lead to harsh competition and low profits. But with this strategy they keep the entrant out of the market because prices are low and the potential entrant would make losses after entry. Profits of the incumbents are lower under entry deterrence than under entry accommodation. If the entrant's fixed costs are so high that entry is blockaded in the sense of Bain (1956) the incumbents choose sharply differentiated products. They do this to reduce competition between themselves since there is no fear of entry.

If the firms collude the scenario is the same under very high and very low entry costs. But if entrant's costs are in the middle range entry might be accommodated. The reason is that incumbents no longer select similar qualities and compete harshly on price but differentiate themselves to increase profits. Therefore entry is profitable.

If we look at the airline industry one might guess that there is some form of collusion going on between the established airlines. They are in the same market for a long time and if one airline deviates from high prices this can be punished easily.[22] This collusive behaviour might not only be in prices but also in other dimensions like different flying times or different qualities. As was shown by Donnenfeld and Weber (1995) this collusion might make entry easier for a potential entrant. Price competition is not very aggressive and the entrant can recover its costs easily. This leads to the conclusion that entry might have been successfully deterred if established airlines acted more as rivals and not as colluders.

Since consumer welfare is higher under entry this has also some interesting policy implications. It might be welfare enhancing to have some form of collusion to make entry easier. Therefore if a form of collusion is observed courts should not necessarily take legal steps against the airlines although their intention was to keep prices high.

[22] This punishment can also be carried out in different routes. See Borenstein (1995).

Minimum Quality Standards

Ronnen (1991) analysed a model which is similar to that of Shaked and Sutton (1982). Firms choose their qualities simultaneously and in equilibrium both choose different qualities. But Ronnen (1991) extended the Shaked and Sutton (1982) model by analysing the consequences of a minimum quality standard. If this minimum quality standard is not way above the quality that the low quality firm would have produced in an unregulated equilibrium, both firms are still active in the market and the low quality firm exactly meets the standard. But also the high quality firm raises its quality although it would satisfy the standard in the absence of regulation. The intuition is that after introduction of the standard, products are more similar which intensifies price competition in the next stage. To alleviate this, the high quality seller chooses a higher quality. However, product qualities are closer to each other because the standard limits the differentiation range. This results in lower prices per unit of quality. Therefore, if the standard is not set too high consumers are better off. Those consumers who already participated in the market select higher qualities while some consumers who did not participate before are now buying the good. The consequences on firm's profits are different. The high quality seller is always worse off. Because competition is intensified price per unit of quality decreases and this hurts the seller more than the gain because of higher demand. Instead of this, the low quality seller gains if the standard is not set too high. Because of the regulation the seller can commit to a higher quality which would be impossible for him without the standard. His profits increase. The overall effect on welfare is also positive. The gain in consumer rents and the profit increase of the low quality seller more than offsets the loss of the high quality seller.

The result of Ronnen (1991) has some interesting implications concerning competition in the airline market. It suggests that the introduction of a minimum quality standard might be desirable from the point of view of society. Since this standard increases quality of both airlines and intensifies price competition passengers would always gain from such a regulation. But governments should be careful in the exact value of the standard. If it is set too high the low cost carriers are no longer profitable because they would lose their own business (low quality, low prices) and would exit the market. This would thwart the consequences of the deregulation. But setting a mild standard can enhance welfare.

Price Discrimination of the Airlines

In the preceding analysis and in the extensions airlines are restricted to uniform pricing. This means that they can only produce one quality and demand one price. But in reality all airlines engage in many tricky forms of price discrimination to segment the passengers. In this subsection I explain how the model can be extended in such a way that each airline is able to produce not only one quality level but many different qualities.

In economic analysis this behaviour can well be depicted by second degree price discrimination. An airline produces different qualities e.g. for business and economy class, offers different flexibilities in booking changes and so on, each for a different price designed for different types of passengers.

Such a competitive price discrimination model in a simultaneous move framework is analysed by Champsaur and Rochet (1989). They show that there will always be a gap between the product lines of the two firms. So one firm will be a high-quality firm and the other firm will be a low one. The reason is that both firms profit from a large product range but to avoid harsh competition they differentiate themselves.

I extend their analysis to a sequential move game (Reisinger (2004)). As a result the incumbent chooses to be the high quality firm. Some consumers are switching from the traditional airline to the low cost carrier and many new passengers are flying now with the low cost carrier. Yet, the effects on consumer rents are ambiguous. Leisure travellers win but it might be possible that business travellers lose after entry. The intuition is the following. Since business travellers have a high preference for quality, for them the low cost carrier is a bad substitute to the established airline. Before entry the monopolistic traditional airline produced many qualities also low ones. To reduce competition after entry this airline produces now only a smaller range of qualities. Since the marginal passenger is indifferent between both airlines his chosen quality is lower than under monopoly. Therefore the business traveller's contract might be more expensive than in monopoly because her incentive to choose the contract designed for the leisure travellers is reduced.

6. Conclusion

In this paper an airline market with an established airline, the incumbent, and a low cost airline, the entrant, is analysed. The incumbent chooses its quality first and therefore has a first mover advantage. It selects that market segment which is more profitable, namely the business travellers. The entrant has to take another market segment, the leisure travellers, to avoid unprofitable price competition. In duopoly some passengers switch from the established airline to the low cost carrier and many new passengers are flying with the low cost carrier.

Profits of the incumbent are decreasing with entry. Thus the incumbent might consider a predation strategy to force the entrant to leave the market. One possible strategy is to build an own low cost airline. This new airline competes very aggressively with the entrant. If the entrant has high operating costs it is probably not possible for him to recover his costs and he will become insolvent. In this case the predation strategy of the established airline was successful. But if the entrant survives the predation, the low cost carrier of the established airline is in competition with its own traditional business. The incumbent's prices and profits are low because of harsh competition between the entrant and its own low cost carrier. Thus it might pay for the established airline to withdraw its low cost airline to lessen price competition.

This model can explain some features that are observable in the airline market after deregulation and provide an intuition why some low cost carriers of traditional companies failed. But it does not mean that traditional companies only build low cost airlines for predatory motives. Since the airline industry faces a crisis nowadays it might be a good idea to develop new business strategies. One might be to separate their business and offer different products. In a recent paper by Lindstädt and Fauser (2003), they argue that it is both more efficient and more effective for an established airline to segment its products, e.g. in a premium product and a low cost product where the first one is designed for the business travellers and the other one for the leisure travellers. They argue that economies of scale which were traditionally considered to be very important in the airline industry do not automatically lead to cost advantages.

An attempt to follow such a strategy is to build a low cost airline not only for predation reasons but also to be prepared for future developments. So it would be interesting to see how e.g. Germanwings, the low cost carrier of Lufthansa, maintains its hold on the market.

Acknowledgements

I would like to thank David Gillen whose comments and suggestions highly improved the quality of the paper and Hans-Martin Niemeier for his patience.

References

Bain, J. (1956), *Barriers to New Competition*, Cambridge University Press, Cambridge.

Borenstein, S. (1992), The Evolution of U.S. Airline Competition, *Journal of Economic Perspectives*, Vol. 6, pp. 45-73.

Champsaur, P. and Rochet, J.-C. (1989), Multiproduct Duopolists, *Econometrica*, Vol. 57, pp. 533-557.

Donnenfeld, S. and Weber, S. (1995), Limit Qualities and Entry Deterrence, *RAND Journal of Economics*, Vol. 29, pp. 113-130.

Dresner, M., Lin J.S.C. and Windle, R. (1996), The Impact of Low Cost Carriers on Airport and Route Competition, *Journal of Transport Economics and Policy*, Vol. 30, pp. 309-328.

Gibbons, R. (1992), *Game Theory for Applied Economists*, Princeton University Press, Princeton.

Gilroy, B.M., Lukas, E. and Volpert, T. (2004), The European 'No-Frills'-Aviation Market: Current and Future Developments, forthcoming in: Forsyth, P., Gillen, D., Mayer, O.G. and Niemeier, H.-M. (eds), *Competition versus Predation in Aviation Markets: A Survey of Experience on North America, Europe and Australia*, Ashgate Publishers.

Goolsbee, A. and Syverson C. (2004), *How Do Incumbents Respond to the Threat of Entry? Evidence from the Major Airlines*, Working Paper, University of Chicago.

Januszewski, S. and M. Lederman (2003), *Entry Patterns of Low-Cost Airlines*, Working Paper, Massachusetts Institute of Technology.

Judd, K.L. (1985), Credible Spatial Preemption, *RAND Journal of Economics*, Vol. 16, pp. 153-166.

Knorr, A. and Arndt, A. (2004), Successful Entry Strategies on the Deregulated US Domestic Market – the Case of Southwest Airlines, in: Forsyth, P., Gillen, D., Mayer, O.G. and Niemeier, H.-M. (eds), *Competition versus Predation in Aviation Markets: A Survey of Experience on North America, Europe and Australia*, Ashgate Publishers.

Lawton, T.C. (2002), *Cleared for Take Off. Structure and Strategy in the Low Fare Airline Business*, Ashgate Burlington.

Lindstädt, H. and Fauser, B. (2003) Separation or Integration? Can Network Carriers create distinct business streams on one integrated production platform?, *Journal of Air Transport Management*, forthcoming.
Reisinger, M. (2004),Vertical Product Differentiation, *Market Entry, and Welfare*, Working Paper, University of Munich.
Rochet, J.-C. and Chonet, P. (1998), Ironing, Sweeping, and Multidimensional Screening, Econometrica, Vol. 66, pp. 783-826.
Rochet, J.-C. and Stole, L. (2003), The Economics of Multidimensional Screening in: *Advances in Economic Theory*, M. Dewatripont, L. Hansen and S. Turnovsky (eds), Cambridge University Press.
Ronnen, U. (1991), Minimum Quality Standards, Fixed Costs, and Competition, *RAND Journal of Economics*, Vol. 22, pp. 490-504.
Selten, R. (1965), Spieltheoretische Behandlung eines Oligopolmodells mit Nachfrageträgheit, *Zeitschrift für die Gesamte Staatswissenschaft*, Vol. 12, pp. 301-324.
Shaked, A. and Sutton, J. (1982), Relaxing Price Competition through Product Differentiation, *Review of Economic Studies*, Vol. 49, pp. 3-13.

Chapter 15

A Capacity Lock-In Rule

David Starkie

Introduction

The object of this chapter is to revisit and argue the merits of a proposed rule, referred to as the Capacity Lock-in Rule (CLR), for deterring predatory behaviour in the airline industry. The Rule itself limits conduct in an *ex-ante* context by focusing on the response of incumbents after entrants have withdrawn from the market.

The next section summarizes the events leading up to the formulation of the proposal and this is followed by a summary of its salient characteristics. There was a lengthy exchange of correspondence between Professor Alfred Kahn and myself about the proposal and because this exchange clarified a number of issues and tested the arguments, it is also included. Finally, I re-visit my original proposal in the light of the various points made by Professor Kahn and with the benefit of having reflected further on the issue.

The Origins of the CLR

In April 1998 the US Department of Transportation published proposed guidelines on predatory conduct in the US domestic aviation industry. The Department had come to the conclusion that unfair exclusionary practices had been a key reason why in the US market new low fare carriers had not been able to gain entry into concentrated hubs to compete with incumbents. It was of a view that a not untypical response by the incumbent to entry was the flooding of the market with additional low fare capacity.

To counter such behaviour by the incumbent, the DoT proposed to examine conduct on a case-by-case basis and institute proceedings when it considered conduct to have been unreasonable. The examination was to take into account how the incumbent had reacted to entry and, in particular, whether the incumbent's response had produced 'lower local revenues than would an alternative reasonable response'. In this respect, the DoT's proposal constituted a significant departure from the usual approach to predatory behaviour, which is to focus on the relationship between prices and costs. The DoT's proposal went beyond these norms by requiring consideration not only of marginal production costs but also of opportunity costs (the net revenues foregone as a result of the incumbent's reaction).

The DoT invited comments on its proposal. In response I submitted a counter-proposal, the CLR, and this was subsequently published in the *European Competition Law Review.* The published paper is attached here as an *Annex.*

The CLR Concept

The essence of the counter proposal was that in the event of an entrant subsequently exiting the market, the post-entry capacity (but not the price(s)) of the incumbent services was to be temporarily frozen or locked in, The basic features of the proposed Rule were that:

- it penalizes only those incumbent airlines responding to entry by offering *excessive* capacity in relation to market demand; because it increases the cost of predation (i.e. extends the period during which the incumbent has to commit to excess capacity), it makes predation less likely to occur.
- An unconstrained competitive response by the incumbent is permitted during the entry phase.
- Post-exit, the incumbent is free to fine tune prices, but the core price(s) will be driven by the extent of the capacity supplied to the market and it is this quantum that would be locked-in.
- The competition authorities need only monitor post-exit capacity supplied in the market (not prices (yields)), which is a much easier task.
- There is no need to try and identify predation as such; the complications of market definition and power, or calculating average variable costs, is avoided.

Professor Kahn's Views

Professor Alfred Kahn (AK), former Chairman of the US Civil Aeronautics Board (CAB) is well known for his abiding interest in antitrust matters[1] and he testified before hearings relating to the DOT's proposals.[2] I had previously worked with him on related matters in New Zealand and therefore I sent him a copy of my submission responding to the US DOT's consultation. Kahn responded with his observations and this led to a further exchange of letters. This exchange helped to clarify a number of issues relating to both the US DOT's proposal and the CLR and it also helped to test the arguments with regards to the latter. The original correspondence is therefore reproduced here in chronological order and with very little editing.

[1] See, for example, A.E. Kahn (1991), Thinking About Predation – A Personal Diary, *Review of Industrial Organization*, 6: 137-146.
[2] Published as A.E. Kahn (1999), Comments on Excusionary Airline Pricing, *Journal of Air Transport Management*, 1-12.

AK to DS 17 August 1998

I take the liberty of sending you a copy of a memo that I sent to my fellow CAB members on this subject twenty years ago (!), which suggests the (Baumol) solution that you propose here.[3]

[Regarding the DOT's proposal] I share your uneasiness about the administrability of – and therefore the ability of airlines to comply with – a standard that requires knowledge of a reasonably available, more profitable response than the one actually made.

I do not think, however, that you (or I) can take refuge in the plausible defence of your proposed alternative that it would be only predatory behaviour (defined in the DOT way or with reference to some measure of cost) that would be taxed by your policy. Suppose the responsive price cut by the incumbent is indeed profit-maximizing in the new circumstances, and not dependent for its justification on the expectation of driving the challenger out: your proposal would nevertheless tax the incumbent in that case, in the event the entrant leaves the market, for the period during which it would be forced to stick with its previous offerings. So the airlines could still claim that your proposal would penalize responses that were not predatory – i.e. not premised on or justified by the expectation that they would have the effect of driving the challengers out – if the challengers nevertheless exited, for whatever combination of reasons.

So your relatively easy defence that the only adverse effect of your proposal on incumbents would be in cases in which they 'took short-term losses' – presumably with reference to some cost standard (including, in the event it increases capacity, the capital or opportunity costs of the additional equipment) – does not hold up.

Observe, incidentally, that the only basis on which I was able to make essentially the same suggestions twenty years ago was in a situation in which we were still regulating the industry. It's not clear to me how the DOT could interpret its present statutory authority as permitting it to impose the safeguard that both you and I favour: as in the case of the antitrust laws, I presume, they could act only in the event of a finding of predatory behaviour, some taking of losses (if only compared with some available, alternative more profitable courses of action – the difficulty in identifying which is, as you clearly point out, the flaw in the proposed DOT standards, even though in principle, I think we both agree, they are sound): if you were correct in contending that your proposed remedy would penalize only incumbents who violate that unenforceable standard, it would indeed be a perfect substitute; but if my reasoning is correct – that it would also penalize carriers that have attempted to respond in the most profitable way available to them in the new circumstances – then it seems to me it is not the felicitous substitute for the unadministrable standard that you suggest.

---

[3] W.J. Baumol (1979), Quasi-Permanence of Price Reductions: A Policy for Prevention of Predatory Pricing, *The Yale Law Journal*, 89,1,1-26. Baumol's essential idea was that the incumbent is committed to a post entry price: i.e. (s)he cannot increase price for some time in the event that the entrant withdraws. The chief problem with this approach in the current context is that there is not a single price to observe.

DS to AK 1 September 1998

It was also very reassuring to know that we have similar views on the predation matter and intriguing to know that we have thought along similar lines with respect to how the problem might be handled. With respect to your specific comments, first it might be useful for me to emphasize that the response that I propose is to lock-in the incumbent's reaction with respect to capacity only. In other words, I would not seek to directly influence the prices that the incumbent would charge during the period after the entrant leaves the market. The requirement would be that the incumbent maintains his offerings of frequency and seats. My reasons for focusing on capacity are as follows:

- to specify both capacity and price (fares) is probably an unnecessary over-specification; it is essentially the capacity available in the market at a particular time that determines the fares and seats available at each fare;
- the Department of Transportation's proposed remedies were essentially focused upon capacity;
- It is easy to monitor and regulate capacity (this is essentially published information) but it is difficult to do the same with fares.

Let me turn, however, to your essential point and one on which I need to convince you that you are wrong! You suggest that my lock-in proposal would tax the incumbent in circumstances where the incumbent's reactions to entry are profit-maximizing and, as such, are not dependent upon driving the entrant out. The consequence of this would be that an airline has been penalized, even though its response was not premised on an expectation of driving out the entrant.

I admit to not being sure why, in the circumstances you put forward, the incumbent airline would be penalized by being required to maintain its offerings for a period of time after exit of the entrant. If the incumbent's offerings, revised upon entry, was indeed a profit-maximizing response, then it is not entirely clear to me why this should not continue to be a profitable position post exit. Although, given the fact that the retirement of the entrant will have reduced total capacity available to the market, which will probably mean that the incumbent is faced with the beneficial effect of an increase in demand for its services and, therefore, might in unconstrained circumstances wish to increase capacity still further to take advantage of this (and thus regain a profit-maximizing position) I cannot foresee how these circumstances can lead to the incumbent facing a loss-making position. I suppose, strictly speaking, one can argue that the incumbent is taxed in so far as there is an opportunity cost in not being able to move from a profitable to a profit-maximizing position, I am not sure that this is what you had in mind and, if it is, I might be inclined to argue that it is a sacrifice worth making to achieve broader aims. Do I begin to convince you?

AK to DS 28 September 1998

I am not wholly satisfied by your proposal that the continuing restraint on the incumbent relate only to the frequency and seats, for a couple of reasons. First of all, you can force it to offer the same number of seats, but if you don't control the price – and of course there's no way of forcing travellers to purchase those seats, regardless of the price – the incumbent will be in a position in effect to go back to its pre-competitive-entry policies of (a) limiting the number of discount seats and (b) charging whatever it can get away with for the unrestricted ones.

I don't suggest that your remedy wouldn't do any good. Where the incumbent has responded by actually increasing its number of flights or total seats, it would presumably do better – after departure of a challenger – to sell the seats for whatever price is necessary: but even in that circumstance, its profit-maximizing price for those seats will presumably be markedly higher after its predation has been successful than before. And to the extent that its response was merely sharply to increase the availability of discount seats on an unchanged total number of flights, it will be free to go right back to the policy that attracted the entry in the first place.

While I do concede you were absolutely correct in pointing out to me that if the competitive response of the incumbent was indeed profit-maximizing without regard to the possibility of its driving the challenger out of the market, forcing it to adhere to those prices and volume of offerings could not be said to force losses upon it; but, as you recognize, such a rule would force it to continue with a policy that was no longer profit-maximizing once the challenger had left. I agree with you, however, that the imposition of such a sacrifice on an innocent incumbent might still be a small price to pay for the additional protection that such a requirement would provide against predatory tactics by incumbents who were not innocent!

DS to AK 29 October 1998

I agree with you that my proposal to lock-in incumbent's capacity offerings does not capture the response that increases the availability of discounted seats on an unchanged total number of flights (and seat capacity). But, in most circumstances this, surely, would be considered a rather timid response and one that perhaps we could live with regardless of its consequences for the entrant.

We might expect a carrier to optimize the equipment used on a particular route to expected demand; the computer inventory systems will provide the incumbent with historic information to enable fine judgements to be made in this regard. The carrier will also take into account its marketing strategy and will usually aim to provide for those requiring full flexibility, including last minute availability of a seat, at first and undiscounted coach fares. Other seats might be priced at a discount and, using yield management systems, the carrier will aim to maximize his yield by progressively increasing the discount nearer to departure time so that ideally no seat is left unsold. The choice of equipment (seats offered) will also reflect this yield management process.

In this light let us consider the proposition when the incumbent's response to entry is to increase the availability of discounted seats on an unchanged total number of flights and seat capacity. If the availability of discounted seats is increased, this might be at the expense of passengers who want the benefits associated with a full fare ticket and, as a consequence, there might be a diversion of these passengers to the entrant. Apart from that, the increase in the availability of discounted seats, by reducing the average ticket yield, might be expected to expand the market. But, if the equipment is configured to the pre-entry levels of demand, the incumbent will find it difficult to accommodate the increase within the constraint (set by the lock-in rule) of existing capacity. Consequently, I think that we are dealing with an ineffective response strategy on the part of the incumbent.

The exception would be in circumstances where the incumbent has adopted the entry deterring strategy of operating with excess capacity [prior to entry], in which case he would be in a position to increase the availability of discounted seats in an effective way, should entry take place. If entry did take place it would, of course, signal that the entry deterring strategy had failed.

The response at which the DoT rules are aimed is one where the incumbent increases his capacity offerings in the market in response to entry; this is the characteristic of an aggressive (but not necessarily inappropriate) response which warrants close scrutiny. Indeed, in your submission [to the DOT] you remark:

> It is the expansion of capacity on the challenged routes that both the DoT rules and I would identify as the critical signal that the competitive responses is not merely defensive but aggressively predatory or exclusionary (p. 19).

I would therefore, say, let us not throw out the baby with the bath water – the lock-in rule is not without its limitations but does it not focus on the essence of the problem.

AK to DS 18 August 1999

Thanks very much for the reprint of your [ECLR] article commenting on the US DoT's proposals. Re-reading your article reminded me that you may well have offered the best possible formula for its doing so.

I know that I have in the past expressed support for the notion of requiring incumbent carriers to continue for some period of time whatever their previous competitive responses were, in the event that they had succeeded in driving entrants out; but I had until now been deterred by my concern that a requirement that they maintain such prices opened the door to pervasive re-regulation: what if, for example, they maintained the prices but deteriorated the quality of their service offerings in other respects? To this concern, it seems to me you provide the effective answer: require them merely to maintain their new levels of capacity. I think your explanations in footnotes 16 and 17 are persuasive. [These footnotes can be read in the Annex to this Chapter.]

Concluding Thoughts

The footnotes Kahn refers to latterly seek to emphasize a number of points made in the correspondence with him. The important one is the relationship between capacity and prices (fares) and the argument that a threat to lock-in capacity does not necessarily constrain post-entry fare offerings. But it is the capacity available in the market at a particular time that determines average price (fare yield).

Consequently if the incumbent constrains his capacity offerings during the entry phase prospectively to allow for profitable operations with the same capacity post-exit, there will be a limit on the extent to which the incumbent can increase the number of discounted seats in response to entry.

The correspondence also clarifies another point: the characteristics of the capacity that is locked-in. The ECLR paper refers to a requirement imposed on the incumbent 'to operate with at least the same capacity (including frequency of service) as it did immediately prior to the withdrawal of the entrant'. This inclusion of service frequency limits the extent to which an incumbent is able to manipulate capacity during the lock-in period. For example, it would not be possible under a CLR expressed in this way, to timetable fewer daily flights by utilizing larger aircraft in order to maintain capacity. Some variation in flight equipment and in the timing of flights would remain a possibility but it would be too prescriptive and introduce too much rigidity if the rule was extended to capture such responses.

Thus the CLR is not a watertight limitation on predatory behaviour. To take another example, it would probably do little or nothing to address an entry deterring strategy wherein the incumbent placed excessive capacity into the market prior to entry.[4] On the other hand I would argue that it is a focused rule and would do much to address the issue that was of particular concern to the US DOT namely the dumping of capacity onto a route(s) by an incumbent in response to entry. And it would do this with a light touch, without having to engage in the messy detail of trying to identify predatory behaviour. It is a proposed rule that leaves untouched competitive but responsible behaviour and leaves accommodating strategies undisturbed; but it does punish those incumbents that engage in the exclusionary practice of flooding the market with excess capacity in response to entry.

[4] I am grateful to William Morrison for this point at the Workshop.

Annex

European Competition Law Review
Vol 20: Issue 5 MAY 1999

The US Department of Transportation's Statement on Predatory Conduct in the Airline Industry: An Alternative Proposal

David Starkie

Introduction

In April 1998, the US Department of Transportation (DoT) issued a controversial proposal for an enforcement policy on predatory conduct in US domestic aviation.[5] Behind the proposal is a growing concern in the US that the considerable reduction in real air fares achieved since de-regulation of the industry in the late 1970s was coming to an end. In the year to February 1997, for example, average fares rose 9%[6] and, whereas previously, decreases in discounted fares more than offset increases in unrestricted full fares, this was no longer the case. The US DoT's view is that this reversal of the previous trend in air fares is due, in large measure, to the difficulties new entrant carriers[7] face when competing with established carriers. New entrants face particular difficulties in penetrating the hub-based networks of the major carriers; the organizational structures of such networks make entry inherently difficult[8] but the barriers are often re-enforced by the 'exclusionary conduct' of the incumbents. It is the latter issue that is the focus of the DoT's proposal.

---

[5] Economics-Plus Limited, Level 2, Berkeley Square House, Berkeley Square, Mayfair, London W1X 6EA. Tel: +44 (0)207 887 6144, E-mail: economicsplus@aol.com. This paper is based on a submission made to the US DoT in response to a consultation process. I am grateful to Professor Alfred Kahn and Frank Berardino for comments on the submission. I remain responsible for all the views expressed. Docket No. OST-98-3713: Statement of the DoT's Enforcement Policy Regarding Unfair Exclusionary Conduct in the Air Transport Industry.

[6] American Express Travel Services.

[7] Defined as independent airlines that have started jet services within the last ten years.

[8] There are sunk costs in establishing the network and entry on a scale necessary for effective competition is sometimes precluded by airport capacity constraints.

The Problem in Outline

Following de-regulation in the late 1970s, the US domestic air carrier industry reorganized itself around a large number of hub-based networks.[9] Previously most journeys between US city-pairs were direct;[10] after de-regulation these were routed increasingly via an intermediate hub, which became a point of (on-line) transfer between flights. The result was that most air passengers benefited from a more extensive, more frequent service and competition between air carriers over their respective hubs led to the inherited economies of a hub-and-spoke network being passed-on to passengers in lower fares.[11] There was and continues to be, however, an important exception to the latter. Because many hub airports are dominated by a single airline, passengers commencing their journeys at a hub pay significantly higher fares[12] than passengers commencing journeys at non-hub airports (from which alternative routings via competing hubs are often possible). Over time, hub dominance has grown; carriers have developed their hubs, a degree of consolidation between established air carriers has produced more single-carrier hubs, and the major carriers have also grown adept at exploiting their local dominance.[13] The DoT noted in its proposal document, for example, that in local hub markets major carriers have focused on high-fare services leaving much of the demand for services at discounted fares not served.

Such conditions should in principle attract competing carriers into the hubs to provide for the unserved discount fare market and to provide more competitive offerings to existing passengers paying fare premiums. These conditions should be particularly attractive to carriers with a low cost base, the archetypal example of which is Southwest Airlines. Southwest, originally a regional based air-carrier has grown substantially on the basis of a strategy that emphasizes discount fares. It has been successful in developing a presence at an increasing number of hubs, whilst other new entrants have entered various local markets, albeit on a more limited scale than Southwest. Overall, however, the impact of the new entrants has been limited, as recent increases in fare levels suggest.

---

[9] Although hubbing in the US pre-dates de-regulation of the airline industry, after 1978 there was an acceleration of its use.

[10] In 1978, domestic on-line connections accounted for 25% and interline connections 23%, of total enplaned passengers

[11] Combining passengers with many different final destinations on the same flight into a hub, enables the use of larger aircraft or, alternatively, higher load factors to be achieved (although there will be a fall in the average distance flown per flight sector).

[12] After adjusting for length of flight and route traffic density.

[13] A useful review is provided in P.P. Belobaba and J. Van Acker, Airline Market Concentration: an analysis of US origin-destination markets, *Journal of Air Transport Management*, 1, 5-14, 1994.

The view of the DoT is that this limited impact is a consequence of predatory behaviour by the incumbent carriers.[14] The hub-dominating incumbent views competition by the new entrant carrier as a threat, in spite of which, in some instances, it chooses to co-exist with the low-fare entrant and concentrate on developing traffic feeding into its network. On the other hand, a not untypical response by the incumbent is to seek to 'flood' the market with additional low-fare capacity. This strategy is often reinforced by the selective enhancement of frequent flyer miles for passengers and of commissions for travel agents. The result is that the entrant finds it difficult to establish break-even load factors and is thus forced to withdraw from the local market. After the entrant withdraws, the incumbent then drops the added capacity and raises fares. The major carrier thus accepts lower profits (or losses) in the short run, in order to secure a higher long-term return from its dominance of the hub. The strategy also benefits the major carriers prospectively because it signals to potential entrants that the major carrier will react aggressively to entry, and thus it acts as a deterrent.

The DoT's Proposal

To counter such behaviour by incumbent carriers, the DoT proposes to examine conduct on a case-by-case basis (after receiving formal and informal complaints but also taking the initiative when appropriate). It will consider whether a response to entry is consistent with behaviour in other competitive markets, especially those in which Southwest is a competitor.[15] In particular, when the incumbent sells a large number of seats at very low fares, or carries a large number of local passengers at the entrant's new low fares such that the entrant's total seat capacity, or passengers carried, is exceeded, the DoT will (unless there are strong arguments to the contrary) institute enforcement proceedings to determine whether the carrier has engaged in unfair exclusionary practices. In addition, the DoT will analyse other types of conduct, such as bonus frequent flyer awards or travel agent commission overrides, to see if incumbents appear to target entrants unfairly.

Notwithstanding this approach, the DoT does not intend to discourage major carriers from competing aggressively against new entrants in their hub markets. The Department suggests that a major carrier could, for example, match the new entrant's low fares on a restricted basis without significantly increasing capacity. It also comments that, conceivably, a major carrier could both lower its fares and add

[14] The following summarizes the DoT's stylization of the pattern of events. There are contrary views that regard the stylization as an inaccurate portrayal of the response to entry and challenge the basic concept of predation behaviour. Other commentators contend that this broad picture is based on a few isolated incidents, whilst others question whether the response of incumbents to entry has been the major reason for entrants' departure from local markets or for the failure of entrants in recent years.

[15] Southwest is used as a bench-mark because its size and profitability are judged to make it a difficult target for exclusionary practices. Its presence on, or potential to enter, a route has been shown to impact significantly on yields (see K. Richards, The Effects of Southwest Airlines on US Airline Markets, in B.S. McMullen, *Research in Transportation Economics Vol. 4*, JAI Press, 1996).

capacity without dilution of its existing revenue base[16] (presumably by price discrimination). And it is willing to concede that a new entrant's service might fail 'for legitimate competitive reasons'; in other words, if the entrant withdraws from the market this should not be construed to indicate predatory behaviour by the incumbent carrier.

It is clear that implementation of these proposals will require difficult judgements. This is evident in the tension in the DoT's Statement with regards to the addition of capacity by the incumbent carrier and in its reference to conceivable circumstances in which adding capacity in response to entry is a legitimate response. Indeed, the DoT's proposal recognizes that 'a reasonable alternative' response by the incumbent could incorporate an element of revenue dilution; thus a great deal could hang-on the interpretation of 'reasonable' conduct. The proposals also constitute a significant departure from established norms for judging predatory behaviour, which have focused previously upon whether an incumbent, in responding to entry, has continued to cover costs. The DoT's proposal goes beyond these norms because an examination of whether the response by the incumbent has had the result of producing 'lower local revenues than would an alternative reasonable response' requires consideration not only of costs but also of revenues forgone. It is recognition, therefore, that net revenues sacrificed by a course of action are an (opportunity) cost to the firm of undertaking that action.[17]

The proposals have been severely criticized by the major airlines (as might be expected) and by a number of distinguished academic experts. The thrust of the more considered criticisms is that the proposals will have the effect of stymieing the reactions of the incumbents to entry; where lower fares (following entry) stimulate an expansion of the market, the incumbent will be inhibited from adding capacity and in the longer term the outcome could be higher rather than lower fares. These criticisms are arguable and have been robustly challenged. But there are those who, whilst strongly supporting what the DoT is attempting to accomplish, nevertheless doubt the 'administrability' of the proposals. In particular, it is questioned whether the major carriers can be expected to know when they are violating the measures that, if exceeded, lead to enforcement procedures. These measures require the incumbent to know the entrant's total seat capacity and the total number of passengers carried.

Back to Basics

What is required is a relatively simple rule that has the effect of making predatory behaviour unprofitable (or less profitable) without imposing significant economic costs on the industry. Let us start the quest by considering why conduct in the

---

[16] Referred to in the docket as 'self diversion of revenue'.

[17] Professor Kahn considers the DoT proposal superior to the Areeda-Turner test because it adds to the marginal production costs, upon which the Areeda-Turner is based, the opportunity costs of the incumbent's response. This is particularly pertinent when the incumbent adds capacity to the contested route because this capacity will, more than likely, have been directed from other markets and thus the incumbent will have sacrificed revenue elsewhere.

airline industry has failed to match the original ideas of contestability theory. When this theory was developed in the late 1970s and early 1980s, it was thought that the market for air travel was a textbook example of a contestable market, the essential idea of which is the ability to challenge for a market rather than necessarily compete within it. This ability to challenge or contest depends upon the ease with which a firm can enter *and* exit from a market without incurring a cost which, in turn, depends particularly upon whether capital is mobile or sunk – that is, irretrievably committed to producing a particular product. Whether sunk costs exist in an industry is crucial to the argument. Sunk costs are costs facing a potential entrant that do not have to be paid *once more* by the incumbent; consequently, to the potential entrant, they constitute a barrier to entry. On the other hand, if all capital is saleable and reusable in alternative markets without loss (other than that corresponding to normal depreciation in use), a potential entrant to an industry can then replicate, without penalty, the cost and output vectors of the incumbent firm(s). Consequently, an industry without sunk costs – even a natural monopoly industry – is said to be a perfectly contestable industry; the possibility of entry by rival firms is a constant threat so that prices and levels of output are what one would expect to find under conditions of perfect competition.

These early claims became contentious; in particular, the speed at which an incumbent could respond to entry was viewed as critical. A number of commentators argued that, to secure a competitive market outcome, it was also necessary to assume that the entrant could undercut the incumbent, capture the market and (if need be) exit before the incumbent could adjust its price.[18] If this was possible (and there were no sunk costs) then the threat of entry alone would encourage the incumbent to post competitive prices absent entry. However, the air carrier industry appears to be a good example of an industry in which sunk costs are low but incumbents are able to adjust prices almost instantaneously, and capacity almost as quickly, when entry occurs. As the DoT's Statement records: 'Compared to firms in other industries, a major carrier can price-discriminate to a much greater extent, adjust prices much faster, and shift resources between markets much more readily … Air carriers have access to comprehensive "real time" information on their competitors' activities and can thus respond to competitors initiatives more precisely and swiftly than firms in other industries.'[19] Airline capital might be 'capital on wings' but the machinery of adjustment is too well oiled; adjustments can be made too cheaply and too quickly for competitive threats to work effectively.

An Alternative Approach

But if it is here that the root of the problem lies, then perhaps the policy response should be designed accordingly. An alternative approach is to *lock-in* the incumbent's response to entry so that if the entrant subsequently withdraws from a route, the incumbent is required to continue to serve the route for a specified period

---

[18] See, for example, W.A. Brock, Contestable Markets and the Theory of Industry Structure: a review article, *Journal of Political Economy, 91*, 1055-1066, 1983.

[19] The cost of adjusting prices (menu costs) is also negligible.

without a reduction in the existing levels of service. During this specified period, the incumbent would be required to operate with at least the same *capacity* (including frequency of service) as it did immediately prior to the withdrawal of the entrant.[20] The lock-in thus builds in a response lag. This has the effect of inducing the incumbent to post prices that one would expect to see in a competitive market. This is because if the incumbent's response to entry is to increase capacity to such an extent that losses are incurred, the incumbent would not be able to start recovering those losses immediately the entrant withdraws from the market. It thus prolongs the period over which an incumbent could incur losses, with the result that this form of predatory behaviour becomes more expensive to pursue and therefore is less likely to occur.[21]

The incentive compatible feature of the lock-in is that it imposes a potential but explicit penalty on predatory behaviour because it penalizes *only* those incumbent carriers who react to entry by introducing levels of capacity that result in the carrier incurring losses in the short term. Any incumbent carrier operating a route that, post entry, continues to make a positive contribution to the bottom line will not be disadvantaged by the lock-in.[22] By, in effect, 'taxing' only predatory behaviour, it increases the risks to the potential predator of adopting such behaviour. The proposed rule, therefore, does not eliminate predatory behaviour in response to entry but it is likely to reduce its occurrence. It enables the incumbent to continue to respond to entry by adjusting capacity but it reduces the likelihood that such adjustments will result in below-cost operations that are unreasonable or unfair to the entrant.

The proposed rule requires a relatively light-handed form of regulatory intervention. The regulator is not required to make difficult judgements regarding the precise cost structure faced by the incumbent or (as the DoT's Statement requires) judge what would be 'a reasonable alternative response' by the incumbent in circumstances where major carriers are not to be discouraged from competing aggressively against new entrants in their hub markets. Instead, the regulator is required only to audit the incumbent's level of service on a route during the lock-in period. This should be relatively easy to achieve: the regulator can match the published *OAG* schedules with actual flight information.

---

[20] The reasons for focusing on capacity are threefold. First, to specify both capacity and prices (fares) is probably unnecessary; the capacity available in the market at a particular time largely determines the seats available at each fare. Second, the DoT's proposed remedies are essentially focused upon the problem of additional capacity. And, third, it is easy to monitor and regulate capacity (this is essentially published information) but it is very difficult to do the same with yields.

[21] Professor Kahn has suggested that the lock-in does not capture a response that increases the availability of discounted seats on an unchanged total number of flights (and seat capacity). I feel that such a response is constrained in its effectiveness because the equipment used on a particular route is likely to have been optimized to the level of expected demand (prior to entry). Consequently, if the incumbent's response to entry is to increase the number of discounted seats on existing flights, this is probably at the expense of passengers wanting full-fare unrestricted tickets. This might lead to a diversion of these passengers to the entrant.

[22] Except in so far as it would have chosen to seek super normal profits during the lock-in period.

The proposed rule is likely to be more effective in reducing predatory responses the longer the lock-in period chosen because the greater will be the adverse consequences to those choosing to behave in a predatory fashion. In international aviation, schedules are usually set on a biannual basis and co-ordinated at scheduling conferences held twice a year. In this context, a six-month lock-in period suggests itself. For US domestic aviation the same scheduling procedures do not apply. Schedule changes are more frequent and a three-month (90 day) period might be more appropriate. However, the issue of an appropriate lock-in period merits further investigation taking into account seasonal factors in demand (which will affect certain markets more than others) and whether there are systematic changes during the course of a year in the price of factor inputs, such as aviation fuel.

Refinements

Seasonal changes, in both demand and the price of inputs purchased by carriers, raise the issue of whether, particularly if a long lock-in period is chosen, there might be a release clause. The object of the lock-in is to reduce predatory behaviour by penalizing the incumbent carrier who adds excess capacity and makes short-term losses in response to entry. In contrast, the incumbent carrier who responds to entry by increasing capacity and reducing fare yields only to levels expected in a competitive market (i.e. a profitable operation is sustained) would not be penalized by the lock-in, unless there was an adverse change in the underlying economic circumstances. If, for example, there was a significant reduction in market demand caused by exogenous factors, or a significant increase in the price of aviation fuel, one would expect air carriers to make corresponding adjustments to capacity (and prices). The lock-in would prevent such adjustment and therefore could lead to possible losses even when the incumbent was pursuing a break-even strategy in response to entry.

There may be differing views regarding the 'fairness' of such an outcome. One argument is that such an eventuality should be treated as another business risk, which the incumbent carrier allows for in his post-entry strategy. There is, after all, the possibility that changing circumstances may lead to an outcome favourable to the incumbent during the lock-in period, demand may increase or the price of factor inputs may fall. If the risks in this context can be viewed as symmetrical (i.e. equal probabilities of demand and factor prices increasing or decreasing) then the presumption must be that the lock-in period does not penalize the incumbent who responds to entry by adopting a legitimate competitive response. If, on the other hand, these risks are not viewed as symmetrical, such that there is a greater likelihood of demand falling or the price of inputs increasing, there is an argument for allowing for this in the lock-in rule. For example, the incumbent could be permitted to alter capacity if it could show that it had been adversely affected by a change in external circumstances. However, a *de minimus* approach might be judged appropriate (thus acknowledging that the exogenous changes may also be favourable to the incumbent): only when adverse demand or cost impacts exceeded a specified threshold would the incumbent carrier be allowed to reduce capacity during the lock-in period. To simplify the administration, and to minimize delay in altering capacity in such instances, the incumbent could make changes without prior authorization, but such changes would have to be filed and a selection of them subject to audit, with penalties for violation.

Summary and Conclusions

The US DoT's Statement on predatory conduct in the airline industry is the latest attempt to address a problem that is concentrating the minds of regulators world-wide. Pinning down predation is recognized as a difficult task and very few actions against predation in the airline industry have been concluded in favour of the aggrieved party. The DoT's proposals are a focused attempt to provide guidelines on the issue and to specify criteria for the introduction of enforcement proceedings. Whether such enforcement proceedings would be more successful than in the past is questionable. The DoT recognizes the fine line between a pro-competitive response to entry and one that is intended to drive the entrant from the market, and much will depend upon its judgement, especially regarding what is a 'reasonable' response in the circumstances of a particular case. The established airlines have strongly opposed the proposals and independent critics have questioned whether the criteria used to trigger enforcement proceedings are workable.

The DoT has focused its proposals on a particular response to entry by the incumbent carrier – that is, a response that floods the market with low fare capacity. With this in mind, this paper advances an alternative approach to this problem. It takes as its starting-point a weakness evident in the application of contestability theory to the aviation industry, namely the absence of effective response lags which theory has shown are needed if prices and levels of output are to be kept at competitive levels regardless of the number of firms competing *in* the market. The proposal put forward in this paper is to introduce a lock-in rule so that if the entrant does withdraw from a route, the incumbent is required to continue to serve the route for a specified period with pre-exit levels of capacity. It is argued that this approach imposes a potential but explicit penalty on predatory behaviour because it penalizes only the incumbent carriers who have reacted to entry by choosing to operate with excess capacity. Incumbent carriers who react pro-competitively, possibly by increasing capacity but continuing to operate profitably, would not be penalized. A further advantage of the proposal is that it focuses the regulatory activities on monitoring and compliance; the regulator is not required to try to define in each case whether the response to entry has been predatory. The proposal appears, therefore, to meet the test of a relatively simple rule that has the effect of making predatory behaviour unprofitable (or less profitable) without imposing significant costs on the airline industry.

Acknowledgment

Professor Alfred Kahn retains the copyright for his contributions to this chapter, and they appear here with his kind permission.

Chapter 16

Wide Angle Preying in Land Modes? The Problem of Missing Modes

Marc Gaudry

1. Narrow Angle Predation at the Margin of Unit Cost

Assignment: Including Missing Modes in the Analysis

Our task in this book is to start a search for clues to the existence of predation outside of the air mode. Without loss of generality, we kick it off by excluding the maritime mode from our inquiry and focus the search on land modes. Even then, the task appears hardly manageable as, excluding known intramodal criminal activities typically carried out to limit or control access within particular land modes (taxis, urban buses, etc.), very few examples of predation, if any, rapidly come to mind. Fewer still come to mind if we search for intermodal and multimodal cases.

We argue that our mandated task would be made easier if the narrow standard price/quality/quantity based definition of predation were enlarged to include the market for all regulation, as firms' actions aim as much at modifying the framework (market structure and regulations) of their activities as they target their competitors within a given framework. In effect, we argue that a critical dimension of successful predation, that of murdered modes, cannot be handled within the confines of the predation literature. Surely, missing modes should matter.

This relevant enlargement of a narrow definition merely allows the effects of firms' actions to include, as matters in public economics, the crucial framework issues that impact firms' markets as narrowly defined. We will therefore argue for a wide angle view of the problem that goes beyond the precise but narrow standard reference analysis.

Indeed, if demand analysis can handle the problems of new modes, such as Personal Rapid Transit (PRT) in cities, it can handle the questions asked about the revival of 'murdered' modes and the literature on predation should adjust to include these limit cases, lest it be accused of irrelevance at the limit: observed zeroes should be explained. It may be the time to remember the missing modes.

Is the Standard Intramodal Reference Analysis Clear and Clean?

In transport analysis as elsewhere, 'reference' predation tends to be defined analytically as a producer's bearing of short run losses to eliminate competitors in order to obtain higher profits in the long run. In this definition, the emphasis on the short run then requires the determination of variable unit costs, often distinguishing, under fixed capital conditions, between the positive ('marginal') and negative ('avoidable') variations at the margin; the emphasis is also per force on the conceptualization, at least of 'intended' long term profits, implicating (under variable capital conditions), some representation of long term cost and price futuribles.

Both of these determinations pose non trivial issues of computational credibility, and more, if the intents of higher profits have to be demonstrated. Consider two land mode cases where even perfect calculations of avoidable costs in standard fashion may miss the presence of predation entirely.

Intramodal Predation without the Cost Margin: the Hollander Squeeze

Consider complementary products, such as return trips by intercity bus between towns like Oxford and London, where the market is of sufficient size to maintain more than one firm. As is well known, the structure of demand in such markets (as in air markets between two large enough cities, e.g. Montreal and Toronto) is characterized by sharp morning and evening peaks. In these cases, a challenger will generally position his new supply very close, or even within, the peaks. But as the new challenger has few services at peak times and the incumbent has many, small differences in ticket price can drive users from the challenger's service to the predominant incumbent's because travellers often know when they leave home in the morning but not when they will come back in the evening.

For this reason, we often note that return ticket discounts are extremely small: return tickets cost almost the same as two one-way tickets. A dominant incumbent wishing to evict a challenger could do so with relatively small price rebates (perhaps as low as say 10%) on return tickets because the latter's lost demand spreads among the former's services. To survive in this market, the challenger might consider offering much lower (return) prices – contributing at least something to his fixed costs – but such prices might not ensure his survival. In effect, his survival requires the incumbent's good will. If the squeezed challenger makes normal profits, the incumbent's good will should allow him to maintain some excess profits while keeping up the appearance of competition and discouraging other newcomers.

Note that this squeezing of the challenger has nothing to do with the incumbent's classical tactic of simply offering excess capacity and frequency, called a 'dilution' of the relevant market. A squeeze is loss-free predation: it reduces revenues somewhat, but without incurring losses at the margin.

Intermodal Predation under Unfair Competitive Conditions: VIA Rail vs Intercity Buses

Another difficulty of the standard definition of predation is that it does formally deal with the fact that the marginal cost very much depends on where it is calculated. In conditions of unfair competition, this pretense at independence between marginal and other costs can kill.

Consider the establishment of unavoidable or marginal costs of mandated subsidized intercity passenger train services, as exists in Canada with VIA Rail. This railway currently receives a yearly operating subsidy of approximately $160 million slowly decreasing over the last years from a peak of about $600 million 15 years ago. Such operating subsidies skew the framework necessary to define predation meaningfully. By contrast, the intercity buses are not subsidized directly, or even indirectly if one takes into account road and bridge damages caused by their kilometer axle loadings and compares them to their specific registration and automotive fuel taxes.

Now is VIA Rail guilty of predation if it establishes, as in the past, a special fare – say for students – that is almost equal to its avoidable expenses for carrying such a passenger and noticeably below the intercity bus fare (that must make some contribution to fixed bus costs)? Some may then simultaneously (i) accept that the competition between these modes is unfair; (ii) reject the claim that the student fare is predatory.

More generally, does the concept of predation make sense in intermodal cases that violate equal playing field rules? As subsidies affect output levels and marginal costs depend on output levels, predation is not a very useful concept in such situations, except if any predation so defined must somehow have a bad effect on market structure and consequently on competition.

In view of the massive subsidies received by many land modes, the link between economic efficiency and the marginal resource use test of anti-predation legislation may reside only in this feedback loop. Of course, changes in market structure and regulation are often the very predatory objective of firms: strikes, for instance, may aim at increasing factor rents through political pressures generated by holding the population to ransom to eliminate competing firms or modes. But is there a body count of eliminated competing modes?

2. Can a Wider Angle Locate the Cadavers of Former Preys and even Revive Them?

A More Complete View of Predation Effects

Assume that predation is the manipulation by a producer:

a) *of the quantity, quality or price of his product*. This means that a strike of a regulated monopoly can count as an act of predation even if output falls to zero;
b) *in order to increase economic rents*. The aim of monopoly need not be higher profits, but a quiet life and gold plating;
c) *by market or regulatory elimination*. Market access of others can be more effective than own price or service change in yielding excess rents;
d) *of producers of close substitutes*. In transport, the suppression of intermodal competition can be as effective as the elimination of intramodal competitors.

In cases of intermodal predation as defined above, it is naturally difficult to determine whether preyed upon modes that are now absent from the market could be revived and recover durably if the effects of past predation to eliminate them (typically through regulation) were removed. We will argue that, in many countries, at least two land modes should be allowed to come out of the cemetery and be granted a chance at revival: intercity buses and collective taxis.

Scheduled Intercity Buses as Best Intercity Public Land Mode

Does this header statement really need demonstration? In North America, intercity buses still reach about 10 times as many towns and cities as passenger trains ever did. Although the demand for their services has been falling slowly, and there is an argument for their deregulation, it is obvious that their absence in Germany and France, for instance, is entirely due to the desire to protect monopolistic railways driven by strong unions pretending notably to defend 'public service', as if public authorities could not subsidize demand instead of supply for socially desirable services …

We pretend that, in many countries where licences have not been available for intercity bus services, such intercity bus services could function without subsidies. Regulations effectively forbidding their existence should consequently be abolished, as the cadaver could revive. Q.E.D.

Collective Taxis or Jitneys

The word 'jitney' is the early 20th century Boston Irish slang term for the nickel, or $0.05 coin: collective taxis charged a fixed fare of five cents ... Collective taxis were outlawed in most North American cities by pressures from the transit companies before the First World War. But they still live happily in many South American cities and could be useful in intercity markets if allowed to operate again. In Santiago de Chile, collective taxis have successfully survived all forms of reorganization of public transit from 1986 onwards. They are expanding in some unexpected places: in Casablanca, 'white taxis', recently allowed to operate within the urban area, are increasing in numbers and competing (as 'white sharks') with public transit bus lines (Zhiri Oualalou, 2004).

The larger issue raised here is that of competition within urban (and intercity) transit markets and of whether the benefits of network coordination in transit networks are compatible with competition with collective taxis. Experience in many places suggests that they are.

Some authors have recently generalized this discussion by pointing out that many of the objections people may have to such competition can be handled by tighter application of 'curb rights' not to dogs misbehaving on sidewalks but to competing bus firms (Klein et al., 1997). It should also be remembered that lighter vehicles should pay lower axle weight damage (Kulakowski, 2002) charges than heavier buses: in consequence, proper infrastructure pricing would contribute to their reanimation.

From the perspective of the some preys, there is hope that the reestablishment of former intermodal competition in this area would durably revive the cadavers. and analysts should deal with the 'observed zeroes' of missing modes.

3. References

Klein, D., Binyam, R. and A.T. Moore (1997), Curb Rights, *The Brookings Institution*, June.

Kulakowski, B., El-Gindy, M., Hoskins, A. and Seokyoung (2002), *Transit Bus and Motor Coach Axle Weight Study*, Final Report, No. PTI-2003-01, Pennsylvania Transportation Institute, The Penn State University, sponsored by Penn DOT.

Zhiri Oualalou, T. (2004), 'Les "requins blancs" de Casablanca et la sécurité', Paper presented at the *ATEC-ITS France International* Conference, Issy-les-Moulineaux, France.

Chapter 17

An Economic Framework for Assessing Predation in Air Services: Markets, Barriers to Entry, Market Power and Tests

Colin A. Gannon

'... Competition in the real world is inevitably imperfect ... naturally monopolistic or oligopolistic (in) most airline markets.'
Professor Alfred E. Kahn (Former Chairman, US CAB), 1988

1. Introduction

1.1 Predation Defined

It is useful at the outset to set down a working definition of predation for this chapter. Predatory behaviour is '... foregoing of maximum current profits in order to eliminate competitors or deter or delay entry, so that greater profits can be earned in the longer run', Dodgson et al. (1990).

A predator's objective is to earn higher profits (in present value terms) by taking actions that weaken the viability of a rival/entrant and influence market structure[1] so as to increase its market power, using prices that are 'too low' (or quality that is 'too high') initially, and supra-competitive later.[2] Note that under this definition losses are lower economic profits and not necessarily financial losses.

1.2 Scope

In this chapter predation in air service markets is examined within the context of a country's domestic trunk routes drawing on recent experience, including from Australia. An economic framework is constructed to improve the general

[1] This includes acquisition, which may have a lower cost than the losses from current predation.

[2] Game-theoretic multi-period models of reputation, signalling, and financing, have strengthened greatly the rationality of predation as a long term strategy, i.e. incorporating 'conjectured' reactions of entrant and incumbent. See Kate and Niels (2002).

understanding of predation in two ways. In Part I, Section 2 provides a brief review of salient features of air services, including market definition, and a perspective on the relationships between market barriers, market power, and predation; Section 3 analyses specific barriers to resolve whether they are independent sources of impediments to entry or whether they are a 'derivative' based upon other more fundamental barriers or factors and, if so, how that may matter. Section 4 gives a brief discussion of market power. In Part II, Section 5 outlines the challenges of formulating an economic test for predation and Section 6 argues how the 'most favoured' economic loss test using an avoidable cost datum should be strengthened; finally, a summary of the main points is presented in Section 7. The two parts may be seen as key ingredients for examining predation: the necessary economic conditions for its rational adoption and evidence for its economic presence.[3]

It warrants reiterating here that predatory conduct is a highly controversial legal and economic issue and allegations are typically viewed with suspicion, especially by the courts.[4] There are two main reasons for such skepticism. First, it is difficult to distinguish definitely aggressive pro-competitive actions from unlawful anti-competitive conduct – even 'pricing temporarily below cost' can be pro-competitive.[5] This has been generally tackled by endeavouring to establish anti-competitive *intent* as the 'inferential purpose'. The second reason is that small firms have a natural incentive to plead 'unfair competition' and seek court *protection* from large incumbents.[6] Understandably, these reasons pose a very substantial challenge for proof and underscore the need for a rigorous analytical economic framework.[7] This chapter is directed at contributing to such a framework.

---

[3] The scope of this chapter is 'Stage 1' of what is conventionally seen as a two-stage analysis of predation, where the second stage assesses whether predation actually took place, usually based on some form of empirical price-cost comparison and evidence of intent with strategic purpose.

[4] Very few cases of predation in any industry have been proven over the last decade or so. One exception in air services is the recent case upheld against Lufthansa-Germania (2002). See Hüschelrath (2003).

[5] Acceptance of legitimate business competition was the key issue in summary dismissal (July 2003) in the American Airlines case. *United States* v. *AMR Corp* (2003).

[6] See McGee (1980).

[7] Historically, especially in the United States, the 'business rationality' of predation was subject to strong criticism, led by the 'Chicago School' of economists who dismissed its existence. See Bork (1978). Despite recent strong counterargument [see Bolten et al. (2000)], the 'Chicago' view has a continuing legacy in legislation and the US courts.

Part I: The Context of Predation

2. Preamble, Preview and Perspective

2.1 Revealed Operation of Deregulated Domestic Air Service Markets

The understanding of how air service markets operate has undergone considerable reassessment and improvement over the last couple of decades. This has emerged from observation of evolving market outcomes as the heavy economic regulation of many markets was removed (notably in North America, Europe, Australasia and, more recently, Asia) and as the ownership of many airlines (and airports) was shifted from the public to the private sector (for example, in Europe, Australasia, and Asia). During the period leading up to and early after deregulation of domestic aviation in the United States (circa 1978), the airline business and air service markets were widely regarded as 'highly contestable'. The key 'plant' – aircraft – were (to say the least) mobile[8] and also could be readily leased. Specialized inputs such as maintenance, ground handling and reservation systems could be contracted out, and access to technology and 'know-how' was largely equal. Sunk costs[9] were believed to be modest and scale economies were recognized to be limited. The initial evidence from unregulated markets in Texas and California seemed to confirm the theory. The conclusion drawn was that entry *and* exit for air service markets would be 'free and easy' – more or less the same as in other transport service industries, such as road haulage, and that the markets would be effectively *competitive or contestable*, i.e. exposed to the discipline of competition from actual entry or the threat of entry.[10] One incumbent would not mean a monopoly. Market power would be small or not exist and predatory conduct would not be possible.

In reality the outcome in the domestic air service markets of countries leading deregulation – USA, Canada, Europe, New Zealand and Australia – has been quite different. Many of the assumptions upon which contestability of markets is based were found wanting.[11] Some *real barriers to entry are now seen to prevail* and the main sources in air service markets are differences or 'asymmetries' between incumbents and entrants in their inputs or products that are unable to be exchanged

---

[8] It was common at the time to refer to aircraft as 'capital on wings'.

[9] 'Sunk costs' are unavoidable costs or expenditures (or expenditure commitments) that cannot be retrieved or revoked; for example, incurred outlays on advertising, training, business planning, regulatory approvals, irreversible contracts (or penalties for termination), and assembly of equipment and facilities.

[10] The susceptibility of an incumbent airline to 'hit and run attacks' hinges on some *time* being required (i.e. lag) before it can respond to entrants. Together with zero sunk costs this is a central condition for perfect contestability. But, with CRS and instant and accurate access to competitors' fare and schedule information, incumbents can react very fast. This collapses the time needed for some adequate 'hit' (i.e. profit) being achieved before the incumbent responds and the entrant then 'runs' (costless exit). An incumbent can also adjust the deployment of existing capacity over its entire network, *selectively* targeting the typically far less extensive set of routes served by a new entrant.

[11] For excellent discussions see Levine (1987), Kahn (1988), Borenstein (1992), and Morrison and Winston (1995).

in a market, i.e. they are *'non-tradable' differences*. Most importantly, these various barriers have been confronted by the 'unbundling' of the attributes of the traditional 'full service airline' (FSA) with the formation of a basic air service business model centred primarily on a 'no frills' single class of service, a homogeneous fleet, enterprise-based labour contracts, point-to-point service, low operating costs, and much lower fares, commonly described as a 'low cost carrier' (LCC).[12] Undoubtedly, this has been the most significant corporate development to emerge from deregulation and it has proven the most effective strategy to deal with many barriers to entry.

2.2 Competition and Barriers to Entry

Typically domestic air service markets are concentrated: there are only a few operators *in* the market; the discipline of competition comes as much, if not more, from the threat of competition *from potential entrants*, as from incumbents, especially if there are only a few (say, 2-3) airlines in the market. Therefore, in air service markets *new entry*, or its likelihood, is very important. If entry is inhibited, impeded, limited, or controlled in some way, then the greater the ability of *individual* firms already in the market to influence *market* prices, i.e. exert market power. Whether, and to what extent, that power is actually *exercised* depends on structural and legal conditions, and the ways in which the firms in the same market strategically interact. Under these circumstances, anti-competitive conduct, such as predation, is of special policy concern since it damages overall social welfare.[13]

2.3 The Core Nexus: Barriers to Entry, Market Power and Predation

Barriers to entry underpin market power. A barrier to entry corresponds to an advantage that an incumbent has in securing an input or product attribute that cannot be matched by an entrant at the same cost.[14] A major reason for such advantages is that entry requires the commitment of *irrecoverable resources*, i.e. *sunk costs*. An example is advertising expenditure; another is expenditure assembling a workforce; another is expenditure establishing a brand and reputation. Thus, entry involves the entrant at a disadvantage – at least for some time – and this renders *the entrant most vulnerable, i.e. at most risk, during the period of becoming established. It follows that 'aggressive competition' from an incumbent is most likely to occur at this time.* The higher the sunk costs the greater the initial disadvantage of the entrant.

---

[12] Other acronyms for this business model are used such as 'value-based airline' (VBA).

[13] A major (ab)use of market power is to set prices above costs and extract above normal profits and/or rents (or allow inefficient costs). Rents may show up as transfers, higher payments to employees (especially unions with monopsony power), or to suppliers of scarce 'essential' inputs who also have market power, for example, airports. This is an abuse since it comes at the expense of a 'deadweight loss' to society.

[14] For example, an incumbent airline may have a highly skilled aircraft scheduling manager, but the highest 'value' of this person to other airlines/entrants, if not the incumbent, may be represented fairly closely by her offered remuneration.

But the initial (and potentially prolonged) disadvantage of the entrant is partly under its own control since it will differ across *entry strategies and targeted market segments*. In particular, in the presence of a full service airline (FSA) incumbent, barriers are likely to be fewer, and less of a handicap, for entry into the low fare/basic service market segment than into the higher fare/full service segment. Sunk costs to the entrant are lower and the product superiority of the incumbent is valued far less.[15] Hence, the equivalent cost disadvantage of this narrower set of barriers is less (and possibly negated entirely and then some) by an entrant who can achieve lower operating costs, attract passengers, and then sustain lower fares.

The approach taken to barriers here is to treat each one separately as a positive advantage to an incumbent. Of course, not all incumbents enjoy such barriers and many incumbents face their own 'barriers', especially 'entry' to reorganization and industrial relations reform. These place the incumbents themselves at some relative cost disadvantages. Moreover, some barriers can be avoided by the entrant offering a different type of service or adopting a different business model. The low cost carrier (LCC) business model is centred upon lower costs and lower fares. These stem from the basic 'no frills' level of service[16] and higher labour productivity (from more flexible working practices, incentive-based awards, etc.) and higher capital productivity and utilization from a single aircraft type and quicker aircraft turnarounds. Lower costs for an entrant may also result from differential and more favourable airport user charges, use of lower cost airports, and financial assistance arrangement from local/regional governments. The low fare market segment also tends to be relatively price elastic and quality of service inelastic; incumbents have less market power in this segment. Hence, it may be possible for the entrant to expand demand by more than capacity – with more limited impact on an incumbent. It is therefore not surprising that most new entrants into air service markets adopt a 'pure' low fare-low cost/leisure segment business model – at least initially.

Entry requires a strategy response by the incumbent that *departs from its otherwise profit maximizing one,*[17] *i.e. it will be costly*; it only makes sense if the

---

[15] For example, many if not most leisure travellers have a lower value of time than business travellers. Hence, barriers such as airport access may be able to be overcome (at the cost of adopting an inferior product) by substituting secondary airports. (These involve higher ground travel time for passengers but also lower landing charges to airlines; the latter will be passed on to passengers if the air service market is competitive.) Although access to landing slots and terminal gates has not been an insurmountable problem in Australasia, it has been a formidable barrier in USA and Europe and as a result the substitution of secondary airports has been a core component of many LCC entry strategies (notably by Ryanair and Southwest).

[16] These features do have some value to passengers (even leisure travellers) and hence their elimination may have a modest affect on demand, especially if the money fares of the entrant and incumbent are the same.

[17] The incumbent may also have *market share* as a separate objective; this tends to emerge along with profits, in oligopolistic markets, particularly air service markets. This is in part because of perceived economies of scope through network size and connectivity and the influence of capacity share (especially frequency share) on demand facing an individual airline. Size in this sense does matter!

benefits of forcing the entrant to contract or leave exceed the costs,[18] and the market structure sought can be sustained for some time.[19]

Several factors shape the possible scenarios. But two major elements are:

- the size of the disadvantage (in cost or demand) facing the entrant, i.e. the '*height* of the barrier',[20] and
- the size of any *additional* cost that the incumbent can impose on the entrant, i.e. the '*depth* of the response damage'.

In the discussion so far no mention has been made of the use of *predation* by the incumbent. What predation generally does is increase the imposed damage from competitive response *and* the cost of the response to the incumbent. Of course, some firms may enter and fail because of the initial cost disadvantage due to one or more barriers to entry *alone*.

2.4 Height of Barriers and Incentive for Predation

Generally, predation is likely to be rational (setting aside the risk of its detection and successful prosecution as unlawful conduct) under three conditions:

a) First, there must be *substantial absolute barriers to entry*; without these, the losses incurred during predation could not be more than recouped; this is commonly recognized;
b) Second, since barriers to entry (and degree of risk aversion) *differ* across entrants,[21] the commercial case for, and likelihood of, an incumbent(s) resorting to predation is stronger for a new entrant, if the entry barriers (and perceived risk) faced by that particular entrant, while still substantial, are *relatively low*!; this does *not* appear to be commonly recognized in treatments of predation to date;

---

[18] It is common for the term 'recoupment' of current losses to be used but this term is somewhat 'sloppy'. Current certain costs/losses are chosen to be incurred – and 'sunk' – in the expectation that uncertain expected *future* profits are *sufficiently higher* to outweigh the current certain losses, when each is measured in *present value certainty-equivalent* terms by appropriate discounting. Interestingly, in some jurisdictions, for example, the United States, evidence of 'recoupment' is required. An argument for this is that it demonstrates *actual exercise* of market power; *potential* for recoupment is somewhat redundant as it stems from market power. Of course, actual recoupment is 'after the event' and a less efficient deterrent.

[19] The number of firms already in the market shapes the incumbents' response and the latest entrant's barriers. The effective height of barriers may be less with more incumbents but the risk of sustainable entry given the limited size of the market may be less.

[20] Most barriers are a matter of degree, but some, such as statutory legal barriers are absolute, i.e. the height is 'infinite'. These were the very 'stuff' of government economic regulation of air services.

[21] For example, experience with entry in Australia after deregulation (circa 1990) shows that entrants *differed* very significantly in their character and, as a consequence, the *absolute height* of the barriers they faced. For the early (three) failures the barriers faced by these entrants were significantly higher than those faced more recently by entrant Virgin Blue.

c) Third, the 'pool' of prospective entrants, for which the barriers to entry are *relatively* low, must be *small.* This combined with the first two conditions, may make predation worth serious consideration when sustainable entry by an entrant appears likely.[22]

This argument warrants reiterating. When the entry barriers faced by a *particular* entrant are absolutely high, the chances of its sustainable entry will be low. In this case, escalation of the competitive response by an incumbent all the way to predation is likely to be *unnecessary and redundant – and not worth the legal risk.* But, when the entry barriers relevant to another particular entrant are less substantial yet still significant, an extraordinarily aggressive response is necessary to undermine viable entry.[23] *In short, the stronger the entrant, the stronger is the rational response by the incumbent.*[24]

This is illustrated in Figure 17.1, in which the height of entry barriers as cost disadvantages are represented vertically along OY, and the entry form of business model is shown horizontally from LCC to FSA. The curve MT depicts the threshold minimum level of disadvantage that can be expected to be overcome for viable and sustainable entry, given the incumbents. The threshold is higher the closer an entrant moves to a FSA model. The curves HF, ME, NS and LR are 'disadvantage or barrier height profiles' that correspond to *specific* potential entrants. For the entrant represented by HF the barriers it faces at entry are too high and it cannot survive, whatever its entry business model. If it enters, an incumbent has little incentive[25] to mount any competitive response. But, the entrant represented by NS can expect to survive if it adopts a business model/level service up to W; its chances are highest as a 'pure LCC'.

But an incumbent might force exit of this entrant if it pushes NS higher to the levels of MT or above by an aggressive response; for entrant LR this may 'require' predation. As a viable entrant becomes established its disadvantage profile can be expected to fall; for example, from NS to UV.

2.5 Defining the Market – An Economic and Legal Perspective

Barriers to entry, market power, and assessment of anti-competitive conduct, such as predation, all require interpretation in the context of a well-defined market. For air services the key dimensions of the most appropriate boundaries for the relevant market are *geographic* (notably a city-pair, route or national trunk network), *functional* (transport modes, including train, car and bus services) and

---

[22] The pool of strong entrant contenders can be made much smaller by restrictions on foreign ownership and the legal barriers to entry into domestic markets by foreign airlines (including cabotage).

[23] Response actions and their outcomes are risky; choice over alternative actions needs to take into account the conditional probabilities of 'win' and 'lose' outcomes, including degree of risk aversion.

[24] To date there is limited (USA) and mixed findings of rigorous empirical studies of incumbent responses to entry.

[25] A competitive response may bring forward the timing of exit and may be profitable.

product/service types (quality of service, direct/indirect carriage, and fare conditions, such as flexibility).

In its most basic terms a market is represented by 'where', with 'what', and 'how' buyers and sellers connect and exchange transactions take place. But a market needs to encompass *interdependencies* in consumers' and producers' choices and decisions.

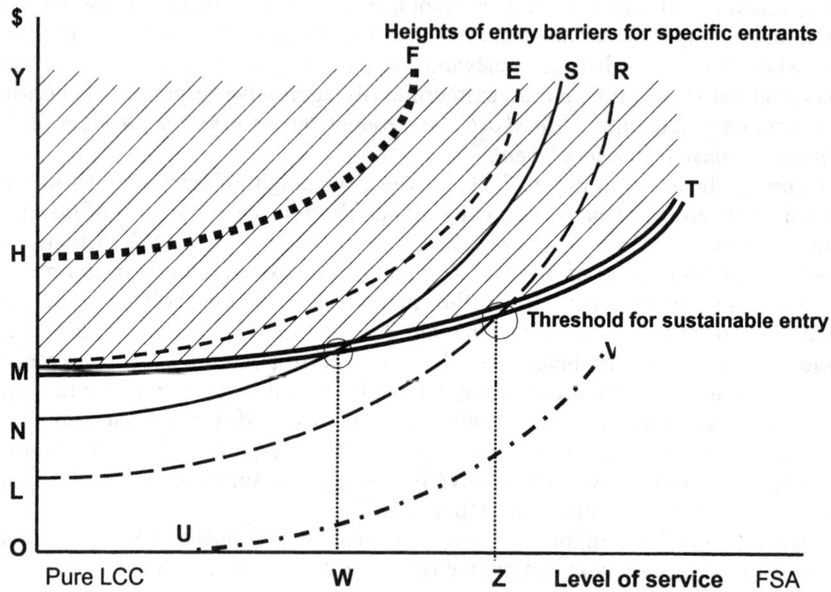

Figure 17.1 Sustainability of Entry and Incumbent Response

Accordingly, in many jurisdictions the combination of economic and legal arguments draws on *substitution* as the central defining characteristic of a market,[26] based upon all substantial substitution possibilities in both demand *and* supply.[27]

With regard to product demand, it is more the *price and attributes* of the various modes or qualities of travel service (notably, speed, schedule convenience, frequency delay, reliability, comfort, amenities, safety, change flexibility, and so

---

[26] In Australia, the High Court case of Queensland Wire (1989) has been highly influential on this matter, although a (predation) case appealed to the High Court adopted a narrower market definition. Boral (2003).

[27] Complementarity among products generally reinforces the case for a broad market definition.

on), and the willingness of passengers to pay for these various attributes, than the type of passenger or trip purpose that shape demand.[28]

Also, FSAs *selectively* target all market segments ('price conscious', 'urgent', 'high service'). Therefore, the relevant market should include *all airline product segments*. This also has the virtue of setting aside the costing 'quagmire' of treating fare types/seats as the unit of analysis.

Also, on the demand side, all passengers (and especially business) prefer non-stop/direct service[29] from A to B to indirect service over an intermediate point C. But if the price of the indirect service is lower (more low fare seats are made available) some passengers will switch to the indirect service. The non-stop and indirect services are clearly close substitutes (closer for passengers with lower values of time). Also for some passengers (especially leisure travellers) the choice of the destination B depends on the total cost of 'consuming the final leisure package' at various destinations. This includes the money fares and travel time to alternative destinations offering comparable leisure type activities. If the air fare from A to D was lowered compared to the fare to B, a significant number of travellers would switch their travel to D, i.e. there are *destinations on a network that are substitutes* for B. A narrow market definition that was limited to only A-B air services would overlook these relevant substitutes. Moreover, recognizing the importance of indirect services reveals the demand-side relevance of all practical indirect routings for example, including A-C-B. For the range of travel distances on the relevant air route, other surface transport modes may not be close substitutes, and do not significantly affect the demand for air travel faced by airlines – except at the margin (for example, private car for some large family groups). Indeed, for many trips, the lowest *air* fares correspond to lowest generalized *travel* cost for which a value of travel time is imputed and included – and hence surface modes are not highly competitive with air. For many countries, including Canada, USA and Australia, the market power of an airline is not, by and large, affected substantially by surface modes, except for 'short' trips.

On the supply side, air services are interdependent through the use of common aircraft and crew over a network; they are substitutable in terms of the ability of the airline to reassign capacity operated on one route/city-pair (D-E) to another route (A-B).[30]

---

[28] Markets may be subdivided into *segments* of passenger and trip purpose types. There may be a positive correlation between, say, higher qualities of service and business trip purpose. But these qualities are also income elastic – some passengers travelling to a holiday destination chose a business class or economy unrestricted fare; business and leisure passengers are heterogeneous.

[29] Direct service normally encompasses non-stop service and service under the same flight number but which may set down at an intermediate (part transit) point. To clearly distinguish non-stop flights from flights with a stop, the term direct is not used here and indirect is used to refer to flights with an intermediate stop (or stops) under the same flight number or not.

[30] On the supply-side, in many cases, airlines can also substitute (and 'integrate') capacity from their international routes. On the demand side, economies of scope for passengers imply possible substitutions involving international connection. However, these considerations are normally secondary and extending the market definition beyond the national boundary is normally not justified.

Thus, while airlines compete on an individual city-pair they also compete across several city-pairs. It follows that in most cases the most rigorous definition of the market is the set of routes and city-pairs that represent the sum of the routes of the parties. If a deregulated environment prevails then this may extend to the entire *national domestic trunk route network of air services*. This stance does *not* require that conduct and its consequences (most notably, pricing, possible losses, and intended exit of the entrant) that are alleged on a specific route(s) must also be proven on 'all' routes, i.e. throughout a broader network.[31] The basis of incorporating the broader market is to systematically recognize the *implications that the relevant network has for the route* of specific focus.

Yet, no market definition is universally accepted. Ultimately, the adoption of the 'relevant market' must be *determined on legislation/case precedent grounds for the relevant jurisdiction*. However, it may be admissible, prudent, and practical to develop analysis based on both a broad (trunk routes) and a narrow (single city-pair) definition.[32]

In Europe, different stances are taken, although the EU's position supports a broad definition.[33] In North America, a 'city-pair market' is normal (for example, in the Canadian WestJet and American Airlines cases).[34] Strictly interpreted, a narrow definition focuses on the city-pair in question – but not on other parts of the network of which the city-pair is an integral part. What is relevant is the *difference* in an alleged predator's costs and revenues, with and without its actions, on a specific route(s), in responding to an entrant. The traditional approach to this question is to assemble the evidence on attributable revenues and costs of the flight as if it were 'free-standing' on the city-pair route. Attributable costs are interpreted as the avoidable costs and these are estimated on the basis that the flight was 'cancelled': the concept of 'avoidability' triggers the hypothetical context as one of 'cancellation'. As elaborated in Section 6 below, this traditional approach has limitations and to correctly identify the economic costs of various resources, including aircraft capacity, operations outside the city-pair in question, i.e. a broader network, needs to be examined.

3. Barriers to Entry in Air Service Markets

3.1 Establishing a Workable Definition of Barriers

The concept of barriers is best defined in relation to the mobility of new resources throughout markets, especially the *mobility* of capital.[35] However, the conventional term 'barriers to entry' is maintained here.

---

[31] For example, an entrant may be 'substantially damaged' if it is forced to withdraw from one of its routes without complete exit, raising the question as to whether the 'withdrawal-forcing conduct' was predatory.

[32] The 'competitive battle' being observed may be on one route but the supply lines and resources available can be drawn from much further afield.

[33] See European Commission (1997).

[34] See Crocioni (2000).

[35] See OECD (2000).

3.2 Types of Barriers

It is useful to recognize two main sources of barriers:

- Structural – which are 'exogenous', and which may be subdivided into
  - 1a. '*natural*': sunk costs of entry such as advertising and the incumbent's brand,[36] and
  - 1b. *institutional:* controls such as government regulation and unequal access to public airport facilities, 'grandfather rights', ownership conditions, and exclusion of foreign airlines.

- Behavioural/Strategic – which are 'endogenous', and which represent incumbents' actions that are specifically designed to thwart or attempt to increase the cost of entry and/or entrant's costs such as vertical integration and long term contracts with key suppliers and distributors travel agents.

These categories are not 'watertight' and there is typically some overlap, especially between natural and strategic.

3.3 Will the Real Barriers to Entry Please Stand Up?

There is considerable debate about whether a particular condition, factor or behaviour is, or is not, a 'real' barrier to entry.[37] The key feature of a barrier is that it drives a wedge between an incumbent(s) and a potential entrant.[38] The incumbent may have established a high market share and be earning supra-normal profits. Post-entry, and its sunk costs, some of these profits may be bid away (and down) by entrants. However, a barrier places a prospective entrant at a disadvantage relative to incumbents. It typically involves *expenditure by a new entrant without any equivalent associated expenditure being imposed on, or*

---

[36] Of course, an incumbent's brand, reputation and customer loyalty may not be all strongly positive. Generally they represent an intrinsic non-transferable advantage and so a barrier. But, some entrants are not de novo companies. For example, in Australia, Virgin Blue brought *at entry* a positive image of the pervasive Virgin (Branson) image and 'brand', especially to low fare passengers.

[37] Some conceptual approaches couch barriers in terms of conditions that impair economic efficiency. This view is important to recognize in a policy formulation context, but it is not incorporated here.

[38] Viewed from the position of the incumbent, barriers give a degree of 'protection' (c.f., a tariff barrier). The incumbent can charge a higher price and offer less than an entrant and still earn supra-normal profits; the higher the barrier the higher the protection and the higher the 'limit price'. It may be even be possible for an incumbent to 'block' entry by setting a 'limit price' such that entry is demonstrably unprofitable. The gap between the limit price and minimum cost is an alternate measure of the height of the barrier.

required by, an incumbent.[39] It is useful to keep in mind that the most important source of such disadvantages is *sunk costs*. These do not have to be incurred by the incumbent (again); i.e. after entry they are no longer a barrier. But, they are a necessary part of the *price of entry*, and when incurred are then sunk, i.e. will not be recoverable – if the entrant exits or not.

It follows that a major source of risk, and hence a determinant of the availability and cost of finance to support new entry, is the extent of the sunk costs. This cost needs to be weighed against the commitment in equity advanced and the prospects for success, taking into account the likely response of the incumbent(s). A sunk cost disadvantage cannot be neutralized without some real countermeasure.[40]

Demand-side asymmetries (for example, in consumer brand loyalty) are reflected as barriers by the expenditure (or lower fare) that an entrant would need to make to overcome resistance by passengers to switch to the entrant's product.

Entry is not achieved immediately – it takes *time*. So does achieving positive cash flow and financial viability. These time aspects of entry[41] have not received much attention as a barrier to entry. The opportunity cost of the time allocated to implementing entry, and the pursuit of its sustainability, is perhaps the 'ultimate example' of a sunk cost. The longer the time required prior to commencing operations: the higher its sunk cost and the sunk cost of the finance charges; the greater the postponement of the commencement of financial recovery of entry costs and; the greater the continuing uncertainty about the market prospects of the entrant – and hence the cost of finance. Entry lags increase risk and enlarge the barriers to entry. Furthermore, time shapes barriers in another way. The longer there is satisfactory experience with the entrant, the less the 'new entry' residual risk and the lower the expected cost of continued finance. Success tends to breed success. The sooner an entrant 'can get points on the board', the sooner the higher cost of capital to the entrant relative to the incumbent will decline.[42]

---

[39] An established firm has already committed resources that give it a 'first-mover advantage'. Indeed, the established firm may have deliberately committed excess capacity to gain a strategic advantage and deter entry in the sense that the entrant realizes (is 'signalled') that the incumbent has made sunk expenditures that make its exit unlikely. See Baumol and Willig (1981) and Dixit (1981).

[40] Stigler (1968), viewed barriers as 'a cost of producing … that must be borne by the firm that seeks to enter but is not borne by the firms already in the industry'. The Stigler view does not regard scale economies per se, for example, as a barrier but only any associated sunk costs.

[41] For example, obtaining an operating certificate, establishing a business plan, and securing investor commitments all take considerable time.

[42] Two to five years is a range commonly forecasted by an LCC in its business plan and prospectus. The present value of financial losses can be substantial over this length of period.

More generally, following actual entry,[43] many though not all,[44] entry barriers associated with a particular new entrant's continued survival are likely to *decline over time*. This poses a challenging conflict in the choice of a response strategy for an incumbent. *Predation will tend to be more effective, and hence more likely, in the early establishment stages of new entry*, especially if the initial entry barriers are relatively modest and indicate a reasonable chance of success.

3.4 Avoidable, Derived, and Primary Barriers

Some factors put forward as barriers to entry are either not strictly barriers at all or actually stem from more fundamental or 'primary' barriers. A careful assessment of the impact of barriers calls for clarification of these distinctions. Examples of the first group are passenger lounges, in-flight entertainment and meals, valet parking, interlining, and distribution channels/agents.[45] For this group procurement is optional at additional *avoidable* costs that may not differ substantially between the incumbent and an entrant.[46] This first group might be termed, '*avoidable barriers*' and, of course, it accounts for many of the package of services (traditionally associated with the relatively complex FSA business model) that can be unbundled and dropped by an entrant in establishing a pure LCC business model. Thus, 'avoidable barriers' are product/quality of service 'advantages' of an incumbent that an entrant can *overcome either by their elimination or selective inclusion in its product mix.*

There is another more subtle set of factors within this first group that are often perceived as 'barriers' but which are also strictly avoidable. These consist of 'scarce resources' that an incumbent airline holds that are not otherwise available to an entrant but are actually transferable or potentially 'tradable' (for example, a peak-period landing slot, a terminal gate, or a top scheduling manager). When such resources can be exchanged they do *not* confer a cost advantage on the incumbent

---

[43] Before actual commencement of operations, sunk costs include air operating certificate, assembly and training of personnel and staff, procurement of aircraft, spares and support equipment and services (CRS, maintenance contracts, sales), establishing terminal space and gate use, posting any security bonds and insurance, and advertising and promotion. For example, in Australia, Virgin Blue initially had built a low cost demountable terminal at SYD and MEL while negotiating lease of main terminal space (later vacated with the collapse of the second incumbent FSA, Ansett); at other airports there was available capacity.

[44] Unequal terms of airport access may persist and a FFP disadvantage by the nature of its incentive structure may be durable.

[45] Even frequency and reliability attributes could be regarded as separable quality of service advantages.

[46] The cost is separable in total (say, for a flight) but generally less between different fare types on the same flight. Video entertainment (headsets) is, and interlining could be, 'sold' separately to individual passengers who need and are willing to pay for it. In Australia, Virgin Blue began offering (June 2003) limited interlining (with United Airlines and REX – a regional operator), basic 'club' lounges on a fee basis, and a booking charge (AUD$30) for extra pitch in front and exit row seats on long flights. These options represent a minor 'blurring' of the pure LCC model and expanded 'entry' into the business-type market segment by overcoming 'avoidable barriers'.

(relative to the entrant) since their market value (or value in some negotiated exchange) represents an *opportunity cost* to the incumbent, and an avoidable cost to the entrant.[47]

In the second group – which depends upon other barriers – the leading two examples are the cost and availability of capital and the response strategy of the incumbent(s). It is useful to distinguish and recognize this second group as '*derived*' barriers. Cost of capital is an important potential 'barrier'. However, it essentially stems from *other* barriers since the main determinant of any *premium* in the cost of capital to an entrant is the judged additional risk exposure with any loan or equity finance. Many factors influence this risk before and after entry (and before the continuation of conditional tranches of finance). Factors include the robustness of the entrant's business plan[48] and macroeconomic conditions.[49] The response strategy of the incumbent as a barrier draws essentially on the primary barriers, except that 'reputation' as a 'ruthless competitor' may also create a barrier itself, especially if there is solid credibility (and problematic successful prosecution).

3.5 An Overview of Specific Barriers

Various types of barriers, with the exception of regulations, are discussed briefly in turn.

Brand, Reputation, and Track Record An entrant has to induce existing passengers to switch or generate new customers, usually by combining promotion of 'more attractive' prices with expenditure on advertising about its product – which typically is 'no frills'. A well-accepted incumbent does not need to incur such expenditures. This imposes a relative cost disadvantage on the entrant.[50] With presence and use, information about the entrant's product becomes more widespread and this price-quality trade-off becomes more informed and reliable. If the entrant brings to entry a brand affiliation (such as Virgin Group) which

---

[47] Another revealing transport example is a taxi licence plate (or medallion). Entry into the taxi industry is usually regulated but plates are generally *tradable*. So a new entrant does not face an entry *barrier* – a cost disadvantage – in the form of the cost of acquiring a plate (risk differences aside), because an incumbent taxi incurs a cost of holding the plate, namely its sale value that is *foregone*, i.e. its opportunity cost; one actual barrier/sunk cost is market knowledge.

[48] For example, underestimation of 'start-up' costs and contingencies can have a debilitating affect by depleting working capital and necessary liquidity as cash flows are trying to be established.

[49] Transport in general and air travel in particular are income elastic (especially for discretionary/leisure travel) and operator margins are fairly small and (in the short run) fixed costs high. Hence cyclical downturns in the economy have a substantial negative financial impact which, depending on the timing of entry, may affect risk for the entrant more than for an incumbent with an established loan 'portfolio'.

[50] The uncertainty associated with the entrant's product and its prospects will not be adequately reflected in imperfect capital markets and hence brand uncertainty is likely to increase the cost of capital to the entrant. But, there is economic debate about the extent of this effect. See Schmalensee (1974).

customers have some prior positive awareness about, then the information gap is less, consumer confidence is higher, and the necessary 'advertising' cost disadvantage is lower. If the incumbent has been established for a long time, brand knowledge will be substantial and pervasive.[51] A 'good' brand represents 'collateral' which also influences favourably the cost of capital. Further, an incumbent with a history of good management, high safety, and sound financial performance, establishes a track record that narrows the risk to investors. This allows a lower cost of debt and equity capital.

To deal with the 'sunk' brand advantages of a well established incumbent an entrant needs to incur promotion expenditure, backed by 'investment' in accumulation of a record of delivering good performance.

Loyalty Programmes A popular frequent flyer programme (FFP) represents an entrenched product advantage for an incumbent since there are economies of scale in continued loyalty to a single brand that cannot be effectively matched by an entrant.

As with other FSA advantages on the demand-side, FFPs can be overcome for many passengers (especially non-business paying their own fare) by a *sufficiently lower money fare; each 'frill' has its 'shadow price'!*

Distribution Channels and Reservation Systems Technological changes in communications, especially the extensive market penetration of the internet, EFT, and electronic ticketing, have reduced the importance of the travel agent channel for much of the domestic low fare/leisure market segment. This has reduced the advantage of established relationships of agents with incumbent airlines, and a distribution advantage can be balanced by paying agents a commission fee.

Computer reservation systems (CRS) are an extremely important input to sales and yield management and viable rapid response pricing strategy. They involve substantial sunk costs, some economies of scale, and require training. But, they can be established internally or contracted out. Entrants wishing to compete up-market can pay the cost for listing in a 'global distribution system' (GDS).

On balance, the barriers to entry associated with reservation systems are indicated to be modest.

Alliances and Code Sharing Alliances and code sharing basically extend the product range and extent of market presence of a member, for example, to improve share of international interlining/feed traffic. They also leverage an airline's FFP. They can represent a barrier to entry, but primarily to the business market segment. The sunk cost of entering an alliance is the 'admission fee', in particular upgrading

---

[51] In Australia, Qantas, for example, has been operating since 1920 and has the distinction of being the oldest airline in the English-speaking world, with the second oldest surviving airline name in the world (KLM Royal Dutch Airlines being a year its senior). Qantas has established distinctive product differentiation – an 'Australian' icon image which it strongly cultivates ('… calling Australia home …' etc.), and a reputation for excellence in safety and reliability. In Qantas' case its 'brand' is a valuable goodwill asset that is inextricably tied to the company as a whole. This advantage cannot be separated and traded.

to FSA type standards. The additional benefits to a LCC (even a hybrid one) can be expected to fall well short of the avoidable costs, which are very substantial.

Network Scale An extensive network of routes is a formidable 'product range' advantage for an entrant to face, even in relation to just a domestic market.[52] These are advantages that cannot easily be balanced, and it takes time to develop an effective network. A well-targeted 'point-to-point niche leisure/low fare' entry strategy reduces their relative importance.

Even at the city-pair level there are substantial sunk station, market intelligence, and other costs as well as risks, for an entrant in attempting to expand fast to approach network benefits.[53] Also, for FSAs a network provides feed and may support complexing with mini-hubs and spokes; this may lower unit costs.

A network allows an incumbent to selectively target aggressive competition at the route(s) where an entrant begins operations. The cost of the response can be spread over its total network with limited impact on the company's balance sheet. It provides more extensive opportunities for profitable price discrimination (particularly in the business segment).[54]

An entrant can minimize an incumbent's network advantage as a barrier by focusing on high density point-to-point services and on prospective customers for whom a network is of low importance.

Absolute Cost Advantages and Vertical Relationships An incumbent may have cost advantages in some key specialized inputs that an entrant cannot purchase on the same terms, for example, aircraft engine heavy maintenance or airframes, and flight simulator training, especially if the markets are thin – *and* there are sunk costs and some economies of scale in producing these services/products.[55] Other areas are crew training and aspects of distribution.[56] This aspect of cost advantage may be seen as part of the more general issue of vertical relationships and restraints. Such absolute cost advantages can represent a barrier.

An entrant can reduce the impact of these barriers by narrowing the range of specialized inputs needed, such as by using a single aircraft type (very commonly a fleet of B737s). Competitive tendering from suppliers may be efficient but, given the transport/aircraft ferrying costs, may involve a cost disadvantage to the entrant. The most effective counterbalance for an entrant is a LCC business model with lower operating costs.[57]

---

52 In addition, a wide network, with its mix of stage lengths and time constraints generally brings some operational efficiencies, especially in aircraft utilization.

53 But there is a strong incentive to do so (witness Ryanair).

54 LCCs can also price discriminate (for example, early vs. late booking); a network widens the scope.

55 This is common for countries for which the market size is not just relatively small but also located distant from the major input markets in the USA and Europe.

56 An incumbent may decide not to offer its internal input services to an entrant – and an entrant may be suspicious of obtaining key inputs from an aggressive competitor.

57 For example, in Australia in the late 1980s, Qantas and Ansett exhibited lower total factor productivity and hence higher operating costs than 'international good FSA practice' would suggest.

Access to Key Infrastructure Of all barriers to entry (except legal prohibition) possibly none is more evident and potentially more binding than unavailability of comparable access to essential airport facilities – landing/take off slots and terminal gates/lounges and apron space.[58] These can act as a crucial vertical restraint.

Landing slots as a differential barrier only applies when slots cannot be traded. Although lack of equal access may only apply during peak hours, it may cascade to attraction, overall scheduling problems, and utilization costs.

By and large, landing slots and terminal gates have not represented a prolonged barrier for entrants. Either access at major airports has been widened or proximate substitute airports have been adopted. However, the timing of adequate access can forestall entry.[59]

Cost of Capital A higher cost of capital for an entrant relative to an incumbent is primarily a 'derived entry barrier' attributable to the expected risk impact of the various barriers noted above.[60] A related issue is the availability of sufficient working capital to finance the sunk costs of entry *and* the operating losses during the establishment phase, taking into *account the most likely strategic responses of the incumbents*. Business plans of entrants with realistic risk management of expected cash flows during the first 2-3 years, entry delay, and other contingencies, can reduce costs. Finally, there tends to be a threshold scale of entry below which the chances of survival will be grim. The scale will depend on the particular market size and expected growth prospects.

3.6 Barriers in Concert

Each of the factors outlined above presents a 'hurdle' for entry and survival. In addition, the barriers are not independent and may be compounding. For example, brand, reputation, and entry to alliances and code sharing,[61] overlap. Assessment of barriers as set out above sharpens the perspective on real barriers and how they

---

[58] Heathrow and CDG are prime examples.

[59] Where major incumbent FSA airlines own terminals/gates unequal access can be a critical disadvantage to an entrant. (This was the case for some airports in Australia in the initial period following deregulation in 1990.) In Europe this problem has been overcome by entrants (especially LCCs) adopting substitute underutilized airports on the fringe of major cities. The location disadvantage, particularly in time, is less important for non-business and non-connecting travellers.

[60] There is debate about the cost of capital, per se, as an entry barrier. The view present here is based upon its derived nature which overcomes most of the points of difference. However, given the weak financial position of many FSAs, their risk exposure and cost of capital may be actually *higher* than for a new entrant/LCC.

[61] Alliances and code sharing 'partnerships' have entry barriers of their own, including operational and quality of service standards; membership comes at a 'fee'.

shape the degree of market power held by an incumbent. This varies significantly across market segments.[62]

It is not easy to measure the extent of this market power for individual barriers, or for all barriers in concert.[63] But, the size of the sunk costs expenditure associated with each barrier would be one leading 'input' measure, and the fare levels/average yields on routes with vs. without the presence of an entrant, would be an 'outcome' measure, provided entry is sustainable over 2-3 years. Also, the money amount by which a LCC fare has to be lower to 'compensate' relevant passengers for each 'frill', would be another measure.[64] The link to market power is discussed next.

4. Market Power

4.1 What is Market Power?

The economic and legal definitions of the concept of 'market power' are usually consistent, although there are several ways to convey the concept. The key to market power is an *ability* of a firm to act in a manner that is different from the behaviour that a *competitive* market would enforce on a firm that faced similar cost and demand conditions.[65] In economic terms, market power exists when a firm can charge a higher price (and/or supply lower quality) than other firms in the same market – and remain profitable for a sustained period.[66] One quantitative economic measure of market power is the *generalized price*[67] *elasticity of demand facing the subject firm* for one or more of its products or services. When this elasticity is

[62] Clearly the degree of market power depends on market segment: FSAs have much more power in the business segment, for which LCCs are generally a weak substitute. However, the business segment extends over a continuum: from small enterprises to large corporate accounts to chief executives, typically with falling price elasticity and rising demand for quality of service.

[63] After entry of a LCC, some FSAs may be at a net disadvantage (for example, in Europe and Australia) and may set up their 'own' LCC as a response strategy, with certain shared costs and economies! However, the success rate of this strategy is not high; some costs, particularly labour, are difficult for the subsidiary LCC to reduce, given its affiliation. In Australia, Qantas has recently set up a LCC subsidiary (JetStar) to compete head-to-head with Virgin Blue – and 'purchased' an ex-Ryanair senior executive to be CEO (a 'tradable' cost advantage).

[64] As attempted for FFP and daily frequency in the Lufthansa-Germania case. See Hüschelrath (2003).

[65] Draws on Kaysen and Turner (1959).

[66] A period of two years is commonly used.

[67] The term 'generalized price' is used to capture price *and* money valuation of non-price features, for example, in seat pitch comfort, cancellation conditions, or delivery requirements; it is very difficult to empirically estimate at the level of the firm, partly because of lack of airline level data. But see Morrison and Winston (1995).

'relatively low'[68] a significant degree of market power is present. It is also important that direct assessment (or measurement) of market power should be oriented to what would prevail *at competitive prices.*[69,70]

However, elasticity alone may be insufficient. Legal interpretations generally give more emphasis to 'supra-competitive prices' and lack of competitive constraints not just in pricing but also, for example, in conditions of supply and distribution. Plainly, market power hinges on very careful definition of the relevant market.[71]

The economic definition recognizes a continuum – *market power is a matter of degree.* Domestic FSAs generally have some market power in their business/first class segments, depending on their competition with other FSAs. They can raise price and *not lose 'many'* of their business passengers to economy class, other FSAs, or to less trips.

The definition of market power should be kept distinct from factors that provide or shape it. The leading determinant of market power is barriers to entry, since they directly reduce or eliminate competitors. Others factors are market share (or an index of concentration such as Herfindahl-Hirschman), and vertical restraints, but these are far less unequivocally linked to market power.[72] Barriers to entry *are directly* linked to market power. The key connection is that an essential *test of market power*, and a sufficient condition for it, is whether the threat of *entry* (or actual sustained entry) will constrain incumbents to behave competitively, and thereby eliminate their supra-normal profits, rents and inefficiencies; if not market power is present.

The question of a direct quantitative measure of market power is typically side-stepped by establishing the existence of its major determinant, namely, barriers to entry.

---

[68] There is no magic threshold number – there is a continuum. But an elasticity of five, for example, means that if the firm in question raises its own price by 5%, it will lose 25% of its sales; a 10% price rise implies a 50% loss. i.e. a relatively small price rise will mean a relatively large loss in sales.

[69] The price prevailing *after* the exercise of market power, 'competition'/'substitutability' will *appear* greater, and the price elasticity measures will be higher – both indicating *less* market power for the firm in question. This is commonly referred to as the 'cellophane fallacy', which arose from criticism of the judgement in a now famous case involving *U.S. v. Du Pont* (1956); 351 U.S. 377.

[70] For business passengers the price elasticity of demand at the appropriate marginal costs of serving them may be, say, only 0.5. But with market power this would indicate that price was below the profit maximizing price (which would involve price elasticity being above 1.0.) If an estimate of marginal costs can be made then the *percentage mark-up in price above marginal cost* (known as the Lerner Index) is a better measure of actual market power. An even more sophisticated related measure is the Landes-Posner Index of market power – but it requires hard-to-come-by estimates of market demand and supply elasticities.

[71] Generally, the narrower the market definition the greater the potential market power, but this depends crucially on the entry barriers to *that* market. *ACCC* v. *Boral* (2003) highlighted the centrality to the court of market definition and market power in predation.

[72] These are mainly applied to estimate changes in market power in merger cases.

4.2 When Does an Incumbent Have Market Power?

The simple answer is when an incumbent 'enjoys' various barriers to entry. A basic inferential test is to assess whether the pattern of conduct of the incumbent would have been the same if it had been operating in a competitive market, especially in terms of pricing, capacity levels, and scheduling.

A number of the factors that represent barriers to entry are also product characteristics that have a substantial effect on passenger *demand*. Thus, for example, the strengths of an incumbent airline in brand, breadth of product range (routes served, airport locations, and service types), and customer base relative to other airlines, and especially entrants, reduces substitution to them (i.e. the airline's own and cross price demand elasticities are relatively low). This translates into the ability to raise fares (and also impose stricter service conditions with lower costs), without substantial loss of sales.

The 'range' of market power of an incumbent is directly related to their breadth of products. i.e. the number of routes and fare types/market segments that they serve, since new entrants invariably must limit offerings to relatively few routes and market segments.

Another indicator of the exercise of market power is an incumbent constraining supply, such as persistent operation of *some* routes, especially non-stop routes, at very high load factors with small quotas on low fares. Price elasticity is lower for non-stop service, partly from self-selection by relatively inelastic business passengers. Also, severe quotas for low fares diverts more elastic passengers to indirect services, builds up density on these routes, and allows the use of larger lower cost (per seat-km) aircraft. In this way total profits from the non-stop and indirect services are more likely to be maximized. Of course, an entrant is likely to target such non-stops, disturb this practice, and initially attract some higher yield traffic, including spill-over back from indirect services: a profitable incumbent 'apple-cart' would be overturned and a strong competitive response from the incumbent could be expected.[73]

Typically, one or a few FSA incumbents have available both a large number of routes and a set of 'full flexibility and high quality of service' market segments (especially high yield business traffic) which are strongly oriented ('captive') to them.[74] These circumstances mean that a FSA incumbent can apply selective and profitable price discrimination to provide a lucrative source of internal funds.[75] These funds would be readily available out of accounting reserves to support aggressive competition against new entrants and could be *selectively targeted* at only the fewer routes where the new entrants typically begin operations.

The incumbent may well be maximizing profits via price discrimination on all routes prior to new entry. New entrants would disturb that discrimination on the few routes initially entered. Nevertheless, the availability of a large number of other routes (and service attributes) is a very valuable asset for the incumbent to

---

[73] Some aspects of this type of scenario are raised in an unpublished paper by Forsyth (2003).

[74] Evidence of this would be where the lowest discount fares drop substantially following LCC entry, but *economy and business* fares remain virtually unchanged or increase.

[75] Price discrimination, per se, does not imply market power; it is commonly used by multi-product firms to recover overhead/joint costs.

draw upon in dealing with new entry – in terms of pricing *and* capacity alternatives. Given a large imbalance between the network sizes of the incumbent(s) and the entrant(s), it is plausible that this 'selective game' could continue 'for as long as it takes' the entrant to give up and exit.

An aggressive competitive response need not extend to predation. As noted above, if the height of the barriers facing the new entrant are high, then matching fares/conditions of the entrant and carefully balancing capacity may result in a skewed split of the total low-fares segment that covers short-run avoidable costs for the incumbent, but falls short of covering financial costs for the new entrant – nd the cash flow, prospects and risks for the entrant remain dim. Resort to predation might be precipitated if the height of the barriers to a particular entrant are *low*, and hence successful sustained entry, 'down the track', is a fair prospect.[76]

Finally, related to exercise of market power is the common response practice by incumbents of matching or undercutting the *money fares* set by an LCC entrant. At issue is interpreting what 'matching' really means. First, matching on lowest/low money fares is typically on a small part of incumbent's service capacity. Even if the incumbent's seat costs are higher than the entrant's, it is likely to have higher average yields. Its market power in the higher fares segments can be expected to generate higher profits, even on routes/services where LCC entrants compete.[77] Second, the 'quality of service' of FSA services is generally higher. The generalized costs for a FSA incumbent and a LCC entrant, when they match money fares, depend on passengers' valuation of the differences in quality between the airlines.[78] However, the extent to which an incumbent is able to match or charge somewhat higher money fares, and still achieve (yield-managed) quotas for the lowest fare segment, indicates market power.[79]

---

[76] It is sometimes argued that the source of funds to finance predation (or aggressive competition) does not matter, and hence cross-subsidization is not necessary. This assumes that capital markets work well and the cost of external funds is not more than the cost of retained earnings. Also risk may be lower in drawing upon a wide range of market segments.

[77] In the low fare segment carriers tend to initially match or slightly undercut each other's fares. However, in Australia, for example, relatively stable periods suggest a fare differential of the order of + 20% of the FSA over the LCC. But, comparison of only lowest fares is vulnerable to 'distortion' by the quota applied to lowest fare seats over a reasonable period. Some work of the US General Accounting Office attempts to measure the market power by premium fare differentials. The work indicated orders of magnitude of 2-10%. But, this should be taken as a very rough indication – it is based on the US system alone, and averages across that system.

[78] This is illustrated in the follow-up to the Lufthansa-Germania Case where the 'Bundeskartellamt' – the German Federal Cartel Office – imposed a (two-year) minimum 'premium' (€35) on the fare that must be charged by Lufthansa above Germania's fare, on the Berlin–Frankfurt route. The premium was based on an estimate of the value of 'auxiliary services' provided by Lufthansa, but not Germania: FFP and higher frequency; benefits of this ex post regulatory approach to sanctions are 'fragile'. See Hüschelrath (2003).

[79] This would be reinforced if, looking at quality of service differences from the supply side, the higher costs of FSA services, when quality adjusted, were equivalent to the lower costs/lower quality of service of the LCC. Persistent higher money fares by the FAS would indicate some market power in the low fare segment as well, depending on the avoidable variable cost of the 'frills' to low fare passengers.

Part II: Improving Economic Tests of Alleged Predatory Conduct

5. Formulation of an Economic Test of Predation

5.1 The Necessary Test Components

Ideally, the development of an economic test of predation should serve to inform the court as to what is expected to be adopted as *the ultimate legal test of predation*. Unfortunately, in many countries there is limited guidance from national competition statutes and/or court case history. This is not entirely surprising as the dividing line between promotion of strong competition on the one hand, and safeguards for anti-competitive conduct on the other, is a 'grey band'. Notwithstanding substantial variation across and within jurisdictions in legal 'tests',[80] there is more universal agreement among economists that a sound economic test, as a *per-se rule*,[81] should explicitly *combine* evidence covering four components:

- possession of a substantial degree of market power;
- intent to substantially damage, or prevent entry of, a competitor, or reduce competition;
- pricing below costs, i.e. deliberately incurring current losses,[82] and
- recoupment, i.e. setting future supra-competitive prices (after withdrawal of the entrant) to secure a net profit gain.[83]

Market power as a necessary condition was covered above. The subject of intent is crucial, but it is inherently subjective. It has no independent economic test; it is heavily a legal issue fraught with difficulty in establishing inferential purpose.[84] Its accepted role in cases differs across jurisdictions: in Australia it is a central and explicit element, but in the United States it is not; its role in Canada and Europe appears problematic.[85] Recoupment may be regarded ex ante as a potential derivative of market power and ex post as evidence of exercise of that

---

80 See Morrison (2003).

81 Consideration of the cost and benefits of 'per se' vs. case-by-case 'rule of reason' approaches, including the probable bias of errors ('Types I and II') to the per se rule, is outside the scope of this chapter. .

82 Pricing above avoidable costs is even sometimes claimed to be predatory since it may block entry of less efficient but still low-price competitors. Edlin (2002). This claim appears weak.

83 Economists acknowledge recoupment as a fourth component. However, it is not commonly accepted by courts, although it is a requirement in the United States. Its status in Australia has been elevated recently by a major predation case appeal decision by the High Court in Boral (2003), where it was tied to the determination of 'substantial market power'. In the 2003 American Airlines Case, the District Court accepted the US Department of Justice's definition that each city-pair was a separate market, and that anticipated recoupment must be expected from the same relevant market!

84 See Posner (2001) for a critical analysis of intent.

85 For a valuable discussion of international differences, also see Morrison (2003).

power and possibly intent. In this Part II the focus is on the third component and, in particular, whether a defendant incurred real *economic* losses as a result of its pricing and capacity responses to an entrant. There is also considerable disquiet about the introduction of this 'economic loss test' component into evidence. Among the concerns the most serious is that there are legitimate exceptions to the 'economic loss test'.[86] Therefore, the loss test by itself is imprecise and inadequate. For this reason it is argued that it should be supplemented by other 'economic rules', especially use of output restrictions or 'strategic and dynamic tests' that attempt to demarcate predation from aggressive competition.[87]

Nevertheless, in general, an economic loss test is useful provided it is used as one component of a more comprehensive test and considered jointly especially with market power. Within this context, the focus here is on the single profitability or loss test that is given prominence in most predation cases,[88] namely, whether the *avoidable costs* are greater than the associated revenue and, in particular, how this test may be improved. The test is generally referred to as 'pricing below average avoidable cost' (or average variable cost).[89]

---

[86] For example, there are circumstances where a firm may consciously price very low (below average avoidable costs) and incur a current loss, and yet by so doing actually improve future market competitiveness! This may arise when an entrant needs to disseminate product information or where production yields 'learning by doing'. In both cases the firm 'invests' by pushing higher output in order to more quickly establish customer awareness or move down the learning curve. Promotion commonly goes so far as to involve setting 'introductory give-away' prices. In these cases the firm's intent is not to force a competitor from the market but to increase sales, in order to improve knowledge – consumers' and/or its own. But, these 'explanations' lack credibility if the incumbent is 'well established'. However, this territory is a legal 'minefield'; evidence and proof of intent are notoriously difficult to secure.

[87] Two leading 'non-cost-based' tests are the 'output restriction rule' proposed by Williamson (1977) and 'quasi-permanence' of price reductions, proposed by Baumol (1979) and (1996). The price-capping feature of the Bauman rule has been proposed by Starkie (1999) as a deterrent device that would attempt to block price rises for recoupment, if an entrant exited. The essential rigidity of the rule renders it not adaptive to inevitable change; also it has the danger of acting as a 'barrier' to new entry by forcing prices to remain low or deterring innocent firms from risking price cutting. Edlin (2002) targets disincentives and proposes fixing the incumbent's price as soon as entry occurs. Leading strategic tests use 'two stages' [Ordover and Willig (1981) and Joskow and Klevorick (1979)] or game-theory [Bolten, et al. (2000)]. Unfortunately, these various rules also have shortcomings in terms of strategic 'gaming', or industry dynamics, or their implementation is complex and their 'point of closure' is difficult to define in practice. Nevertheless, future refinement of such sophisticated approaches is called for to strengthen 'proof'.

[88] Interestingly, the test is relevant to inferential purpose. But it is explicitly *not necessary* in Australia, where the trade practices legislation does not mention predation per se (only anti-competitive conduct with the proscribed purpose '… to take advantage of market power to eliminate or substantially damage a competitor, or … prevent … competition'. The few court decisions available for guidance confirm that establishing market power and intent are crucial.

[89] For multi-product firms total flight revenue is needed (Baumol (1996)).This cost-based loss test was originally proposed by Areeda and Turner (1975) in the form of pricing above average variable cost (as a proxy for short-run marginal cost).

For ease of discussion a simple competitive context of a single FSA incumbent and a LCC entrant is adopted. The predatory action considered is a 'scheduled flight' by the incumbent on one particular route of its network that is operated by the entrant. This context allows the economic issues to be readily analysed and the conclusions easily extended.

There are several issues that must be resolved before any test can be applied; prominent among these are establishing operational and economically meaningful definitions of the relevant market, the appropriate unit of output, the appropriate time period, and the relevant costs. These issues are briefly discussed (Section 5) before in-depth attention is given to how the 'avoidable economic loss' test may be improved (Section 6).

5.2 Market Definition

Approaches to market definition differ internationally (and sometimes within the same jurisdiction), and the economic criteria to guide adoption of the relevant market were discussed in Section 2.5. It was argued that *substitution* between demand for products (destinations, fare types and quality of services) *and* sources of supply of those products (including redeployment of capacity/aircraft from operations on other routes) was the single most important criterion. As emphasized below, substitution of aircraft and crew capacity across routes goes to the heart of cost determination and (arguably) intent.

On these bases, and on balance, the broad set regional or national (network) of routes operated, typically by the incumbent, is the relevant market[90] – or alternatively the sum of the sets of city-pair markets flown. However, a city-pair market definition should not preclude argument and evidence that comes from beyond that narrow market and likewise a regional route network market definition *should not* oblige the cost and revenue analysis to embrace all routes in a whole network.

5.3 Unit of Analysis

In general, the *scheduled flight* is the most suitable unit of analysis for operations, revenues, costs, and losses. This is a flight scheduled to fly at a *given time on a given day* (typically daily over a weekly cycle). It is supportable with available data, has international recognition, and corresponds closely (with route and frequency) to a basic management decision variable for an airline. The justification of the scheduled flight as the most appropriate unit is also strengthened by noting that an alternative, such as route level output, represents *average* outcomes which can mask the performance of individual flights. Assessment should be focused on additional scheduled flights, i.e. at the *margin*. However, sums at the route level may be worth doing, especially if flight data are unavailable: if they reveal negative average profitability then at least some (possibly all) flights are unprofitable.[91]

[90] Whether the network is national or regional depends on the network most relevant to the route(s) on which the alleged predation was conducted.

[91] See, for example, discussion in the Canadian WestJet case.

5.4 Time Period for an Economic Test

Data for profit and loss outcomes is typically compiled on a *monthly basis* and averaging scheduled flight performance outcomes over one month is satisfactory for analysis. This accommodates the random and seasonal variation of demand. With regard to costs, in the long-run *all* costs are avoidable.[92] In the very short run, *few* costs are avoidable. Last minute cancellation of a scheduled flight and consolidation of passengers on later flights would save fuel burn for example, and its cost would be avoidable. Most other costs associated with that individual flight cancelled at short notice on that day would not be. The more advanced the notice, the more costs could be avoided or the inputs productively redeployed.[93] Furthermore, some costs can be adjusted or avoided entirely for the airline at a 'cost penalty'. This may be the case for many labour or servicing contracts or the cost may arise in the form of a profit foregone penalty. Thus, there is an (avoidable!) cost of unanticipated avoidability or input flexibility. This cost may be regarded as a type of 'transaction cost'.

In predation cases a major focus is whether the defendant deliberately incurred a real loss. Identifying the revenues at stake on a scheduled flight is normally straightforward and not especially contentious. Most[94] argument centres on costs. The conventional approach to costs is to look at *financial* costs.[95] The defendant will typically claim them to be low, by asserting that a large proportion of airline expenditures could not be avoided over the period of the case; i.e. they are not attributable to the actions under litigation. In short, the defendant will typically submit argument that many if not most costs are unavoidable, i.e. *sunk.*

It follows that the time period over which each category of cost can be avoided is likely to be highly contentious. It is attempted to be resolved by distinguishing between two situations: *adding* a new flight (or failing to remove a flight) to a route, and *cancelling* an existing flight on a route. This distinction focuses on a forward-looking perspective regarding decisions to add or not remove

---

[92] The airline could go out of business, merge, change ownership, restructure its network and levels of service, appoint a new CEO, adopt a new brand, modify alliances, dispose of some equipment, form vertical relationships, establish a different fleet mix of aircraft, relocate its base(s), and operate under an entirely different business plan. Acquisition of new aircraft is normally (part of) 'long run' decisions. Some of these 'long run' costs are overhead/joint costs and not attributable to a scheduled flight. Also expanding (aircraft) capacity pushes company average variable costs towards average total costs; the gap related to 'common/joint/overhead' costs and any residual fixed costs.

[93] In the Canadian Competition Tribunal (2003) 'WestJet' proceedings, the Tribunal accepted the view that, for an individual flight, an airline should know its profit contribution, and be able to avoid all attributable flight costs within *3 months*. Thus, management decisions could be planned 3 months in advance and confirmed by a rolling 3 month average profitability indicator, adjusted for seasonality and macroeconomic forecasts as appropriate.

[94] Issues such as 'frequent flyer revenue/cost attribution' and 'beyond revenue' are sometimes contentious but normally should not be included or tested on a strict with/without flight basis.

[95] The correct form of costs to measure and, in particular, financial expenditures by the airline or economic costs is addressed below.

capacity. All of the costs associated with adding or maintaining capacity should be regarded as avoidable. With respect to cancellation of an existing scheduled flight, some of the costs will take longer to avoid, either through shedding the cost, or through redeployment. Generally, it is more important to identify those costs that are intrinsically avoidable rather than the precise period over which they can be avoided.

6. Improving the Use of Avoidable Costs in Tests of Predation

6.1 Developing the Fundamental Basis for Avoidable Costs of a Flight

This section presents a rigorous economic approach to guide legitimate classification and accurate estimation of what costs are attributable to, and avoidable with, a scheduled flight. It develops economic argument for expanding and sharpening what constitutes avoidable costs. It may be seen as complementary to the conventional approach to estimating the various categories of airline operating costs. Two parallel strands of argument are developed; each stands alone on its own merits but each is based on the same economic principle of opportunity cost. The section could be re-titled: 'How to Overcome Getting Hung-Up on What is After All Unavoidable'!

It is surprising that argument applying the concept of opportunity costs does not appear to have been brought to bear, in the way developed below, in earlier predation litigation arguments.[96] This may be due to the contrast that can arise between *financial cum accounting* costs and *economic* costs, and the challenges of reconciliation of this contrast in court proceedings. The term 'financial' is used here to denote a money transaction – an expenditure or a receipt – with no distinction made between avoidable and unavoidable transactions. The term 'economic' denotes discretionary real resource transfers (including money) – their disbursement or receipt.[97] Economic transactions are all avoidable.[98] It follows that the financial cost of an action can be small or zero and its economic cost can be large, and vice versa. The limited use of the opportunity cost concept in court proceedings dealing with commercial matters also may be due in part to the fact that it is typical to consider industries in 'equilibrium', where there may be more congruence between financial and economic costs. But, disequilibrium is common and an imbalance of capacity on a route (too little or too much), combined with the nature of airline cost structures, can 'drive a wedge' between financial and economic costs.

---

[96] The role of opportunity cost in the Areeda-Turner test was touched upon by Baumol (1996) with respect to owned-input costs (deemed relevant) and revenues foregone (deemed irrelevant) in judgement of acceptable prices for an 'equally-efficient' rival. The issue of profits foregone in other markets was put forward, but rejected, in the American Airlines case, *United States* v. *AMR Corp.* (2003); this was brought to the author's attention by David Gillen.

[97] The term 'economic' is used here in a narrow sense with reference to the airline, and not its wider social sense which encompasses externalities.

[98] This does not imply that unavoidable costs cannot have impacts that have economic as well as financial dimensions.

6.2 Focusing on Costs through the Opportunity Cost Looking Glass

Four central economic issues arise in assessing an incumbent's conduct using an economic test for predation:

a) the status of an incumbent's existing capacity and, in particular, whether it is in 'excess demand' (scarce) or not;
b) the incumbent's priorities for the allocation of its capacity across its domestic network;
c) valid cost and profit information at the flight/route level, and its role in establishing priorities; and
d) the incumbent's response strategy to competitive entry.

Here the focus is primarily on (a), (b) and (c), although against the background of (d).

At the core of a test for predation is the question as to whether the incumbent consciously pursued a strategy of confronting entry by *selectively* targeting aggressive (low) pricing and (high) capacity at the entrant's routes only. If so, the incumbent is likely to have sacrificed some of its profits elsewhere on other domestic routes, and possibly to have incurred economic losses on these routes. *Profit forgone to confront entry is a real cost* and a key economic issue.

At a general level, an incumbent may defend its response actions, such as redeployment of capacity to, and selectively lowering some fares on, any route, and specifically one of the new entrant's routes, by claiming the 'impracticality' of the company keeping track of all profit opportunities and the difficulty of predicting with precision the outcome of any change/redeployment of existing capacity from one route to another.[99] But a grain of truth does not make an edible loaf. If an incumbent was not vigorously attempting to maximize expected profits for the company, and its shareholders as a whole (without predation), what was the reason?

If an incumbent foregoes outstanding profit opportunities on some routes, then what would be the purpose of chasing demonstrable smaller profits (or potential losses) on a new entrant's route?

Could it be lack of sound commercial judgement, unawareness, or simply uncertainty? After all future demand and demand responses to large fare reductions and capacity decisions are not known with certainty. This is a difficult question to resolve. If the FSA is highly experienced with sophisticated information systems and a proven track record of sound management and generally solid financial performance, then ignorance, incompetence, or weak demand forecasting, strains

[99] There are two key aspects of an FSA's conduct in responding to an entrant. The first is price-matching and seat control behaviour. Here the key issue is whether the FSA'a conduct resulted in predictable losses. Subsequent to additional capacity and lower fares yields, passenger numbers can be expected to increase but yields decline. The second aspect is selective use of flights on some of the entrant's routes to target its response, specifically at the entrant's low fare service type. This involves effects on profitability via 'dilution' from selective yield management on the targeted routes and willingness to substitute on those routes, low fare capacity for business/full economy seats, and possibly increase business fares.

credibility. This is particularly so if unsatisfactory profit outcomes (losses) *persist* on a route(s) for some time *and* if other better profit opportunities are available. When does *continuation* of a business decision negate a presumption of mistake belief?

Could it be lack of capacity? If the FSA has scarce capacity, there is greater pressure, incentive, and opportunity to nail down the most profitable uses/routes/frequencies for deploying that capacity – as well as to assess in detail the cases for investing in, or leasing, *additional* capacity, how much, and *when*. An *active* programme of fleet *expansion* with aircraft being brought on stream is based upon current, as well as future, profit opportunities for that *additional* capacity – not simply efficient replacement of existing capacity. Thus, in a circumstance of scarce capacity, if the incumbent chooses to marginalize the use of some of its scarce capacity, in the face of high profit alternative uses, what would be the purpose? What would be its intent?

Could it be surplus capacity? If the FSA has surplus capacity, does this imply that there are no *domestic routes* where deployment of additional capacity is profitable? The lack of domestic profit opportunities must be based on costing capacity at the *maximum* receipts that would be available from deploying the capacity to *non-domestic use*: on any international route, or leasing, or selling the aircraft. If the capacity is being held for anticipated future use, based upon the costs and benefits of holding surplus capacity rather than purchasing it later, then its 'cost while holding' is its highest *internal* use, i.e. its opportunity costs based upon its highest profit contribution on alternative routes; this requires examination of the FSA's broader network of routes. Finally, could it be strategic? Holding 'surplus' capacity is recognized as a potential way to deter entry, especially in moderately thin markets.

In basic terms, an airline's commercial aircraft decisions are straightforward:

a) employ the aircraft (and other associated inputs) if it can make a positive contribution to profits;
b) deploy the aircraft on that route where the higher profits are earned;
c) employ additional aircraft so long as opportunities remain for expected positive profits;
d) dispose of the aircraft if it cannot be expected to be used to generate profits (based on its disposal value and other complementary input avoidable costs).

Of course, this grossly oversimplifies an airline management's fleet allocation and planning, investment timing, network, and product pricing decision making, but it goes to the very heart of the relationship between the airline's profits and its costs. If an incumbent is undertaking substantial (net) additions to its fleet and domestic capacity, condition (C) holds.

6.3 What should be the Basis for Determining Costs? The Role of Economic Opportunity Cost

The form of costs normally submitted to the court as evidence is accounting and financial. These typically represent expenditures associated with actual market transactions or 'book values'. The expenditures may be past or future, committed or discretionary, avoidable or sunk. The determination of whether they are avoidable or not is a major task for the court since most admissible economic tests involve, inter alia, estimating the profitability or loss of the conduct in question. Avoidability matters since if a cost is unavoidable or sunk, then it cannot be changed, no matter what decisions an airline makes; in a financial sense it may be regarded as 'common' to all company performance outcomes. *The company can only be held responsible for its avoidable costs as they are discretionary and hence accountable in profit outcomes.* Thus, a test for predatory conduct requires that avoidable costs be carefully determined.

From an economic perspective a financial approach can be inadequate and misleading. *Economic costs are opportunity costs*, i.e. what is given up by taking some action. These may correspond to the giving up of one real resource for another (where the money cost acts as an intermediate proxy for the opportunity cost), or more subtly they may correspond to the giving up of one *outcome* for another. i.e. the opportunity cost of outcome A is the giving up or loss of outcome B. The concept of opportunity cost is as profound as is it simple; it is a powerful economic idea and it can be highly relevant to a predation case.

6.4 Testing for Predation under Conditions of Fleet and Capacity Expansion – The First Argument

Circumstances in which an incumbent airline is expanding its capacity are very revealing. The cost of each new aircraft is fully avoidable (prior to order commitments/'sunk' deposits). An incumbent is making forward-looking, 'long-run' decisions over a range of choices, taking into account *all* of the associated costs, and hence treating them as *avoidable.*[100]

Moreover, during an acquisition programme condition (A) above must hold for existing aircraft. It follows, on commercial grounds, that in such circumstances:

- aircraft capacity is scarce and profitable to employ;
- there will exist several existing schedule flights that yield substantial profits and many opportunities for profitable employment of new capacity;
- the value in use of an aircraft exceeds the costs of acquiring, deploying, and operating new aircraft;

[100] There are a couple of caveats that should be noted. First, new additional aircraft represent 'lumpy' additions to capacity and therefore they may not be fully utilized for some time. Second, the complementary inputs (flight crew, maintenance, ground staff, and so on) may not be all 'fully avoidable' costs in the short term. But, on the first, if an incumbent has a relatively large domestic network and existing fleet, the quantum of the new acquisitions and the control of delivery timing can 'smooth' and optimize integration of aircraft into operations. Therefore, each new aircraft would rapidly contribute positive profits. On the second, expansions in complementary inputs are discretionary and their costs are avoidable over a modest time frame; the provisions in any agreements on future employment conditions are actionable at the discretion of the incumbent airline and hence, whatever the costs are (and for labour they may be high) they are fully avoidable. [Baumol (1996) coins the term 'decremental' for avoidable (costs).]

- the costs of adding and operating new capacity are all avoidable;
- the value in use or profits of an existing schedule flight exceeds the total cost of a (new) schedule flight – all of which was avoidable.

In short, when capacity is scarce, and new capacity is being added, the profit value of using existing capacity *exceeds* the total and fully avoidable costs of acquiring and operating new capacity. The opportunity cost in profits foregone of existing capacity is *at least as great* as the total cost of new capacity. Thus, when capacity is scarce, treating *all* schedule flight costs as avoidable is a *conservative* assumption; the real opportunity costs to the airline are *higher and profits lower.*[101]

Note that this holds even if existing aircraft are 'more expensive' to *operate* than new aircraft; new aircraft will only be acquired if they improve (the present value of) profits so that the *total cost* of aircraft ownership/lease *and* operating costs are minimized.

6.5 The Role of Opportunity Costs in the Presence of Unavoidable Costs – The Second Argument

An *economic* loss on a route or schedule flight does not necessarily mean negative financial profits. An economic loss involves earning *lower* profits than could have been obtained from better use of the resources available. Thus, for example, shifting a schedule flight from a route R, where it contributes + $10m in economic profit to the company) to route S, where it contributes + $7m in economic profit) involves an *economic loss* of $10m – $7m = $3m. This arises if aircraft capacity and crew are not used on the most profitable route opportunity available but on a route that provides *lower* profits, which could still be positive. This prospect becomes more acute under circumstances of a (temporally) fixed domestic fleet, and so limited and potentially scarce capacity.

To encompass recognition of this within an avoidable cost-based economic loss test is straightforward provided the presence of more profitable opportunities can be measured *or demonstrated.* As outlined under the 'First Argument' above, the 'cost' of providing a schedule flight on route R by redeploying an aircraft is the *opportunity cost* measured in terms of the profits foregone by its (equivalent) use

[101] In circumstances of *excess* capacity, the highest (profit) value in use of marginal capacity is negative (or zero), based on the disposal value of aircraft and other avoidable complementary input costs. Condition (4) obtains and the best action is to minimize losses by reducing and disposing of capacity.

on its source route S.[102] From an economic standpoint, forgoing $3m in profits that were available is a loss and just the same as incurring an additional $3m in resource costs – both are real costs and *both result in lower company profits*.

Consider an individual schedule flight as a free-standing single aircraft operation. The financial costs of the schedule flight consist of a sum of avoidable and unavoidable costs. Since only avoidable costs are discretionary, it is the economic profits of the schedule flight that matter, and these depend only on revenues and the avoidable costs. So, economic profits are always *higher* than financial profits (revenues less financial costs).[103] Moreover, the greater the share of financial costs that are deemed to be unavoidable, the greater the corresponding economic profits. Indeed, economic profit is always more than financial profit by the extent of the unavoidable costs.[104]

Now consider circumstances where airline capacity is limited to the provision of this one schedule flight, but it can be deployed on only one of two routes, R or S.

In the debate on avoidable vs. unavoidable costs much is commonly made of a correspondence between the feasibility of redeployment of the schedule flight and the avoidability of its financial costs, i.e. it is often argued that redeployment implies avoidability.[105] But, *redeployment is neither necessary nor reliable as a basis to establish that all the flight's associated financial costs are avoidable.* Redeployment may carry with it some unavoidable costs (such as labour contracts that hold for say 12 months). Any unavoidable costs of operating the flight on route R are likely to *remain* unavoidable when the flight resources are redeployed on route S. However, what the potential for redeployment *does* bring into focus is the *opportunity cost* of redeploying resources from Route R to Route S, and thereby the real economic costs.

By focusing on opportunity cum economic cost the dispute over what are unavoidable and what are avoidable costs is actually turned on its head! It is typical in predation cases for the defendant airline (usually the incumbent) to argue vehemently that most of its costs attributable to a flight are unavoidable – as this stance supports a higher level of profits and guards against a loss being found.

---

[102] Note immediately that whatever costs of certain ('mobile') inputs are not avoidable, their alternative *uses* or redeployment opportunities *are*. For example, suppose the financial cost of a labour arrangement that has been put in place for 12 months is $1 million. This becomes an unavoidable cost for that period (setting aside any contract pre-maturity options and penalties). But the *economic* cost in its current use is its opportunity cost, i.e. its value/profit contribution available from its redeployment in its *next-best* use. (If there were no other complementary input avoidable costs this alternative use value/profit would be simply its revenue contribution in that alternative use; otherwise any avoidable additional costs would have to be netted out.) This opportunity cost may be *more or less* than the (sunk) financial cost of $1 million; it could be zero if the next highest net revenue available was zero. At the end of the contract period the cost of the labour for another 12 months is fully avoidable and the decision to incur that cost will be based upon its benefits (net of any spillover costs of non-renewal of the contract) exceeding $1 million.

[103] Setting aside 'creative, albeit legal, tax accounting manipulations'.

[104] Recall that in the 'long-run' all costs are avoidable, so these are 'short-run' economic profits.

[105] This is very evident in the proceedings before the Canadian Competition Tribunal (2003) in the Air Canada-WestJet predation case.

Since redeployment has as its opportunity cost the profits foregone, those *foregone profits are higher with larger unavoidable costs*. Hence, on the route to which the flight is redeployed, the economic costs are higher and the profits are lower for higher unavoidable costs!

Larger unavoidable costs increase, not decrease, the contribution of an additional scheduled flight on a route to an airline's total profitability, whenever total capacity is fixed and scarce, and hence that flight is redeployed from a profitable route.

On the one hand, if aggregate demand conditions are weak (for example, the national economy is in a recession/business-cycle trough or there has been an exogenous shock/disaster with a sudden unexpected drop in air service demand) there may be aggregate excess capacity, with many flights and even routes at marginal or even negative profitability; and it may become unprofitable to operate some capacity at all, taking into account the returns available from lease or disposal alternatives, if any. On the other hand, if an additional flight is made possible by a net addition to the airline's (domestic) capacity through aircraft leasing or acquisition, then plainly all the costs of that flight are avoidable. Moreover, a net additional flight on any specific route corresponds to a marginal increase across the network as a whole (subject to some scheduling and repositioning network constraints). i.e. all individual flights are 'at the margin' and have a cost responsibility given by the total avoidable acquisition cost of the new aircraft. In economic terms, if capacity is being expanded the marginal economic costs of an additional flight are the full financial costs, all of which are avoidable.

If a decision by airline management is made to redeploy capacity from route R to route S, and this decision is motivated by obtaining higher profits on route S, then this implies that revenues (for a schedule flight in total or per flying hour) on route S exceed revenues on route R – accepting that the total flying time of the redeployed capacity and the passenger loadings remain approximately the same. Thus, the influence of costs and, in particular, the role of unavoidable costs, are 'neutralized' since they are common (and fixed) to the operation of the schedule flight on both routes. Moreover, if flying time and the number of passengers did differ somewhat between the flights on the two routes, then the difference in flying time and passenger loading would clearly correspond to a difference in *variable avoidable* costs (fuel, crew time, maintenance, passenger servicing) which could readily be factored into the economic profits calculation.[106]

These fundamental statements of profit outcomes highlight three key conclusions that in general prevail:

- when there is fixed and scarce capacity on an airline's network, the profit on any route R decreases as the extent of any unavoidable costs increases: profits of shifting an additional flight onto route R are given by the difference in revenues less any difference in avoidable costs, between the routes; this

[106] For example, if the supply and demand quantities of flight redeployment did not closely match, and the 'new' flight on route R involved t hours of flying time more than on route S, at a variable cost of \$v per flying hour, and n more ('economy equivalent') passengers at a cost of \$s per passenger, then the profit on route A is less by *approximately* \$(vt + ns) per flight.

difference would be small, except for lumpiness in aircraft assignments on the network;

- if schedule flight/route profits are positive (and expected to persist or expand) across several routes, then there is a commercial case to purchase more capacity (aircraft, crews, etc.); this is a 'long run' decision for which the associated costs are all avoidable and under which financial and economic costs correspond[107] and, across 'all' routes, subject to some lumpiness in capacity assignment, the incremental profits[108] are of the same order of magnitude (say, per ASK); and
- if the profitability of the additional schedule flight capacity on the source route is zero or negative, and so the capacity is demonstrably 'excess' or idle, then the opportunity costs of redeployment are at a minimum[109] and equal to that capacity's opportunity cost in terms of the net receipts value of disposal or leasing (for example, at least for leasing the aircraft component of excess capacity); these represent the 'residual' avoidable and economic costs and could be zero; profits on the route in this case would be flight revenue less the residual economic cost; the alternative route opportunity cost is zero.

It warrants stressing that under domestic air service market circumstances of on-going capacity expansion, the true economic cost of redeploying a flight is actually *greater* than all financial costs taken as avoidable; the full or 'long run' financial cost of operating additional flights is below the true economic costs as measured by profits foregone, and hence provides a conservative estimate of profitability on the route in question. i.e. *unavoidable costs are not pertinent!*

Therefore, the often protracted debate over the extent of avoidability or unavoidability of costs, especially in the context of short term cancellation of a flight[110] is misplaced and, strictly speaking, irrelevant.

This logic is of course highly inimical to the standard line of defence in predation cases that most costs are unavoidable and that profits for the actionable flight in question are positive. Thus, heavy rebuttal of this opportunity cost/profits foregone economic argument can be anticipated by defendants and also probably from the courts. The initial criticism may be that the idea is 'esoteric, academic, impractical', and not considered in, or relevant to, 'normal business practice'. This claim is patently false; it can be refuted by selecting any decision of the defendant airline. Any decision can always be interpreted in opportunity cost terms – assuming there were choices to be made, i.e. alternatives to assess and a real decision required or implied. All companies, implicitly or explicitly, draw on the

---

[107] Ignoring externalities.

[108] If the airline has market power and exercises it, then the profit level will be supra-normal, i.e. revenues will exceed full financial/avoidable costs. Without any market power, supra-normal profits would be reduced to zero and total revenues would match total full financial and economic costs.

[109] It is conceivable that there could be a global scarcity of particular aircraft capacity and that the 'international price' of used capacity could exceed the marginal (i.e. minimum) profit value in use of a company's capacity; opportunity cost to the company becomes the international price.

[110] Canadian Competition Tribunal (2003).

concept every day! Opportunity cost is a cornerstone of rational commercial logic not a true/false hypothesis or an empirical fact.

7. Key Points in Summary

- Barriers to entry provide market power which is a necessary condition for anti-competitive conduct, such as predation.
- Barriers to entry prevail in air service markets and the main sources are: brand, reputation, network scale, absolute cost advantages, and access to key inputs such as airports; certain 'barriers' may actually be 'ladders' for which entrants may have their own advantages.
- Some product and cost advantages held by an incumbent cannot be purchased; they can only be countered by an entrant adopting a different business model (such as LCC) and incurring some up-front sunk costs, including time.
- Incumbents can also form strategic barriers that thwart an entrant's sustainability; these are costly and only make sense if the barriers limit further entry and the future market power benefits from influencing market structure exceed the costs.
- Barriers to entry differ across firms; the higher the barriers faced by a particular entrant, the lower the chances of sustainable entry. Therefore, the stronger the entrant, the stronger and sooner is any rational response by the incumbent.
- The effective height of barriers differs across entry strategies. In particular, barriers are likely to be fewer, less of a handicap, and overcome, for entry into the 'low fare/basic service' market segment, using a LCC business model.
- A predatory response is commonly regarded as one that combines means (market power), motive (intent to drive the entrant from the market) and method (loss-making conduct); however, these are necessary but not sufficient conditions.
- It is difficult to distinguish between a response by an incumbent to a new entrant that is aggressively competitive and one that is anti-competitive.
- There is no definitive test for predation; moreover, competition statutes and court case judgements in most countries provide limited guidance on legal and economic 'tests'; courts take a very cautious approach to ruling business conduct predatory.
- Notwithstanding the legal difficulties, the three components of market power, intent, and an economic test of losses, applied in combination, are generally adopted as a strong basis for assessing the likelihood of conduct being predatory. These are sometimes augmented by additional economic tests, for example, regarding changes in an incumbent's output.
- A test of economic loss remains a key component of assessing alleged predatory conduct.
- Conventional determination of an economic loss is based upon avoidable operating costs of a schedule flight. This inevitably reduces to a technical and

legal dispute over which cost elements are avoidable and which elements are not; this focus can be grossly misguided.

- Protracted debate and technical argument over what are unavoidable and what are avoidable costs of a schedule flight, in many cases, is irrelevant. The normal debate over avoidable costs should be turned on its head! Larger unavoidable costs decrease, not increase, the contribution of an additional schedule flight on a route to an airline's total profitability, whenever total capacity is scarce, and hence that flight is redeployed from another route.

Acknowledgements

The author thanks David Gillen and also Peter Forsyth for valuable comments.

Remarks

This chapter stems from economic background work undertaken by the author in 2003 as part of litigation in *ACCC* v. *Qantas Airways Ltd*, also involving Virgin Blue Holdings Ltd. In the event, the case was subsequently dropped by the ACCC and did not go to trial. Due to confidentiality the chapter does not discuss any details of that particular case; nor should its content be construed to suggest any specific interpretation of any issues in that matter. In addition, the views expressed in this chapter are those of the author and should not in any way be regarded as representing the views of the ACCC or its staff.

References

Areeda, P. and D. Turner (1975), Predatory Pricing and Related Practices under Section 2 of the Sherman Act, *Harvard Law Review*, 88, 1975.

Baumol, W. (1979), Quasi-Permanence of Price Reductions: A Policy for Prevention of Predatory Pricing, *Yale Law Journal*, Vol. 89, No. 1, 1-26.

Baumol, W. (1996), Predation and the Logic of the Average Variable Cost Test, *Journal of Law and Economics*, 39, 49-72.

Baumol, W. and R. Willig (1981), Contestability, Sunk Costs, Entry Barriers, and the Sustainability of Monopoly, *Quarterly Journal of Economics*, 71, 178-183.

Bolten, P., J. Broadley and M. Riordan (2000), Predatory Pricing: Strategic Theory and Legal Policy, *Georgetown Law Journal*, 88, 8, August, 2000.

Boral Besser Masonry Ltd v. *Australian Competition and Consumer Commission* (2003), HCA 5, 7 February, 2003.

Borenstein, S. (1992), The Evolution of the U.S. Airline Industry, *Journal of Economic Perspectives*, 6(2), Spring, 45-73.

Canadian Competition Tribunal (2003), *Commissioner of Competition* v. *Air Canada (WestJet)*, CT 2001-002, Decision 145a, 2003-07-22.

Caves, D.W., L.R. Christensen, and M. Tretheway, (1984), Economies of Density Versus Economies of Scale: Why Trunk and Local Service Airlines Costs Differ, *Rand Journal of Economics*, 15(4), Winter, 471-489.

Crocioni, P. (2002), Defining Airline Markets: A Comparison of U.S. and EU Experiences, *Antitrust Bulletin*, Spring 2000.

Dixit, A. (1980), The Role of Investment in Entry Deterrence, *Economic Journal*, 90, 95-106.

Dodgson, J., Y. Katsoulacos, and R. Pryke (1990), *Predation Behavior in Aviation*, A Report to the Competition Directorate of the European Commission, Luxemburg.

Edlin, A. (2002), Stopping Above-Cost Predatory Pricing, *Yale Law Journal*, 111, January 2002.

European Commission (1997), *Commission Notice on the definition of the relevant market for the purposes of Community competition law*, Official Journal: OJ C 372 on 9/12/1997.

Forsyth, P. (2003), Predation Behavior in Australian Aviation Markets, Paper presented at the HWWA Workshop on Predation in Aviation Markets, Hamburg, February 12, 2003.

Hüschelrath, K. (2003), How to Prey on Predators? – The Lufthansa-Germania Case, Ms. IIO, Wissenschaftliche Hochschule fur Unternehmensführung, Vallengar, Germany.

Joskow, P.L. and A.K. Klevorick (1979), A Framework for Analyzing Predatory Pricing Policy, *Yale Law Journal*, 89 (2), December, 1979.

Kahn, A. (1988), Surprises of Airline Deregulation, *American Economic Review*, 78 (2), 316-322.

Kate, A. and G. Niels (2002), On the Rationality of Predatory Pricing: The Debate between Chicago and Post-Chicago, *Antitrust Bulletin*, Spring, 2002.

Levine, M.E. (1987) Airline Competition in Deregulated Markets: Theory, Firm Strategy, and Public Policy, *Yale Journal of Regulation*, 4, 393-495.

McGee, J. (1980), Predatory Pricing Revisited, *Journal of Law and Economics*, Vol. 23, 238-330.

Morrison, W.G. (2003), Dimensions of Predatory Pricing in Air Service Markets, Paper presented to the 6th Hamburg Aviation Conference, Hamburg, February, 13, 2003.

Morrison, S.A. and S. Winston (1987), Empirical Implications and Tests of the Contestability Hypothesis, *Journal of Law and Economics*, 30, 53-66.

Morrison, S.A. and S. Winston (1995), *The Evolution of the Airline Industry*, Brookings Institution, Washington.

OECD (2000), *Airline Mergers and Alliances*, Report DAFFE/CLP(2000)1.

Ordover, J.A. and R.D. Willig (1981), An Economic Definition of Predation: Pricing and Product Innovation, *Yale Law Journal*, 91 (1).

Oster, C.V. and J.S. Strong (2001), *Predatory Practices in the U.S. Airline Industry*, A Report to the US Department of Transportation, January 15, 2001.

Posner, R.A. (1976), *Antitrust Law: An Economic Perspective, (Second Edition)*, University of Chicago Press, Chicago, 2001.

Queensland Wire Industries v. *Broken Hill Co. Pty. Ltd.* (1989), 167, CLR, 177.

Schmalensee, R. (1974), Brand Loyalty and Barriers to Entry, *Southern Economic Journal*, April, (40), 579-588.

Starkie, D. (1999), The US Department of Transportation's Statement on Predatory Conduct in Airline Industry: An Alternative Proposal, 1999 ECLR 281.

US General Accounting Office (1991), *Airline Competition: Effects of Airline Market Concentration and Barriers to Entry on Air Fares*, Report GAO/RCED-91-101, April, Washington, D.C.

United States v. *AMR Corp (2003)*, No. 01-3202, 335 F. 3d 1109 (10th Cir. July 2003).

Williamson, O. (1977). Predatory Pricing: A Strategic and Welfare Analysis, *Yale Law Journal*, 87, 284-340.